Molecular Inclusion and Molecular Recognition — Clathrates II

Functional Group Assisted Clathrate Formation — Scissor-Like and Roof-Shaped Host Molecules
E. Weber and M. Czugler

The Significance of Molecular Type, Shape and Complementarity in Clathrate Inclusion
I. Goldberg

New Types of Helical Canal Inclusion Networks
R. Bishop and I. Dance

Recent Progress in Molecular Recognition
J. Rebek, Jr.

Reaction Control of Guest Compounds in Host-Guest Complexes
F. Toda

AKADEMIE-VERLAG BERLIN

Topics in Current Chemistry · Vol. 149

Molecular Inclusion and Molecular Recognition — Clathrates II

Molecular Inclusion and Molecular Recognition – Clathrates II

With 130 Figures and 45 Tables

AKADEMIE-VERLAG BERLIN
1988

Die Originalausgabe erscheint im Springer-Verlag
Berlin · Heidelberg · New York · London · Paris · Tokyo
in der Schriftenreihe „Topics in Current Chemistry“,
Volume 149

Vertriebsrechte für die sozialistischen Länder:
Akademie-Verlag Berlin

Vertriebsrechte für alle Staaten mit Ausnahme der sozialistischen Länder:
Springer-Verlag Berlin · Heidelberg · New York · London · Paris · Tokyo

ISBN 3-540-19338-3 Springer-Verlag Berlin · Heidelberg · New York
ISBN 0-387-19338-3 Springer-Verlag New York · Heidelberg · Berlin

ISBN 3-05-500613-5 Akademie-Verlag Berlin

Erschienen im Akademie-Verlag Berlin, DDR-1086 Berlin, Leipziger Straße 3—4
Lizenznummer: 202 · 100/540/88
Printed in the German Democratic Republic
Gesamtherstellung: VEB Druckerei „Thomas Müntzer“, 5820 Bad Langensalza
LSV 1205
Bestellnummer: 763 922 3 (3072/149)
16800

Table of Contents

Preface

This book is the second of a two-volume series that attempts to bring together in-depth investigations in currently important areas of "*molecular inclusion and molecular recognition*" mainly based on *clathrate* formation. The first volume of this series (Volume 140) was published in 1987.

The well-known fact that certain biological compounds (enzymes, carriers, etc.) can bind substrates selectively while markedly enhancing their chemical modification or transporting them through membranes has inspired chemists to design and synthesize molecules capable of mimicking these and other features. Remarkable progress has already been achieved in the preparation of macrocycles and related systems which combine specifically with ions as well as with various uncharged molecules and provide functional properties based on recognition, binding, catalysis, and transport of ions and molecules. The 1987 Nobel Prize for chemistry was awarded to the trio of scientists Pedersen, Cram, and Lehn who did fundamental work on this relatively new field, called "host-guest" or "supramolecular chemistry" [cf. Weber, E., Vögtle, F.: Nachr. Chem. Tech. Lab. *35*, 1149 (1987)].

A *supramolecular* species (*host-guest* compound) results from the interaction of a substrate (the guest) with its receptor molecule (the host). Normally one finds enclosure of the guest molecule in the cavity formed by a host framework (cf. crown complexes, cryptates). The host-guest association is not established by covalent and ionic bonds, but is caused by H-bonds *and/or* van der Waals interactions. With reference to the chemistry of weak intermolecular bonds, supramolecular association has contributed to the fundamental understanding of the elementary interactions on which molecular recognition and binding is based and represents an interface between chemistry and biology.

Within this approach, *clathrates* (for a definition see Chapter 1 in Vol. 140 of this series) and related lattice-type aggregates may be considered as *multi-supramolecular systems* where guest molecules are included in a crystal matrix. They allow a great many applications which have been specified in Vol. 140, first of all the separation of enantiomers by enantioselective recognition and inclusion of racemic guest molecules.

Unfortunately, chemistry of clathrates has not made such rapid progress as the organic complexes until recently. Now new approaches and techniques have amply encouraged the design of selective host molecules and clathrate structures. The aim of this book is to illustrate important developments in these fields including functional group assisted clathrates, helical channel formation, architecture of crystalline clefts, and other molecular design principles given in Chapters 1–4, while Chapter 5 is devoted to the fascinating possibilities of stereocontrolled solid-state reactions using clathrate inclusion. Thus the present book is not a mere sequel of Volume 140 (part I of this topic) published in 1987, but complements the theme.

Both books could only be completed with the assistance of the individual contributing authors. Their efforts cannot be underestimated, and the editor wishes to express his heartfelt appreciation.

Bonn, April 1988 Edwin Weber

The Significance of Molecular Type, Shape and Complementarity in Clathrate Inclusion

Israel Goldberg

School of Chemistry, Sackler Faculty of Exact Sciences, Tel-Aviv University, 69978 Ramat-Aviv, Israel

Table of Contents

1 Introduction

The formation of clathrate inclusion compounds in which guest species are enclosed by channels or cages that occur in a given host lattice was first discovered in the middle of the past century [1], Since then a large number of clathrates and solvates, found mostly by chance as by-products of research in other areas, have been characterized [2-4]. These multicomponent systems attracted however relatively little attention, and a systematic development of their chemistry was very slow. The rapid advancement of host-guest chemistry in the last two decades, since the pioneering work of Pedersen on crown ethers [5], provided a turning point in this respect. Initially, the class of *molecular* inclusion complexes, in which larger host species are able to enclose and bind smaller guest molecules, has been subjected to most intensive investigations aiming at the synthesis of molecular models that can mimic the chemical behaviour of natural systems. The success was overwhelming with respect to the number and complexity of suitable compounds (ligand-substrate complexes, ionophores, molecular receptors) that have been prepared and analyzed. Most extensively studied were macro(poly)cyclic hosts of the crown/cryptate type which turned out to be excellent and often selective complexors of metal ions and amino acid derivatives [6-8]. The presence of O and/or N heteroatoms in these macrocycles made them also suitable for an effective complexation through hydrogen bonding with uncharged guest molecules containing proton donating sites such as acidic CH, NH or OH [9-11]. Most recent investigations of models for molecular separation and transport processes relate to carefully designed organic systems which involve encapsulation of apolar guests within apolar host molecules [12-14].

As a natural development of the successful molecular inclusion concept, which involved electrostatic [6-11] as well as van der Waals [12-14] forces between the interacting host and guest entities, an increasing interest has been shown in the systematic study of *lattice-inclusion* type systems. A considerable effort has been devoted to the design of new hosts for the formation of stable crystalline clathrates and the improvement of selective complexations with potential guests. Suitable examples of clathrates studied in recent years include hosts such as Dianin's compounds [15a], perhydrotriphenylene [15b], cyclotriveratrylene [15c], triphenylmethane [15d], hexakis-(arylthio) and -(arylthiomethyl)benzenes [15e], tri-o-thymotide (TOT) [15f], and choleic acids [15g] (cf. Fig. 1 in Ch. 1 of Vol. 140). Selected series of such clathrate inclusion systems have particularly been useful in research of photochemical reactions in the solid state [16,17] and of selective molecular complexation that is central to biological phenomena [18,19].

This report surveys the *structural* aspects of clathrates formed with a series of other host systems, illustrating the conceptual progress made over the last few years in our ability to achieve a better controlled and more selective lattice-type complexation of uncharged molecules. It also emphasizes an important observation that in molecular separations the approach of lattice inclusion can often be as effective and easy to apply as that of molecular inclusion; both phenomena utilize a similar variety of binding forces by which host and guest constituents are held together in a structured way. Most of the examples presented below were structurally analyzed in this laboratory, and they refer to systems based on organic host lattices only. They reflect on the author's general interest in structural aspects of host-guest chemistry.

They reflect further on the ideas originally proposed by Prof. Fumio Toda of Ehime University [20], Prof. Harold Hart of Michigan State University [21] and Prof. Edwin Weber of Bonn University [22] with respect to the design and use of suitable hosts, and then jointly evaluated with the aid of detailed structural interpretations [23-26]. These are complemented by additional citations of the literature, trying to avoid as much as possible an overlap with the contents of other chapters in this volume and with the preceding vol. 140 of this series; the respective drawings were prepared with atomic coordinates retrieved from the Cambridge Crystallographic Data Base.

2 Solvates and Clathrates as Common Phenomena

2.1 General Considerations

A survey of the available literature indicates that crystal packing of molecular entities that are characterized by an irregular shape and a small number of configurational degrees of freedom frequently results in the formation of a host lattice with intermolecular voids. To achieve thermodynamic stability and dense packing (cf. Ch. 2, Sect. 3.1 of this volume), this space in the lattice must be filled by another (guest) component. In inclusion structures with imperfect geometric relationships and lacking specific bindings between the complex constituents, the guest species appear either disordered or thermally smeared within the host lattice. Both, topological as well as functional interactions therefore play an important role in stabilizing clathrates, and the ability to achieve a selective crystallization will depend significantly on the steric compatibility between the complexing partners. Furthermore, if the solvation forces are sufficiently strong in comparison with other intermolecular interactions, the solvating species will be carried out from the solution into the crystal along with the solvated substance. As a result, structural domains that dominate dynamic equilibria in the liquid phase may persist also, at least to some extent, in the solid inclusion system.

2.2 Solvates — The Simplest Example of Lattice Inclusion

At one extreme of the inclusion compounds variety are complexes, usually classified as solvates, in which molecules of the solvent are enclosed within intermolecular voids of the crystal to fill empty space or to complete a coordination sphere around a functional moiety. Quite often, it is difficult to control such a co-crystallization

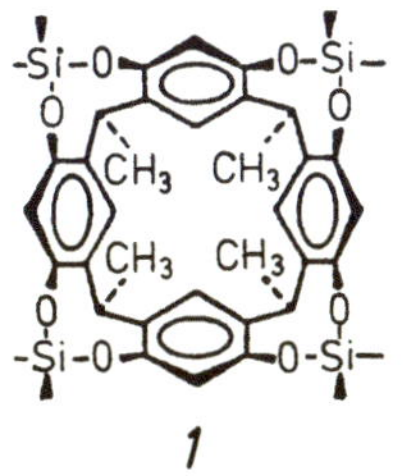

1

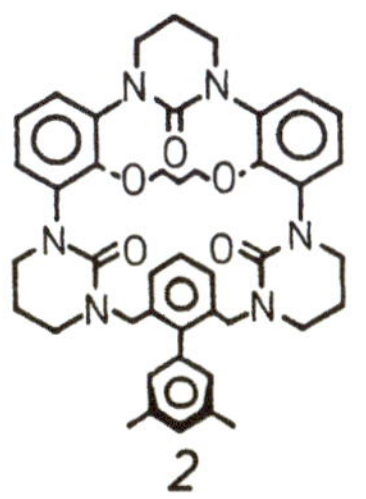

2

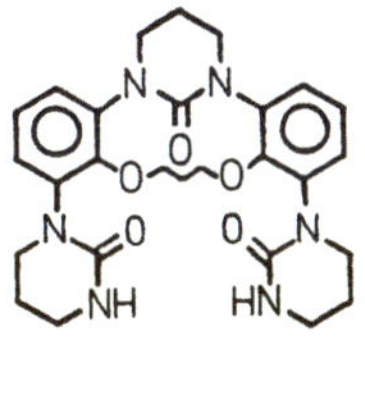

3

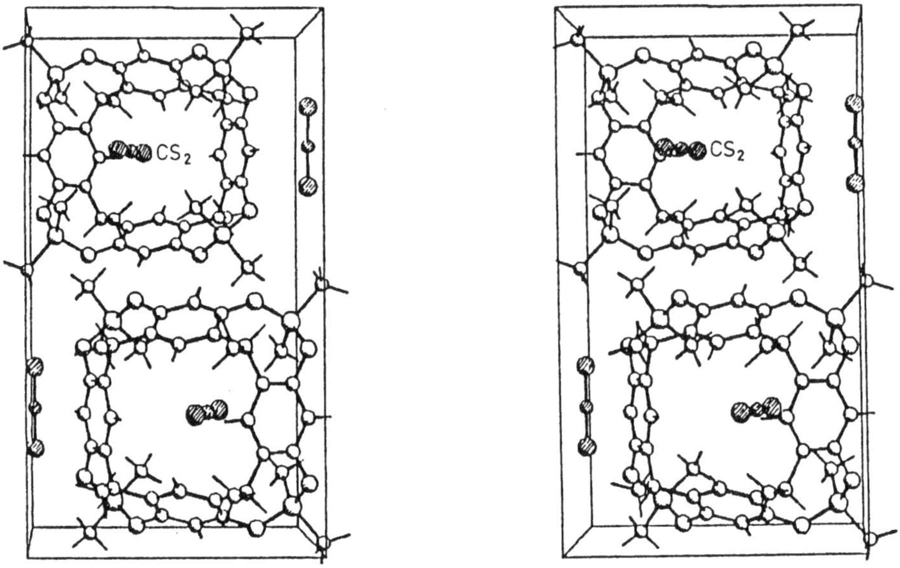

Fig 1. Stereographic projection of the crystal structure of the 2:1 inclusion compound between CS_2 and cavitand *1*. One CS_2 ("guest") molecule is encapsulated within the host cavity, the second CS_2 ("solvent") being located between the complexed entities (taken from Ref. [27])

process and to predict the stoichiometric ratio between host and solvent. This aspect of the solvate formation is illustrated by several interesting examples, the first two of which combine molecular type and lattice type enclathration in the same structure.

Figure 1 shows the crystal structure of the 5,10:12,17:19,24:26,3-tetrakis (dimethylsiladioxa)-1,8,15,22-tetramethyl[1_4]metacyclophane cavitand (*1*) which has an enforced cavity appropriately sized to include only slim linear guests [12b]. This cavitand forms crystals of a 1:1 *molecular* inclusion complex with CS_2, the guest species being almost entirely encapsulated within the host cavity [27]. The crystal structure of the complexed

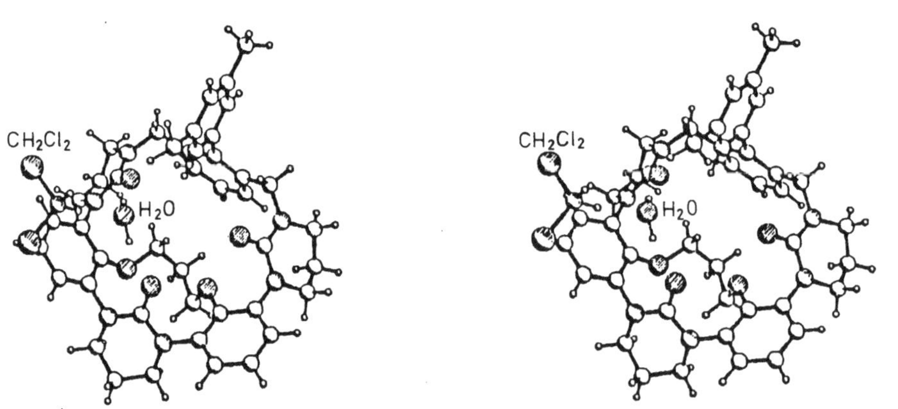

Fig. 2. Stereoview of the inclusion complex between host *2* and water. The methylene chloride moiety from the solvent supplements the coordination sphere around the complexed water [28]

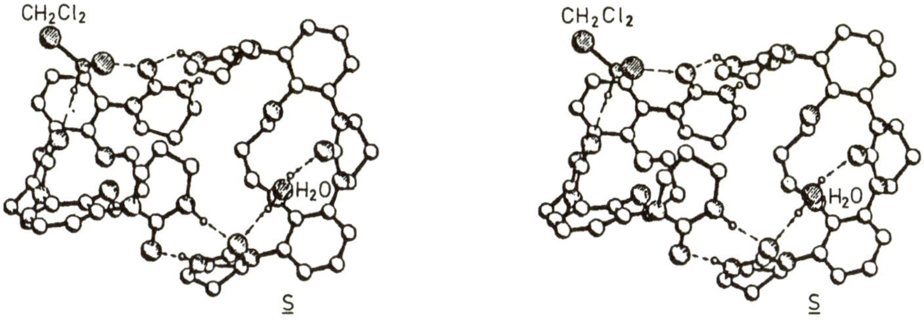

Fig. 3. Stereoview of the inclusion complex between a hydrogen-bonded dimer of host *3* with water and methylene chloride as guests. The crystal structure of this compound contains two additional species, a disordered CH_2Cl_2 and fractional water (donated by "S"). Both lie outside the complex between the bound water and an adjacent host unit (taken from Ref. [28])

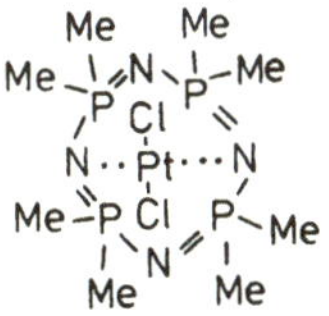

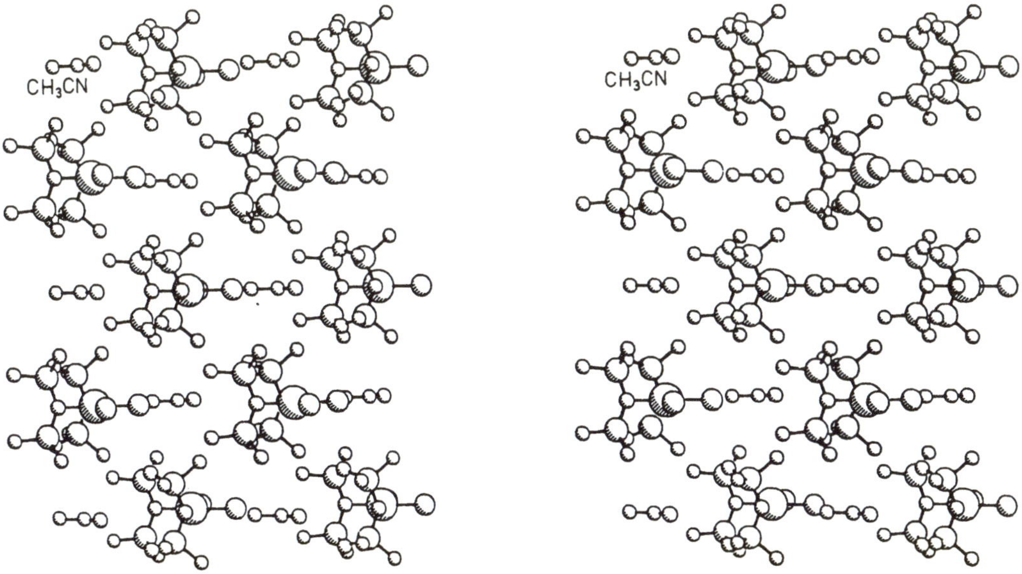

Fig. 4. Stereoview of the intermolecular arrangement in the crystal structure of cis-dichloro-(octa-methyl-cyclotetraphosphazene-N,N'-)Pt(II) · acetonitrile solvate [31a]

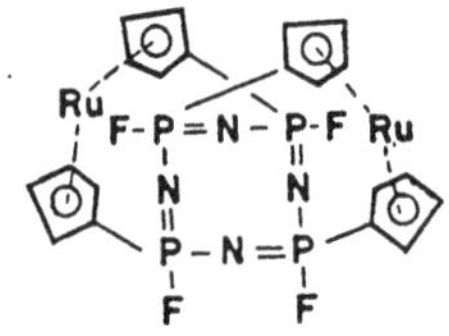

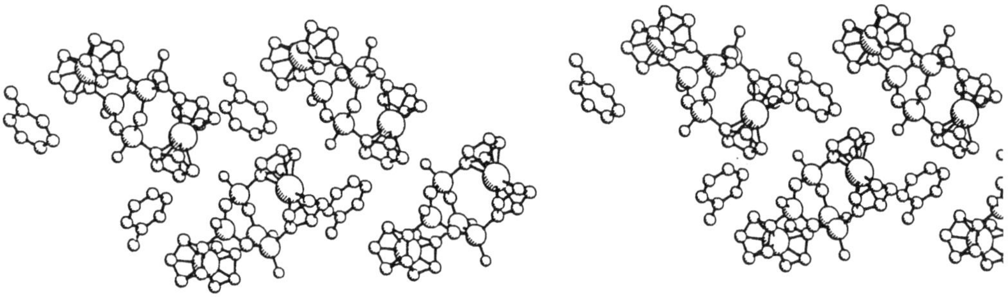

Fig. 5. Stereoview of the intermolecular arrangement in the crystal structure of 2,6:4,8-bis(ruthenocenyl)-2,4,6,8-tetrafluoro-cyclotetraphosphazene · toluene solvate [31h)]

molecule which has a lipophilic exterior is stabilized by weak dispersion forces, and contains voids between neighboring entities. The latter are filled stoichiometrically by a second, solvent-type, molecule of CS_2. The cavitand-encapsulated CS_2 is well ordered in the lattice; the "solvent" CS_2 being less constrained by its crystal environment is, however, slightly disordered even at 128 K.

A similar feature of solvent inclusion has been revealed by crystallographic analyses of complexes involving directional association of water and methylene chloride guests with organic hosts containing only cyclic urea binding sites (*2,3*) [28)]. In these structures (Figs. 2 and 3), the guest species are located on and strongly bound to the surface of the corresponding host containing appropriately sized polar cavities. Additional molecules of the solvent (H_2O, THF or CH_2Cl_2) are located however between adjacent units of the complex, supplementing the coordination sphere of the host-bound water. An attempt to change the packing pattern in one of the structures by several recrystallization experiments from different solvents was unsuccessful.

An enormous variety of solvates associated with many different kinds of compounds is reported in the literature. In most cases this aspect of the structure deserved little attention as it had no effect on other properties of the compound under investigation. Suitable examples include a dihydrate of a diphosphabicyclo[3.3.1]nonane derivative [29)], benzene and chloroform solvates of crown ether complexes with alkylammonium ions [30, 54)], and acetonitrile (Fig. 4) and toluene (Fig. 5) solvates of organometallic derivatives of cyclotetraphosphazene [31)]. In most of these structures the solvent entities are rather loosely held in the lattice (as is reflected in relatively high thermal parameters of the corresponding atoms), and are classified as "solvent of crystallization" or a "space filler" [31a)]. However, if the geometric definition set at the outset is used to describe clathrates as crystalline solids in which "guest" molecules occupy *cavities*, *channels* or *tunnels* in the host lattice [32], many of the solvates can also be considered as clathrate type structures [33].

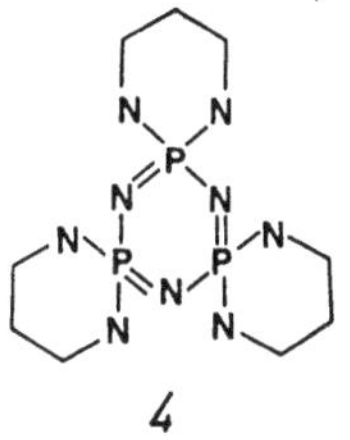

4

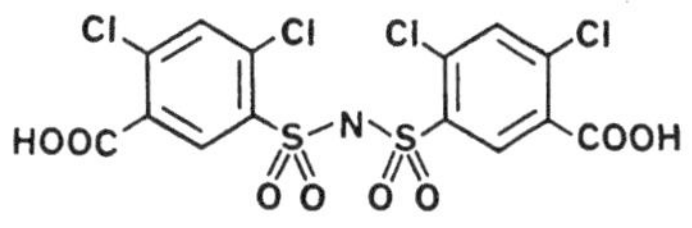

5

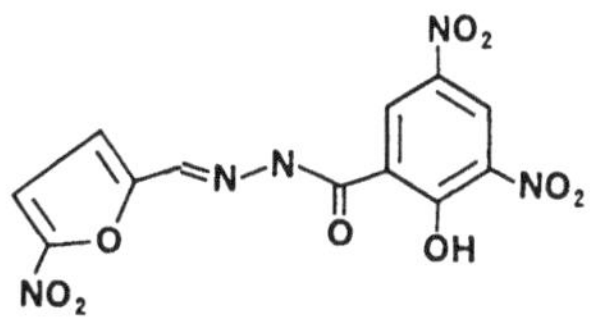

6

In selected cases, the effect of solvation on the crystalline structure formed is, however, considerably more pronounced. For example, the observed packing in the crystal of 2,4,6-tris(1,3-propylenediamine-N,N′-)cyclotriphosphazene (*4*) dihydrate (Fig. 6) is due to strong intermolecular hydrogen bonds between molecules of water and suitable couples of N–H groups on the host moiety [34]. The H_2O species form also continuous H-bonded layers of solvation around the cyclophosphazene derivatives, thus stabilizing the crystal lattice.

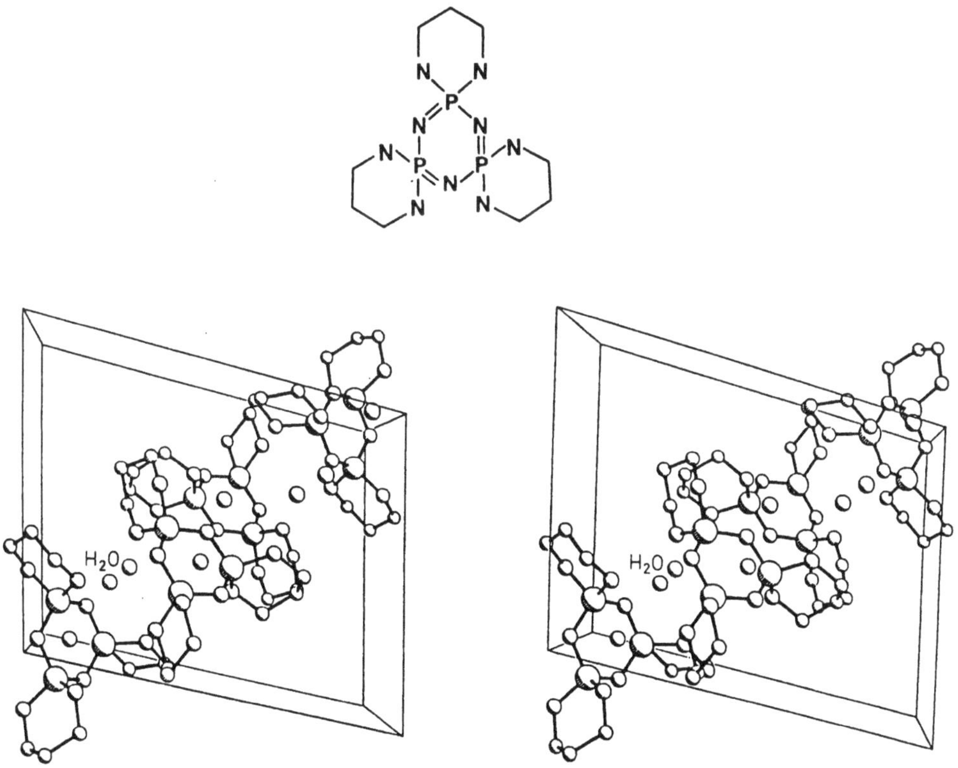

Fig. 6. Stereoscopic illustration of the *4* · 2 H_2O clathrate, showing the water solvated cyclophosphazene [34]

Furthermore, it is often possible to extract from the structural analysis of solid solvates a significant information on solvation patterns and their relation to induced structural polymorphism. An interesting illustration has been provided by crystal structure determinations of solvated 2,4-dichloro-5-carboxy-benzsulfonimide (*5*) [35]. This compound contains a large number of polar functions and potential donors and acceptors of hydrogen bonds and appears to have only a few conformational degrees of freedom associated with soft modes of torsional isomerism. It co-crystallizes with a variety of solvents in different structural forms. The observed modes of crystallization and molecular conformation of the host compound were found to be primarily dependent on the nature of the solvent environment. Thus, from protic media such as water and wet acetic acid layered structures were formed which resemble intercalation type compounds.

In the compound with water, continuous layers of water alternate with bilayers of host molecules, defining two distinct regions in the solid (Fig. 7). Within the bilayers, the structure is stabilized mainly by dipolar interactions between the C–Cl groups turning inward. All the oxygen-containing functions of the host point outward on both sides of the bilayer, and are linked efficiently to the adjacent hydration layers.

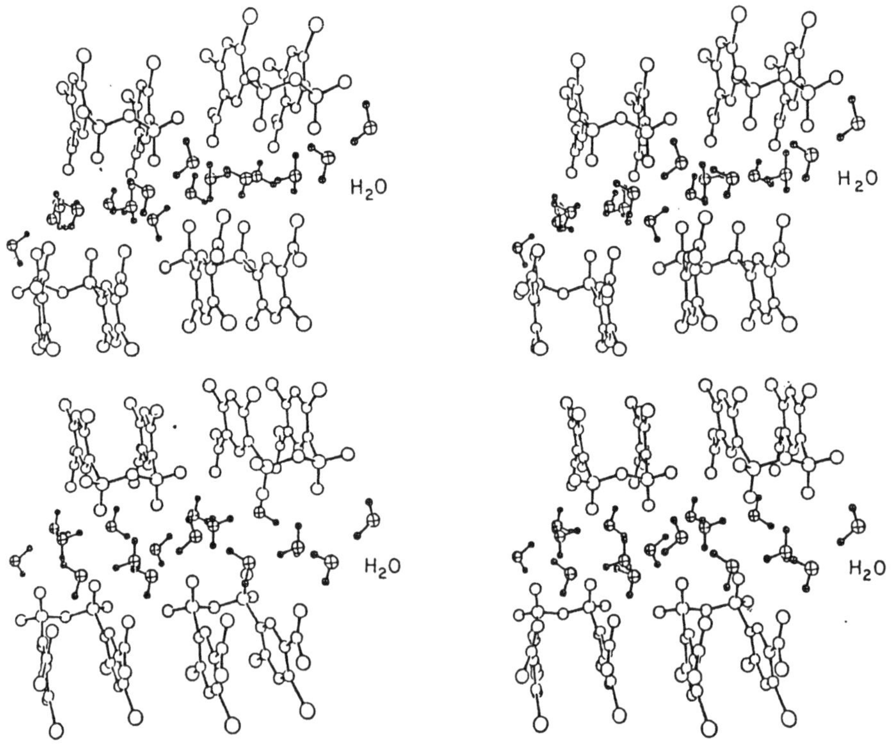

Fig. 7. Stereoview of the crystal structure of water solvated host *5* (folded conformation). The structure is held together by host-host and host-solvent hydrogen bonding interactions. Within the solvation layer there are chains of circular H-bonds between the molecules of water (crystal data: $a = 8.227$, $b = 8.964$, $c = 16.945$ Å, $\alpha = 89.64$, $\beta = 97.51$, $\gamma = 114.28°$, space group $P\bar{1}$; taken from Ref. [35])

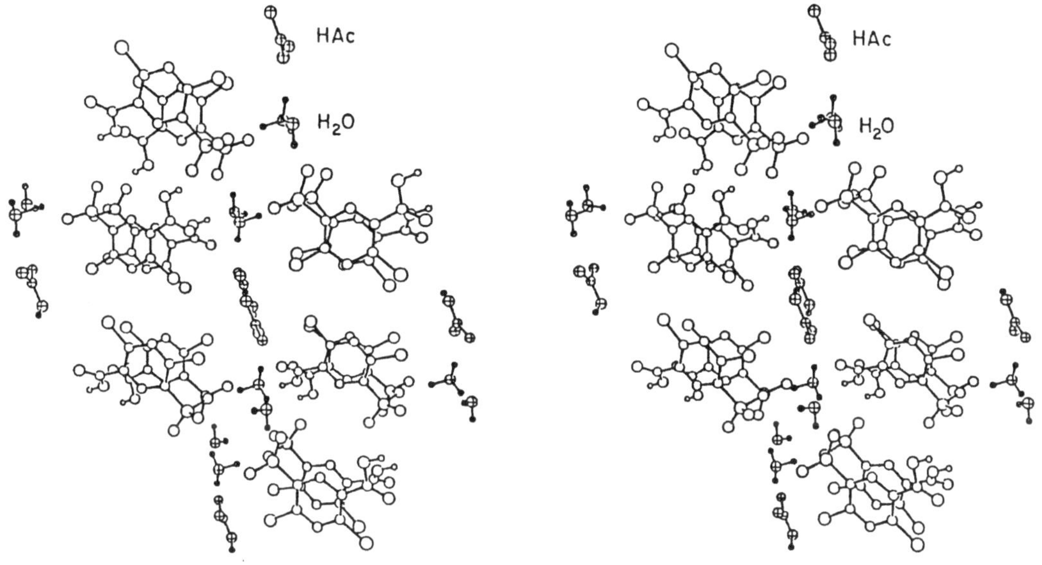

Fig. 8. Stereoscopic illustration of the inclusion compound of host *5* (folded conformation) with acetic acid and 2 mol of water. Host-host and host-water hydrogen bonding interactions stabilize the structure. The solvation layers consist of cyclic carboxy dimers of acetic acid surrounded by water species (crystal data: $a = 7.857, b = 11.379, c = 13.831$ Å, $\alpha = 92.50, \beta = 101.21, \gamma = 101.12°$, space group $P\bar{1}$; taken from Ref. [35])

The water molecules form polymeric chains which include *homodromic* and *antidromic* six-membered hydrogen-bonded circles; each one of the water species is either threefold or fourfold coordinated, donating at least one H-bond to the surface of a neighboring bilayer.

In the crystals containing acetic acid and water the solvent layers are included in between monolayers of the host, as illustrated in Fig. 8. Molecules of the host now interact simultaneously with two neighboring bands of the guest species. The solvent zone itself consists of cyclic dimers of the acetic acid surrounded by water and oxonium ions.

Crystallization of *5* in the open air from an initially aprotic solvent (N,N-dimethylacetamide) led to a non-layered structure which is characterized by a three-dimensional lattice of loosely-packed host species interspaced by channel-type zones accommodating the solvent guest components (Fig. 9).

The appearance of three different structure types was found to be primarily dependent on the nature of the solvent environment. The differences in functionality of the solvating media are clearly reflected in the number of carboxy protons that are exposed by the host to solvent hydrogen bonding; this number increased (from zero to two) as the proton-donating capability of the solvent decreased. Correspondingly, while a folded conformation of the host is induced in solvation environments that are good donors of H-bonds (cf. Figs. 7 and 8), an extended conformation is favored in a solvent which is a poor hydrogen donor (Fig. 9).

Another host, 2-hydroxy-3,5-dinitro-N′-(5-nitrofurfurylidene)benzohydrazide (*6*),

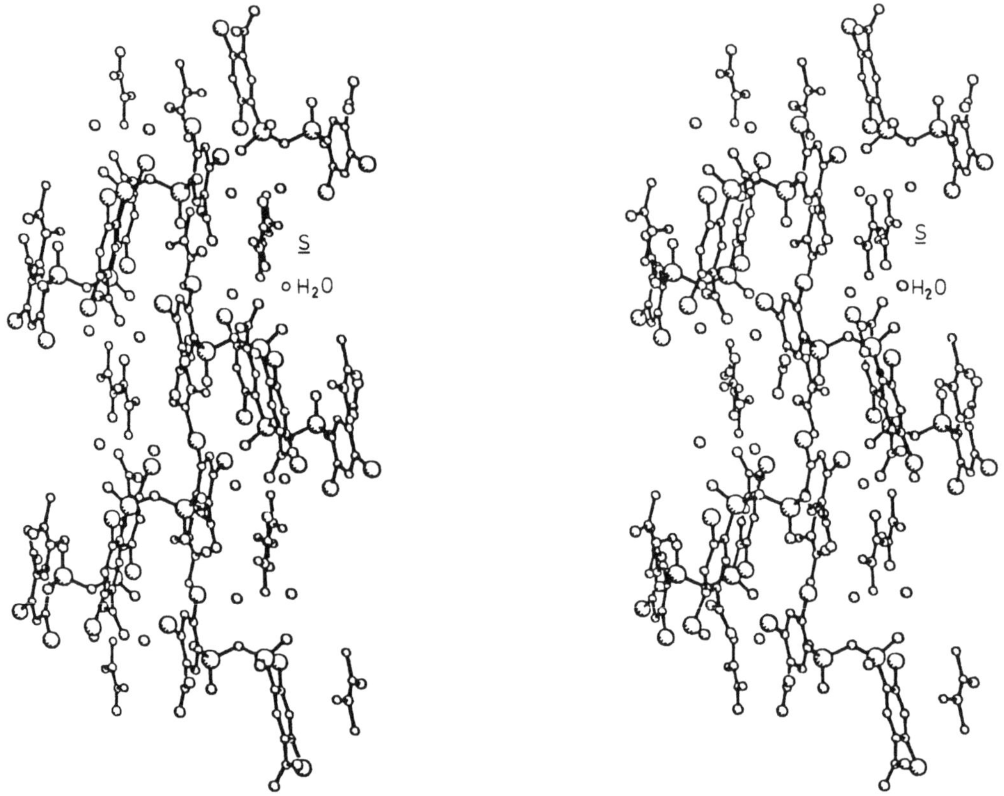

Fig. 9. Stereoview of the inclusion compound between host *5* (extended conformation) and 2 mol of N,N-dimethylacetamide (denoted by "S") + 2 mol of water (isolated circles). In the crystal structure there are hydrogen bonds between the solvent molecules and the —COOH and $>SO_2$ groups of the host (crystal data: $a = 14.838$, $b = 14.818$, $c = 14.500$ Å, $\alpha = \beta = \gamma = 90°$, space group $P2_12_12_1$) [35]

which has a flat shape and contains a large number of polar functions as well as sites available for hydrogen bonding, also reveals a remarkable tendency to form inclusion complexes when crystallized from various solvents. Its inclusion behaviour towards acetonitrile and acetic acid was studied crystallographically (Figs. 10 and 11) [36]. As in the previous structures, the observed structural differences reflect a considerable influence of solvation forces on the host conformation. In both complexes the intermolecular arrangement of the flat host molecules containing highly polar groups was found to be dominated by attractive as well as repulsive dipole-dipole interactions. The hydrogen bonding pattern is different, however, in the crystals obtained from acetonitrile, an aprotic environment, than in those obtained from a protic environment of the acetic acid.

The former structure contains an intramolecular H-bond within the host, which stabilizes its planar conformation, and an N–H ... N link to the guest species. The other structure with acetic acid contains hydrogen-bond stabilized clusters of two hosts and two guests around the crystallographic inversion centers. A distortion of

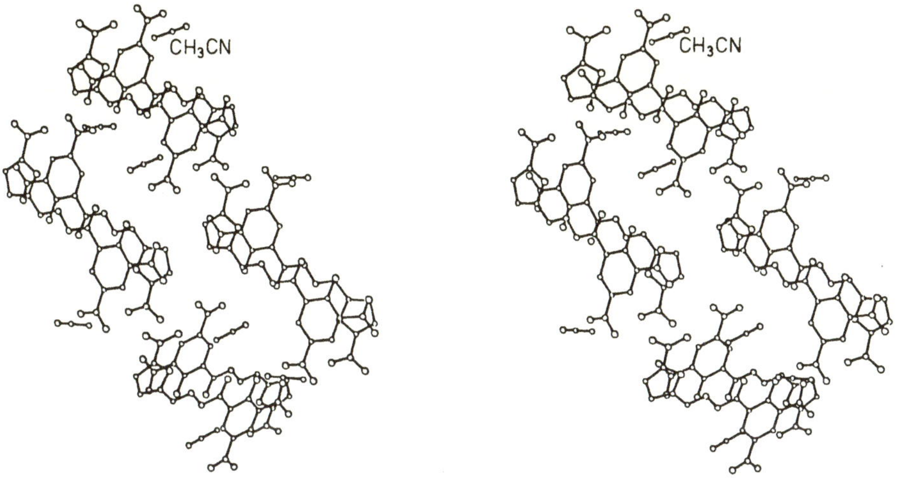

Fig. 10. Stereoview of the inclusion structure formed by host *6* with acetonitrile, showing dipolar pairing of the host molecules. The latter are nearly planar, and contain an intramolecular hydrogen bond. The —NH site of the host is approached by the nitrogen end of acetonitrile, forming an N—H ... N bond (crystal data: $a = 12.977$, $b = 8.021$, $c = 16.830$ Å, $\beta = 106.95°$, space group $P2_1/c$) [36]

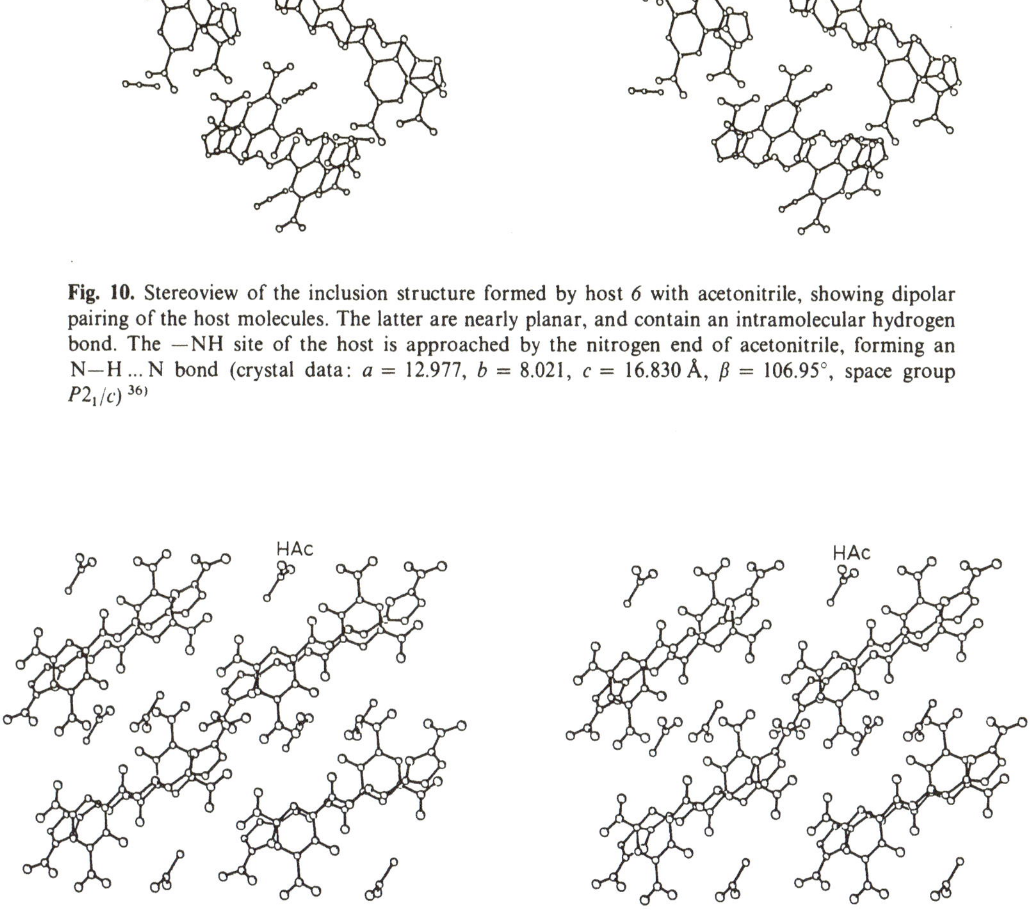

Fig. 11. Stereoview of the 1:1 complex of the dipolar host *6* with acetic acid. Every guest species hydrogen bonds simultaneously to two centrosymmetrically related and dipolarly interacting host entities (crystal data: $a = 8.665$, $b = 8.696$, $c = 12.077$ Å, $\alpha = 90.82$, $\beta = 99.82$, $\gamma = 105.15°$, space group $P\bar{1}$) [36]

the intramolecular H-bonds in favor of intermolecular ones allows significant deviations from planarity of the host by decreasing the torsion angles about bonds which bridge between the substituted benzene and furan rings.

2.3 Channels and Cages in Organic Lattices

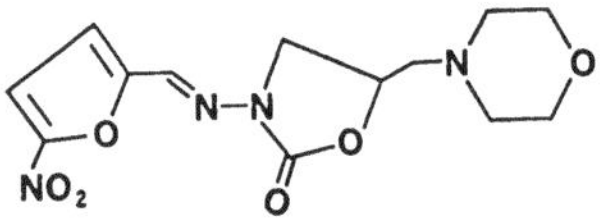

7

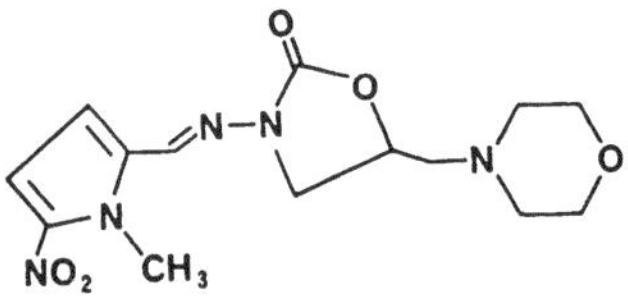

8

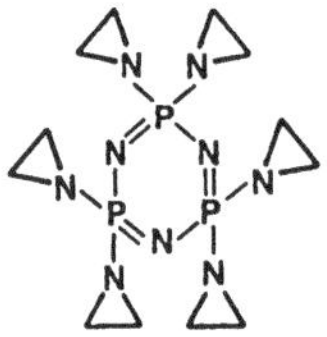

9

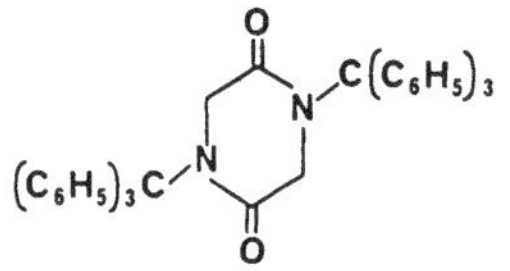

10

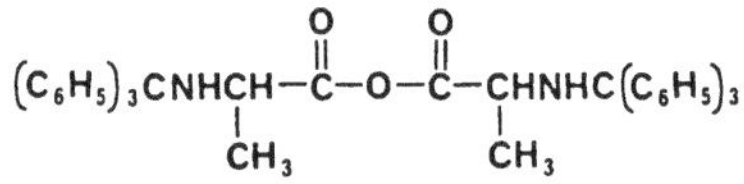

11

Perhaps the most typical examples of solid clathrates involve channel-type inclusion compounds [3,4]. For example, the crystalline complexes formed by furaltadone (*7*) hydrochloride represent such clathrates, in which the host lattice is stabilized by dipolar forces and can accommodate various guest molecules in continuous channels running through the crystal (Fig. 12) [37]. The channels have an approximately cylindrical shape, and are lined with oxygen and chlorine nucleophiles. They turned out to be flexible, expanding upon inclusion of increasingly larger guest species in the order water < acetic acid < propionic acid. The enclathrated species were found translationally disordered in their location in the channel, lacking a sufficiently strong interaction with the channel walls. The complex with acetic acid (Fig. 12b) provided a unique example of a solid structure that contains discrete molecules of the acetic acid guest; however, it tends to deteriorate slowly at ambient room temperature due to diffusion of the acetic acid into the surrounding atmosphere. These examples illustrated again the significance of specific interactions other than hydrogen bonding in the formation of clathrate inclusion compounds.

Nevertheless, as in many previous observations, the clathrate formation by dipolar host compounds could not have been predicted in advance. In fact, there are no channels in the crystal structures of hydrated moxnidazole hydrochloride (*8* · HCl, a closely related species to furaltadone hydrochloride) and of hydrated furaltadone base (Fig. 13) [37]. Rather, the latter two structures are best described as solvates with the H_2O molecules contained in local voids between adjacent moieties of the host.

The difficulties encountered in predicting clathrate formation are further illustrated ▶

Fig. 12a and b. Stereoviews of the channel inclusion structures formed by furaltadone (*7*) hydrochloride: **a** with 2 mol of water (channels are horizontal; crystal data: a = 10.016, b = 16.742, c = 11.241 Å, β = 91.68°, space group $P2_1/c$); **b** with acetic acid (channels are vertical; crystal data: a = 10,212, b = 17.440, c = 11.181 Å, β = 93.06°, space group $P2_1/c$) [37]

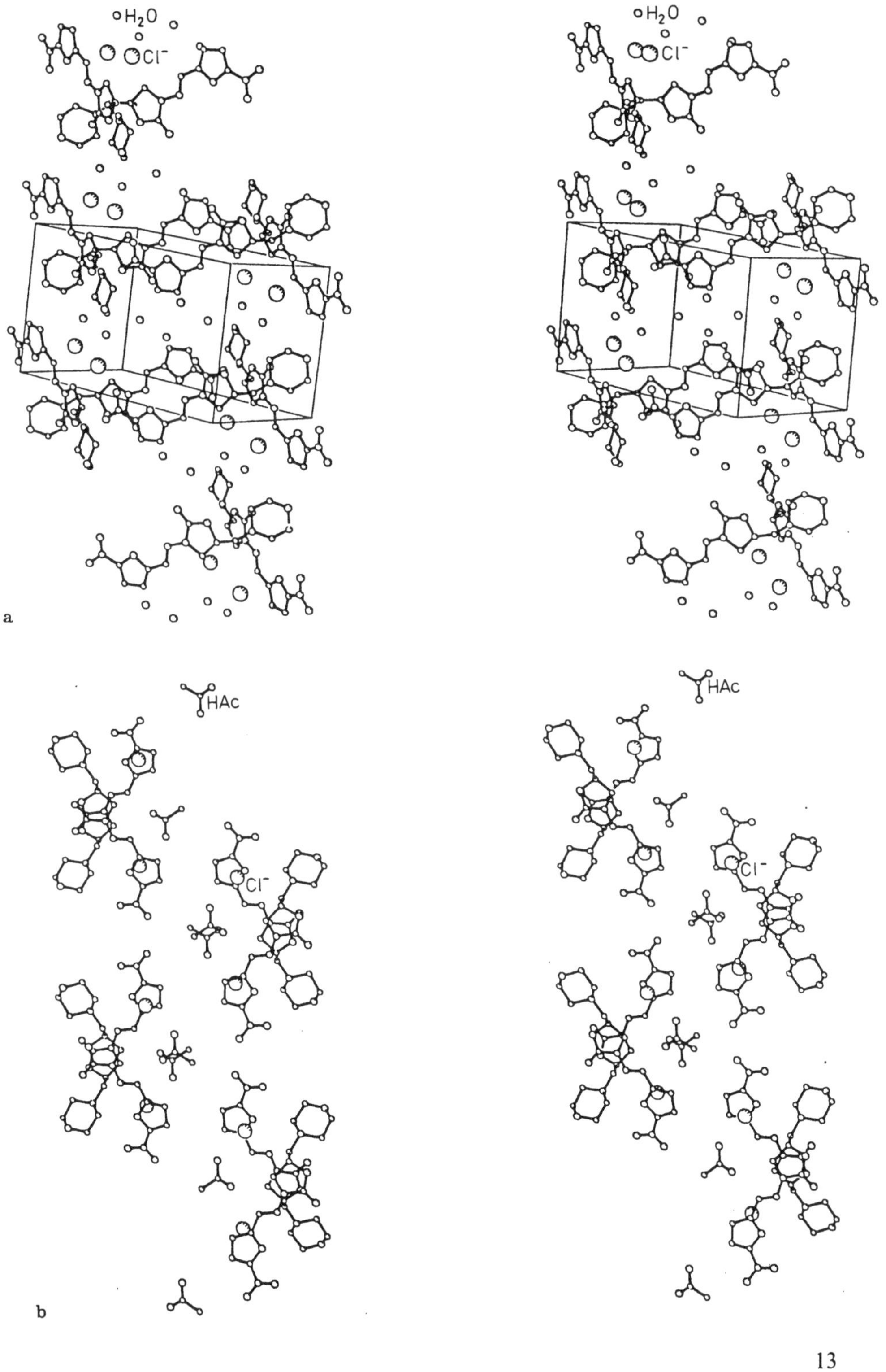
H₂O
Cl⁻
H₂O
Cl⁻
a
HAc
Cl⁻
HAc
Cl⁻
b

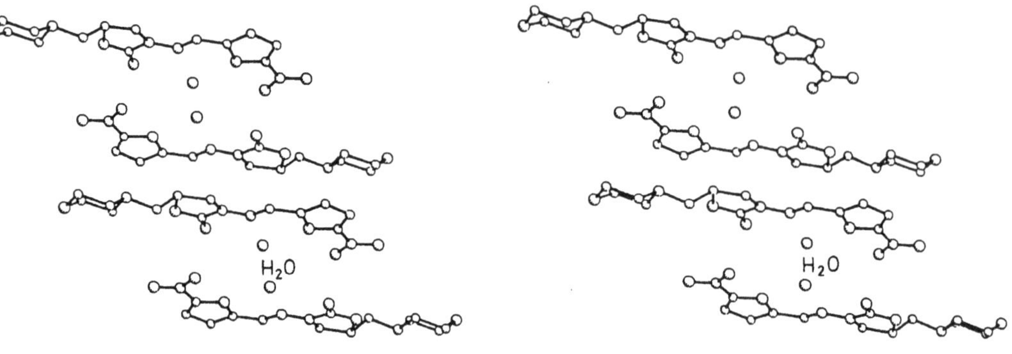

a

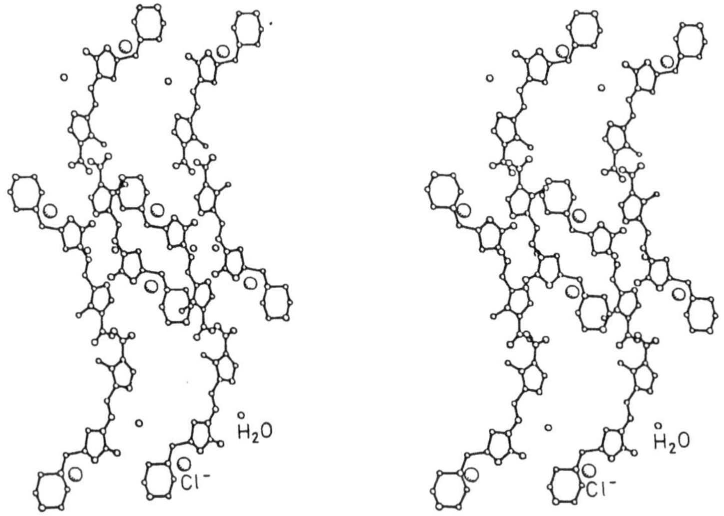

b

Fig. 13a and b. Water inclusion in "local" intermolecular voids in the crystal structures of **a** the *7* · H_2O clathrate (crystal data: a = 6.958, b = 8.100, c = 14.823 Å, α = 96.60, β = 85.24, γ = 104.69°, space group $P\bar{1}$) and **b** the *8* · HCl · H_2O clathrate (crystal data: a = 9.779, b = 26.142, c = 6.977 Å, β = 97.74°, space group $P2_1/c$) [37]

by the inclusion behaviour of 2,2,4,4,6,6-hexakis(1-aziridinyl)-cyclotriphosphazene (*9*), a well known antitumor agent. This compound is extremely soluble in water and no crystals could be obtained from it. Crystallization experiments from *m*-xylene led to an orthorhombic crystal form of unsolvated species, while crystallization from CS_2 led to an unsolvated monoclinic phase. On the other hand, a very stable channel type clathrate of the cyclophosphazene derivative was obtained upon crystallization from benzene (Fig. 14) [38]. The aziridyl substituent has a prominent lone-pair orbital which occupies space within the lattice, and it has been assumed that an efficient packing of these electron pairs (in competition with host-solvent interactions) must be a factor in the crystallization of this compound.

Another way to satisfy this requirement is exhibited in the formation of stable charge transfer complexes with several other solvents [39]. Of a particular interest is

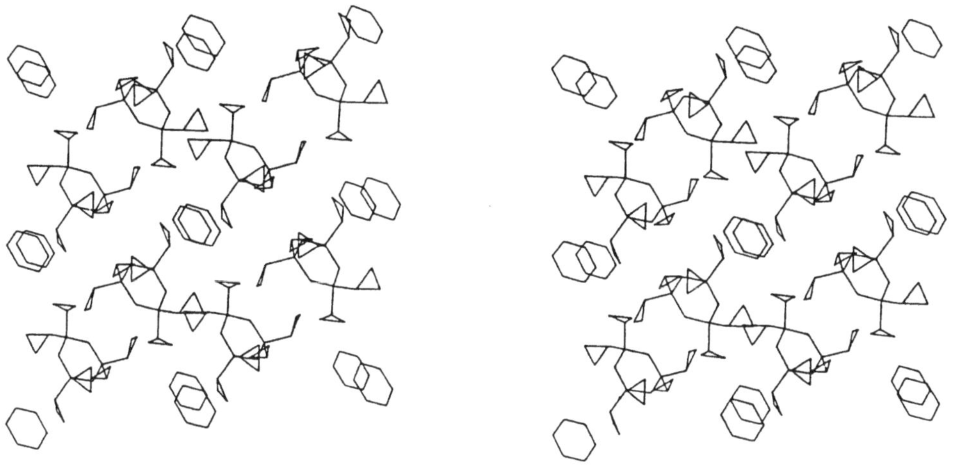

Fig. 14. Stereoview of the 1:1 *9*·benzene clathrate, showing a channel-type arrangement of the benzene guest species [38]

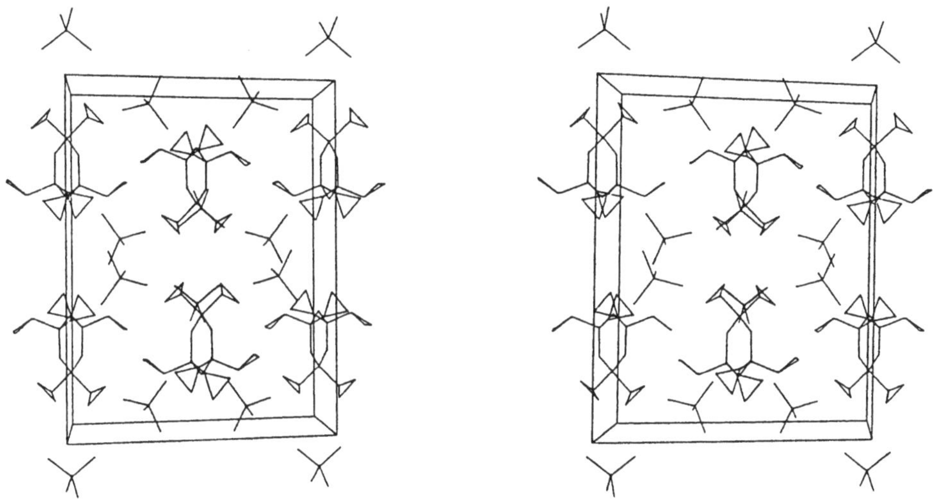

Fig. 15. Stereoview of the 1:3 *9* · CCl_4 "anticlathrate". Each molecule of *9* is surrounded by 13 molecules of CCl_4 [39]

the structure of the complex with carbon tetrachloride (Fig. 15). This structure has been termed an "anticlathrate" since the host lattice in it is essentially built up from the solvent CCl_4 entities, resembling a "bee's nest net". Each molecule of the cyclophosphazene derivative is inserted in a sphere (or rather a monocapped icosahedron) of 13 CCl_4 molecules, and is directly connected to six of them by a halogen bridge.

Different modes of crystallization were observed also for the N,N′-ditrityldiketopiperazine compound (*10*), the molecular surface of which is not oval but has thick ends and a thin center [40]. It was found to crystallize in an unsolvated form from

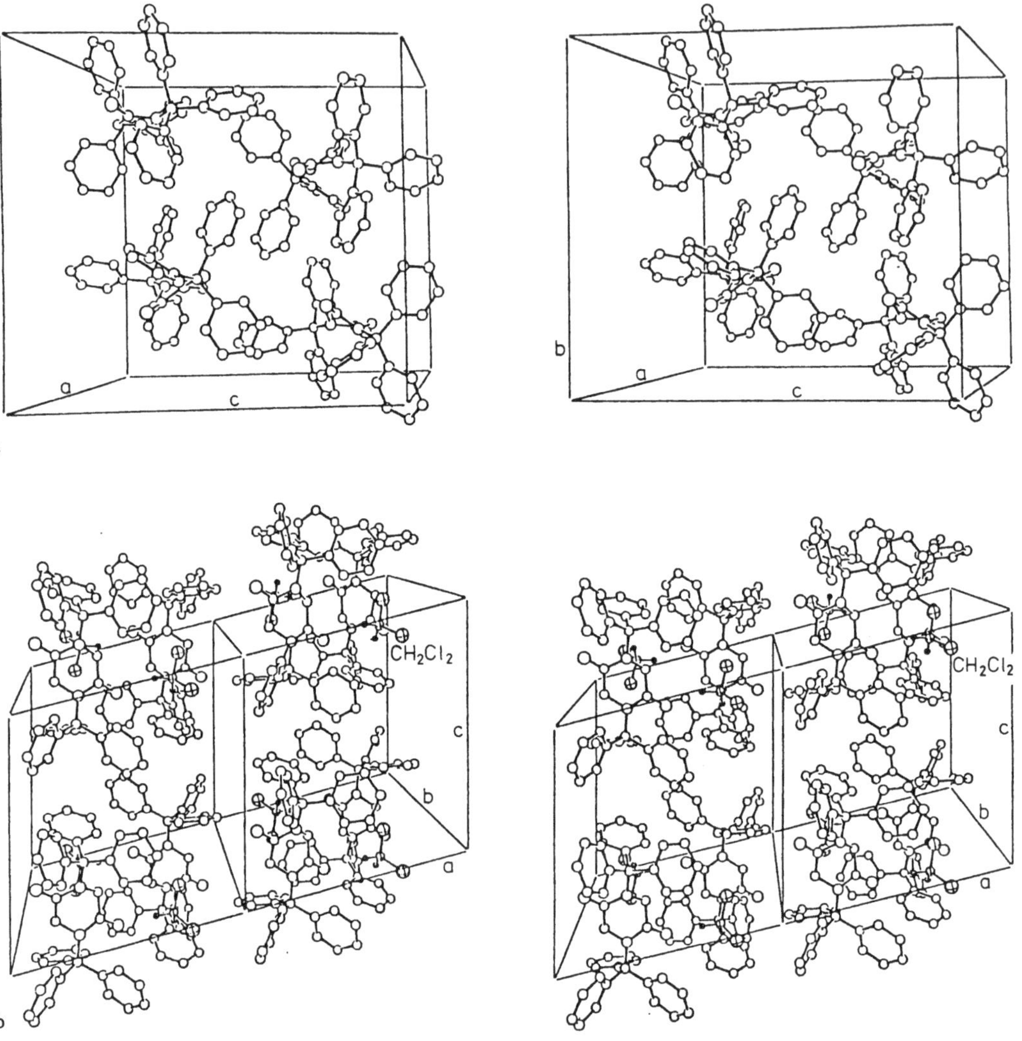

Fig. 16a and b. Stereoviews of **a** the crystal structure of the free host *10* (crystal data: $a = 11.737$, $b = 16.199$, $c = 17.710$ Å, $\beta = 108.37°$, space group $P2_1/c$ and **b** its crystalline 1:1 inclusion complex with methylene chloride (crystal data: $a = 12.193$, $b = 12.378$, $c = 12.670$ Å, $\alpha = 106.74$, $\beta = 104.01$, $\gamma = 90.38°$, space group $P\bar{1}$) (taken from Ref. [40])

ethylacetate and hexane, and also as inclusion complexes with certain small molecules such as methylene chloride. In the crystalline phase the empty space on the molecular surface (between the bulky ditrityl substituents) of this host can be occupied in two different manners; either by suitable fragments of an adjacent molecule, or by other appropriately sized species from the solvent. Indeed, in the structure of the free compound (Fig. 16a)[40], the molecules interpenetrate each other with the trityl groups of one moiety lying above the central ring of an adjacent molecule. In the inclusion complex (Fig. 16b)[40], the host molecules are arranged in layers (aligned with their

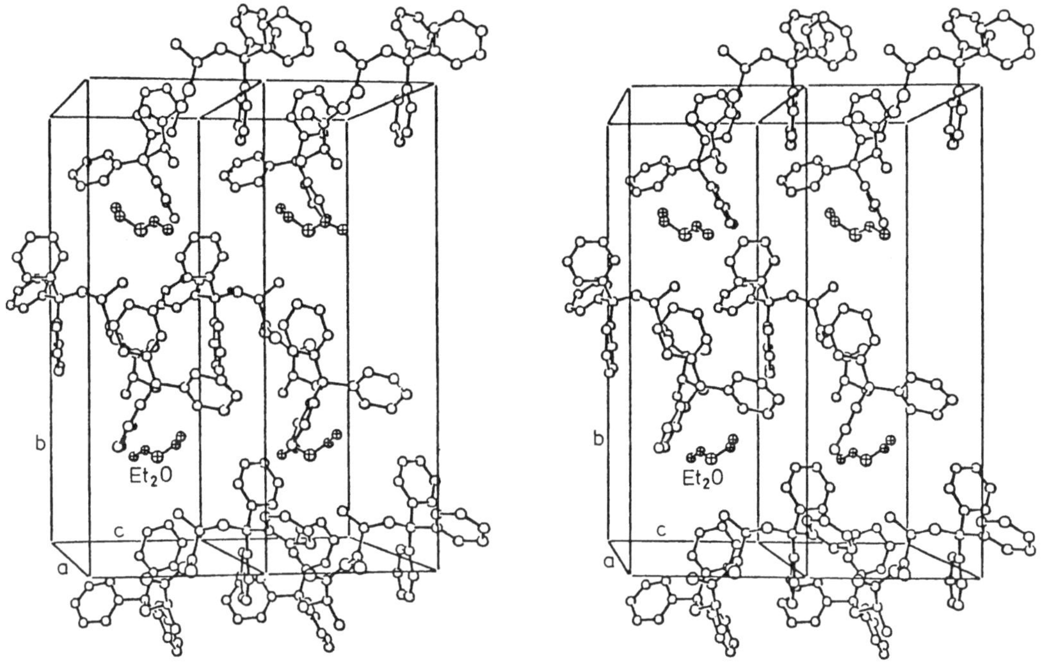

Fig. 17. Stereoview of the inclusion structure formed by host *11* with diethyl ether (crystal data: a = 9.092, b = 24.809, c = 9.135 Å, β = 102.60°, space group $P2_1$) (taken from Ref. [41])

N ... N axes parallel to each other) with the trityl groups in contact and the space between the diketopiperazine rings filled by methylene chloride. In the resulting structure every guest moiety is within pseudo-cage-type voids surrounded by four adjacent hosts, representing lattice inclusion which is not assisted by any specific coordination between host and guest.

The significant role of the triphenylmethyl group as a "spacer" which tends to prevent close packing of the host molecules has also been exhibited by the inclusion behaviour of N-tritylalanine anhydride (*11*) towards neutral guest components [41]. The crystal structure of the 1:1 adduct with diethyl ether is shown in Fig. 17. It can be described as consisting of layers of host molecules interspaced by pseudo-channels occupied by the guest species. These results indicate that the trityl groups, even when attached to a relatively flexible aliphatic fragment of atoms, can act as spacers to prevent close packing of the molecules which then become suitable hosts for crystalline inclusion of various guests (see Sects. 3.2 and 4.1).

Notwithstanding the variety of structural patterns that characterize the crystalline inclusions referred to above, it has recently been shown that the clathrate formation can be induced in a more systematic manner. The use of the clathration phenomenon to store selected guest molecules or to separate one type of guest species from another is the subject of the following sections in this account.

3 Molecular Design of Hosts for the Formation of Inclusion Compounds with Apolar Guests

3.1 Background — Clathrate Inclusion Compounds of Spirotriphosphazenes

Aromatic derivatives of cyclotriphosphazenes, rigid six-membered ring systems built on a framework of alternating P and N atoms, provide one of the more beautiful early examples of hosts that form channel-type clathrates and can be useful for molecular separations [32,42]. Although these clathrate systems were discovered by accident, the conclusions that emerged from their investigation have been extremely helpful for the molecular design of other potential host molecules.

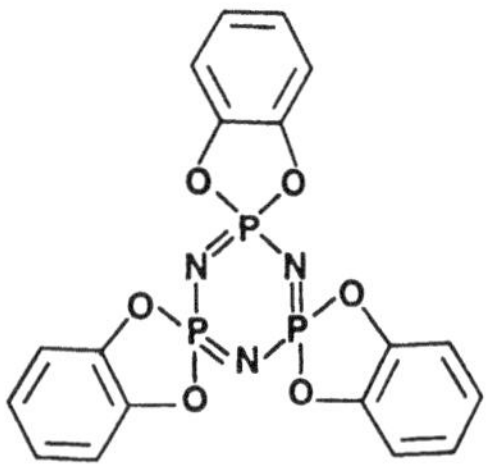

12

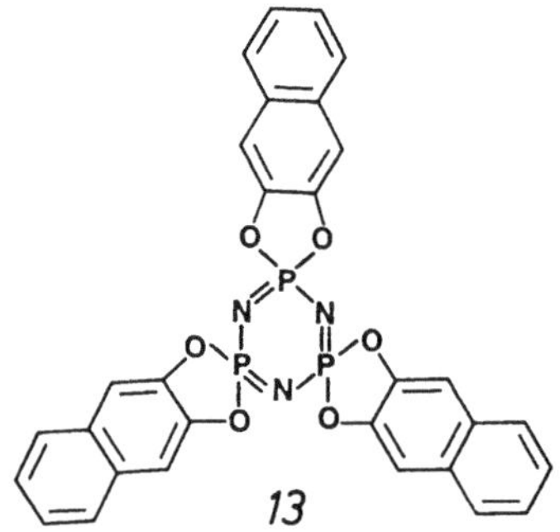

13

The range of inclusion adducts formed by the organophosphazenes is very broad, the guest species varying from aliphatic and aromatic hydrocarbons to ethers, ketones and alcohols [42]. Some of the hosts [e.g., tris(*o*-phenylenedioxy)cyclotriphosphazene (*12*)] form clathrates not only when recrystallized from organic solvents but also

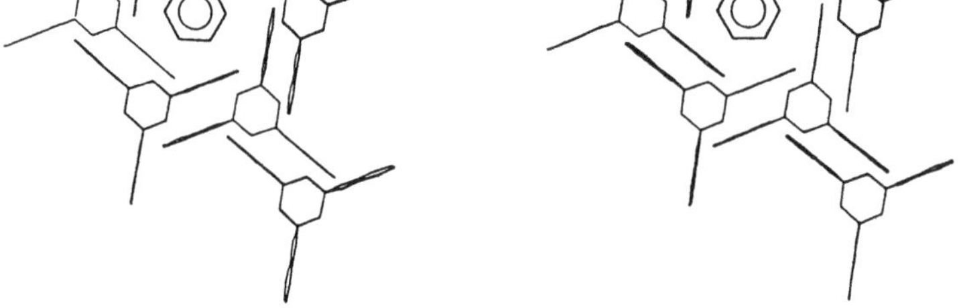

Fig. 18. Stereoview of the channel-type clathrate structures formed by host *12* with benzene (shown schematically) or *o*-xylene

when the pure host is brought in direct contact with the liquid or even the vapor phase of the guest component. Furthermore, a selective absorption into the host lattice of one component from a mixture of potential guests has been observed.

The crystallographic investigation of a representative complex of host *12* with benzene or *o*-xylene revealed a channel-type structure [42] in which the guest species are retained in tunnels that are continuous through the crystals, the cross-sectional diameter of the latter varying from 5.0 Å (at the narrowest point) to 7.0 Å (Fig. 18). As indicated by X-ray diffraction and broad-line ^{1}H NMR spectroscopy [42], the guest species are translationally as well as orientationally disordered within these channels even at −60 °C. The apparent possibility to displace one guest component by another (provided in excess) is consistent with the observed motion of the guests in the lattice.

Similar phenomena characterize the clathration behavior of the larger host tris(2,3-naphthalenedioxy)cyclotriphosphazene (*13*) [42]. Due to the bulkier backbone, this host forms, in relation to the former structure type, a markedly expanded (with channels of diameter 9–10 Å at the narrowest point) and thus less stable structure. Evidently, the driving force for clathration of the substituted cyclotriphosphazene species is associated with their unusual "paddle-wheel" like shape. Van der Waals packing of these molecules in a high symmetry arrangement leaves substantial voids in the lattice which are filled by guest components to lower the enthalpy of the total system. The potential use of this clathration behaviour for separation of hydrocarbons was envisioned already about fifteen years ago [32].

3.2 "Wheel-and-Axle" Type of Host Molecules in Clathrate Inclusion

The basic principle of steric fit has recently been applied to the molecular design of other host systems (see Ch. 1, Sect. 3 of Vol. 140). For example, compounds with a long molecular axis ("axle") and with large and relatively rigid groups at each end ("wheels") are also expected to function as hosts since the large groups acting as spacers prevent the hosts from packing efficiently in the solid. Indeed, well-defined clathrate type complexes were formed between a series of chemically different but similarly shaped hosts and a variety of aromatic guests [21]. Examples of such hosts (all characterized by thick ends and a thin center) include *14*–*19*.

$R{-}(C{\equiv}C)_n{-}R$ n = 2,3 *14*

$R{-}(CH_2)_n{-}R$ n = 4,6 *15*

$R{-}CH_2{-}CH{=}CH{-}CH_2{-}R$ *16*

$R{-}CH{=}N{-}N{=}CH{-}R$ *17*

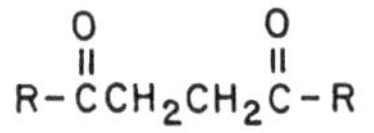

18

R'–CH=N–N=CH–R'

19

R = Ph_3C

$(4\text{-}MeO\text{-}C_6H_4)_3C$

$(4\text{-}Ph\text{-}C_6H_4)Ph_2C$

R' =

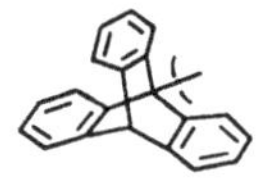

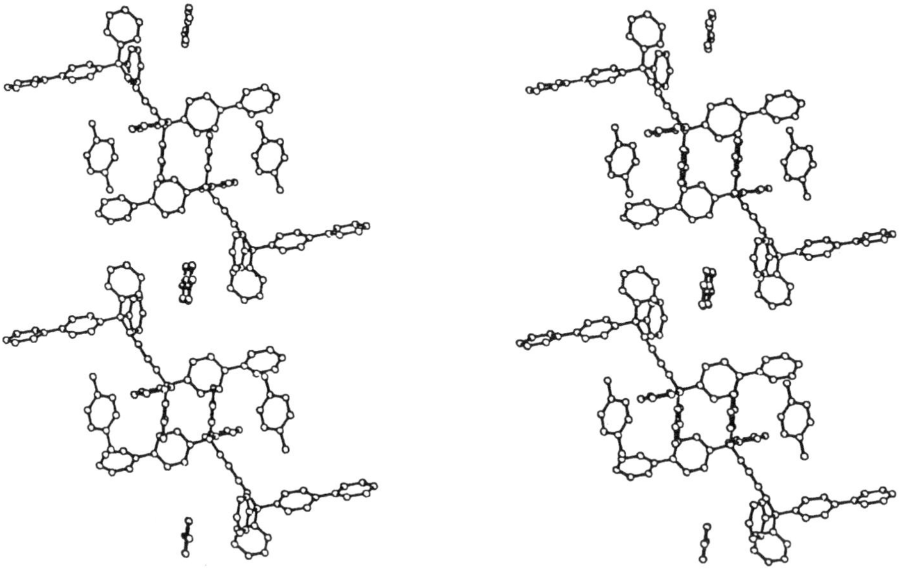

Fig. 19. Stereoview of the 1:1 inclusion compound *14* [R = $(4\text{-Ph-}C_6H_4)Ph_2C$, n = 2] with *p*-xylene [21)]

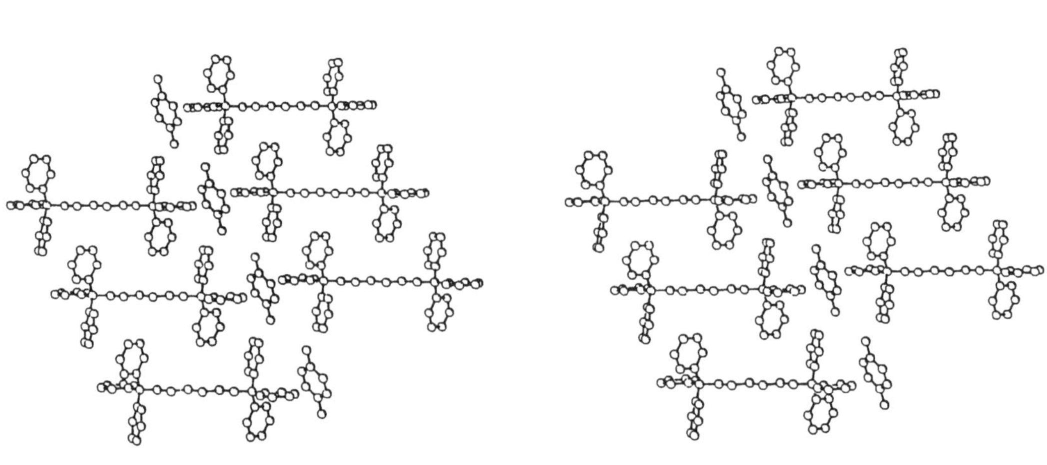

Fig. 20. Stereoview of the 1:1 inclusion compound of *15* (R = Ph_3C, n = 6) with *p*-xylene [21)]

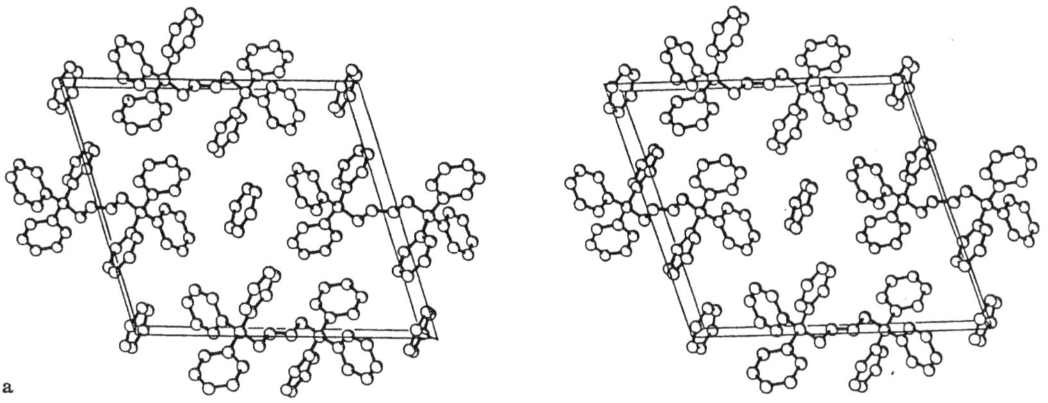

a

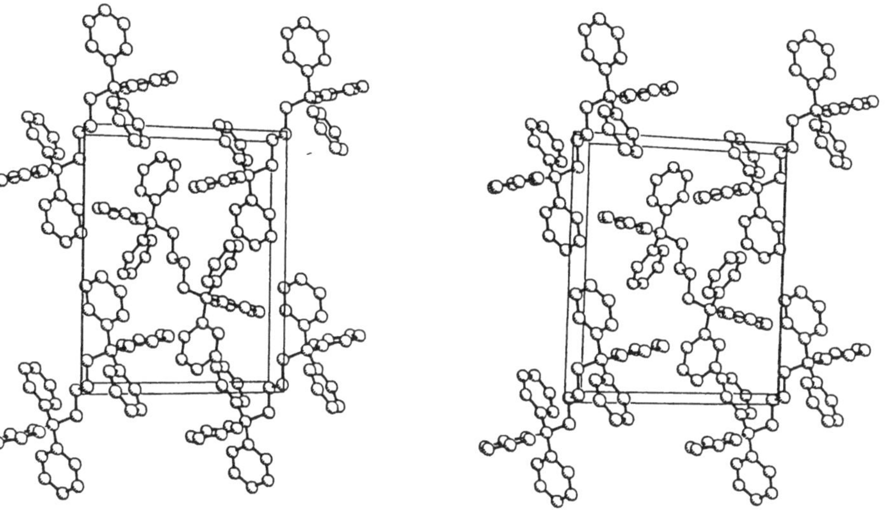

b

Fig. 21 a and b. Stereoviews of **a** the 1:1 *16* (R = Ph_3C) toluene clathrate (the disordered guest methyl is not shown), and **b** the crystal structure of the uncomplexed host (it represents one of a very few examples of a guest-free lattice which could be obtained from this type of compounds) [21]

The detailed structures of several clathrates have been characterized, and a certain degree of selectivity on complexation with different isomers has been detected [21]. Most of these complexes are of the channel type, but some of them have structures which simultaneously qualify for channel and cage type descriptors; representative examples are illustrated in Figs. 19–21. The crystal data of the complexes are summarized in Table 1.

Table 1. Crystal data for structures of inclusion compounds with wheel-and-axle type hosts containing large aromatic end groups [21]

Compound:	(I)	(II)	(III)	(IV)	(V)	(VI)
a (Å)	12.978(3)	16.381(3)	16.685(9)	8.972(4)	14.786(6)	12.250(6)
b	18.799(5)	16.381	7.915(4)	13.754(5)	7.525(4)	14.148(8)
c	9.560(2)	16.381	14.772(7)	8.646(4)	17.063(7)	9.400(4)
α (deg)	102.06(2)	90.0	90.0	103.75(3)	90.0	90.0
β	91.69(2)	90.0	110.98(3)	100.80(3)	107.98(2)	110.67(3)
γ	89.56(2)	90.0	90.0	103.86(2)	90.0	90.0
Z	2	4	2	1	2	2
D_c gcm^{-3}	1.16	1.26	1.16	1.16	1.16	1.18
Space group	$P\bar{1}$	$Pa3$	$P2_1/n$	$P1$	$P2_1/n$	$P2_1/n$

Inclusion compounds:
(I) *14* [R = (4-Ph-C_6H_4)Ph_2C, n = 2] · *p*-xylene (1:1);
(II) *14* [R = (4-MeO-C_6H_4)$_3C$, n = 2] · chloroform (1:1);
(III) *15* (R = Ph_3C, n = 4) · toluene (1:1);
(IV) *15* (R = Ph_3C, n = 6) · *p*-xylene (1:1);
(V) *16* (R = Ph_3C) · toluene (1:1);
(VI) *16* (R = Ph_3C) (uncomplexed host)

3.3 Substituted Allenes as Clathrate Hosts

Further exploitation of the molecular shape feature has led more recently to the design of another series of clathrate hosts by substituting on the allene rigid backbone bulky groups [22]. Representative compounds of this new host family are *20* and *21*. The allene *20* (R = t-butyl) shows an interesting clathration behaviour upon crystallization from various environments, including alicyclic and aromatic compounds, heterocycles, cyclic ketones and cyclohexaneamine [26].

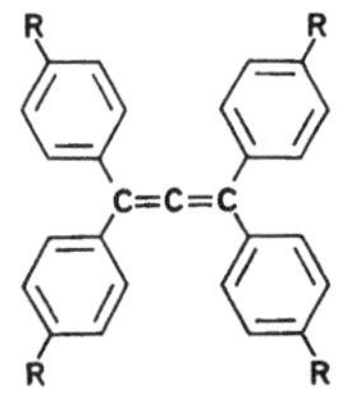

(R=t–Bu)

20

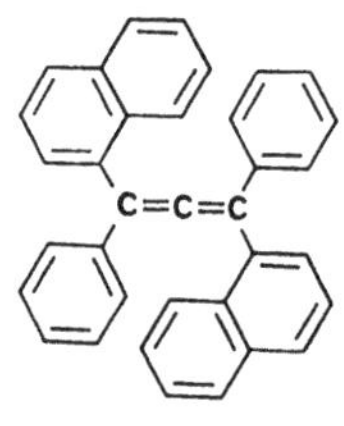

21

The structures of three different inclusion compounds of this host with toluene, cyclohexene and cyclohexane have already been determined by X-ray diffraction (Table 2). They reveal a relatively invariable host lattice with continuous channels that are occupied by the corresponding guest entities (Figs. 22 and 23). These channels have a nearly rectangular shape of van der Waals dimensions 4 × 6.5 Å, well fitting the thickness and the width of the guest ring. In clathrates with cyclohexene and cyclohexane the opening between two neighboring guests along the channels narrows from 6.5 to about 3.4 Å, thus forming a partial enclosure around each guest.

Table 2. Crystal data for the 1:1 inclusion compounds of host *20* [26]

Guests:	Toluene	Cyclohexene	Cyclohexane
a (Å)	12.614(1)	12.395(6)	12.367(9)
b	12.621(1)	12.835(7)	12.719(3)
c	14.317(2)	15.372(3)	15.228(3)
α (deg)	82.12(2)	101.84(4)	101.85(2)
β	74.87(1)	95.95(4)	96.55(4)
γ	72.29(1)	113.57(3)	112.60(3)
Z	2	2	2
D_c gcm^{-3}	1.056	1.014	1.032
Space group	$P\bar{1}$	$P\bar{1}$	$P\bar{1}$

The allene host *20* exhibits high inclusion selectivity in competition experiments using mixtures of solvents [26]. In fact, preliminary results indicate that it can be used for the separation of constitutional isomers, of homologues, and of aliphatic from alicyclic or aromatic compounds (Table 3). Other derivatives of the allene, e.g. *21*, are also functional as host molecules, and are currently being subjected to further investigations [22].

All the inclusion compounds referred to in this section involve uncharged host and guest constituents of low polarity. The guests occupy extramolecular cavities in the lattice, interacting only weakly through dispersion forces with the surrounding molecules of the host. The relative stability of such clathrates depends on the strength of host-host interactions as well as on the degree of steric fit of the guests within the lattice.

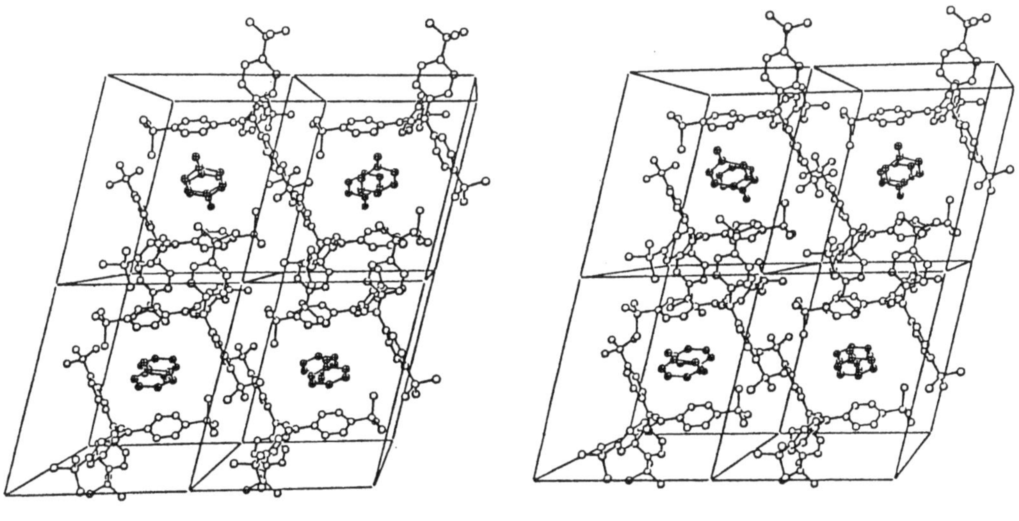

Fig. 22. Stereoview of the 1:1 *20* · toluene clathrate viewed down the channel axis (taken from Ref. [26])

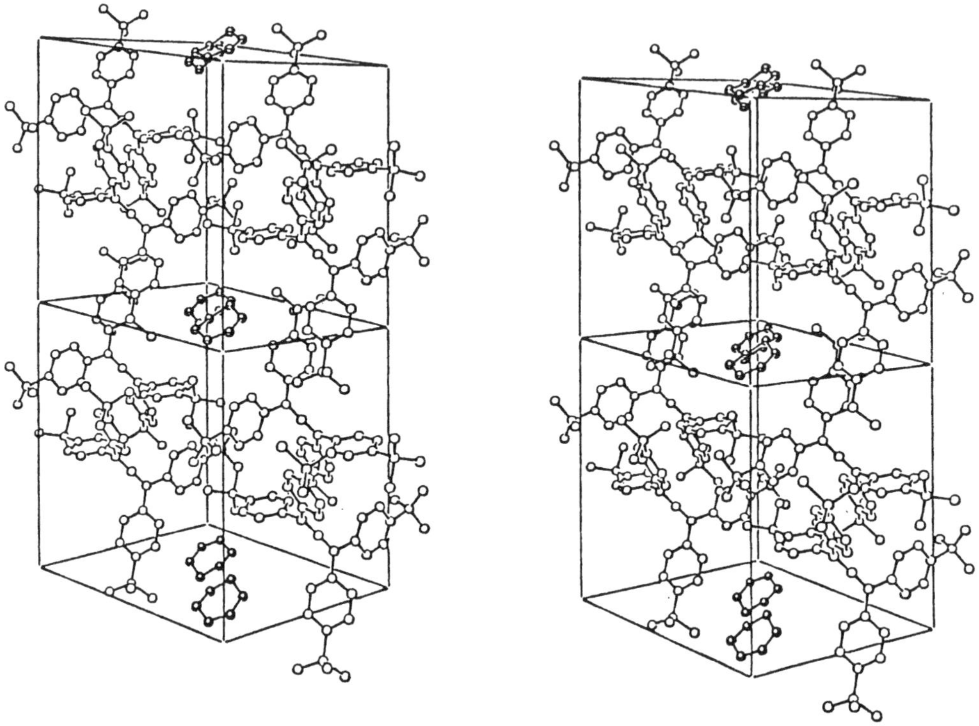

Fig. 23. Stereoview of the 1:1 *20*·cyclohexene clathrate viewed down the channel axis (the inclusion compound of *20* with cyclohexane is isomorphous)[26]

Table 3. Studies of guest binding to host *20* from a two-component solvent mixture[26]

Entry	Equimolar guest mixture (guest I/guest II)	Inclusion compound stoichiometric ratio[a] (host:guest I:guest II)
1	Cyclohexane/methylcyclopentane	1:1:0
2	Benzene/toluene	1:0:1
3	1,3-Dioxolane/1,4-dioxane	1:1:0
4	Cyclohexane/benzene	1:1:0
5	Benzene/pyridine	1:1:0

[a] Determined by NMR integration of the isolated crystals

4 Selective Complexations of Polar Guests by Hosts Containing Functional "Sensors"

On the basis of previous observations it was anticipated that the clathrate formation would be more selective in controlling guest selectivities if a functionality which could form hydrogen-bonds with guest species and add this strengthening feature to that of the molecular shape were included in the host. Several successful attempts which utilized the above concept of coordination assisted clathration with polar guests are detailed below. A similar approach to the problem of lattice inclusion selectivity and guest discrimination has recently been proposed by Weber and coworkers [43,44], and is described in Ch. 2 of this volume (see also Ch. 1, Sect. 3 of Vol. 140).

4.1 Coordinato-Clathrates of N,N′-Ditritylurea

N,N′-Ditritylurea (DTU, *22*) contains bulky end groups as large rigid spacers and an amide moiety as a source for proton-donating and proton-accepting sites. The amide group adds rigidity to the molecular axis, and the hydrogen-bonding feature favors complexation with polar guests. DTU is an excellent host for guests containing varied functionality [23,24,45]. It formed stoichiometric complexes (mostly 1:1) with a variety of amines, amides, alcohols, esters and small molecules with other functions

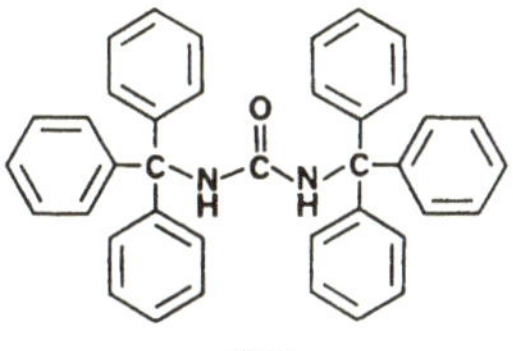

22

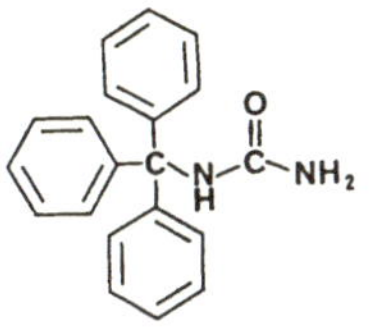

23

Table 4. Crystalline inclusion compounds with host *22* [45]

Guest molecules	
1:1 Complexes:	Diethyl ether, methyl n-propyl ether, diethylamine, N-methyl-1-propanamine, acetone, allyl alcohol, dimethylformamide, propanamide, 2-methylpropanamide, 2,2-dimethylpropanamide, benzamide, dichloromethane, toluene, ethyl N-acetyl-glycinate, -alaninate, -methioninate, and -aspartate, ethyl acetate, tetrahydrofuran
1:2 Complexes:	Methanol, ethanol, 1-propanol, 1-butanol
2:1 Complexes:	N,N,N′,N′-tetramethylsuccinamide
No complexes:	2-butanol, cyclohexanone, camphor, triethylamine, ethyl 4-aminobenzoate, ethyl N-acetyl-prolinate, -serinate, -phenylalaninate, and -glutamate

from solutions of the components in hot ethyl acetate (Table 4). Most of these complexes are stable and can be heated under vacuum well above the boiling points of low-boiling guests without decomposition.

Amide derivatives provide the functionally most compatible guests. The crystal structures of the 1:1 DTU inclusion complexes with propanamide (space group *C*2/*c*) and with ethyl N-acetylglycinate (space group $P\bar{1}$) are illustrated in Figs. 24 and 25, respectively [23]. These structures contain continuous chains of alternating host and guest molecules held together by an extensive network of directional hydrogen bonds between the amide groups. Along each such chain the C=O and N—H groups

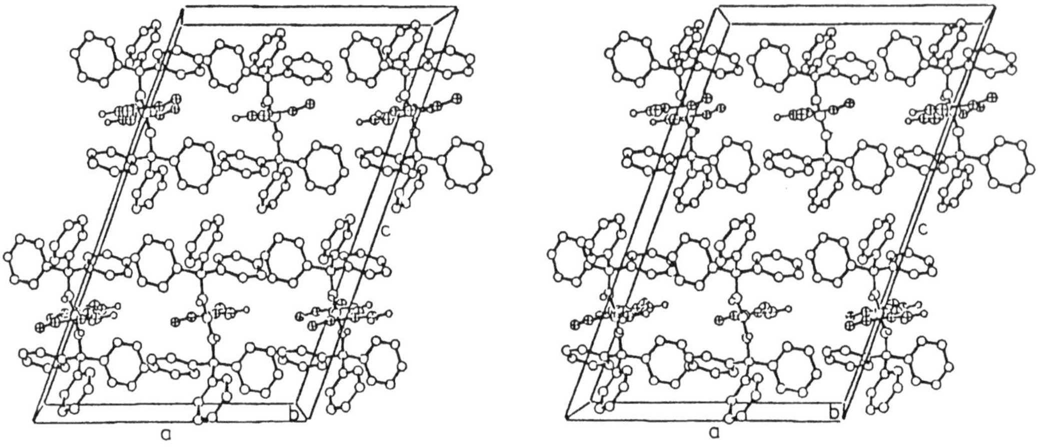

Fig. 24. Stereoview of the 1 : 1 DTU · propanamide clathrate. For clarity, only one orientation of the disordered guest is shown at each site (taken from Ref. [23)])

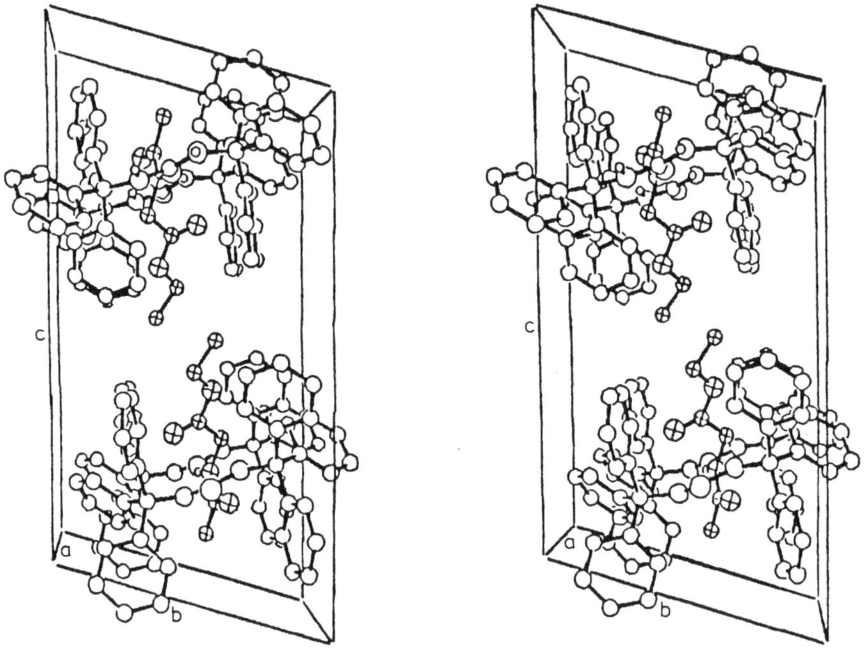

Fig. 25. Stereoscopic illustration of the crystal structure of 1 : 1 DTU · ethyl N-acetylglycinate (taken from Ref. [23)])

point in opposite directions, thus providing complementary sites for an effective H-bonding association. The alkyl residue of the guest lies either above or below the nearly planar binding site. The structure of the propanamide complex has a two-fold rotational symmetry with the C=O bonds of DTU hosts coinciding with crystallographic two-fold axes and the propanamide guests being orientationally disordered about them.

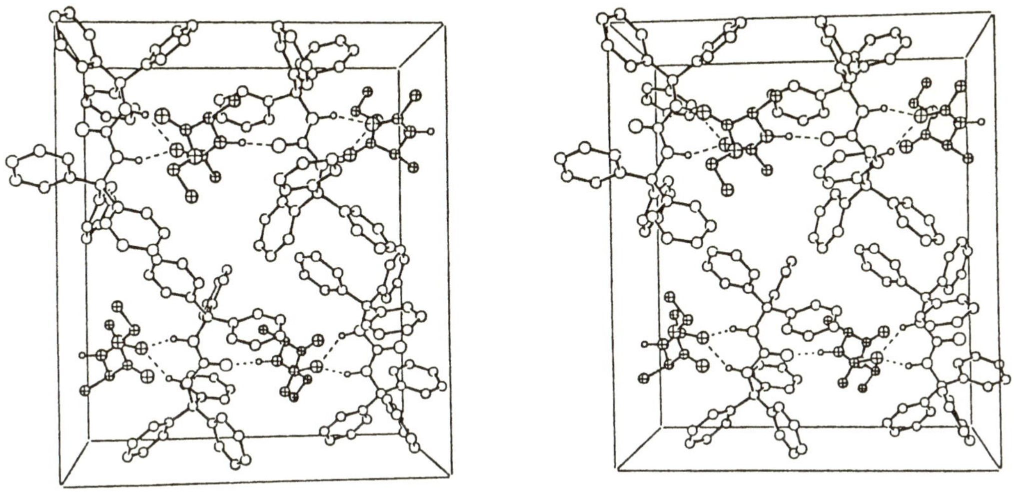

Fig. 26. Stereoview of the crystal structure of 1:1 DTU · ethyl N-acetylalaninate, showing the continuous pattern of hydrogen bonds in the lattice [46]

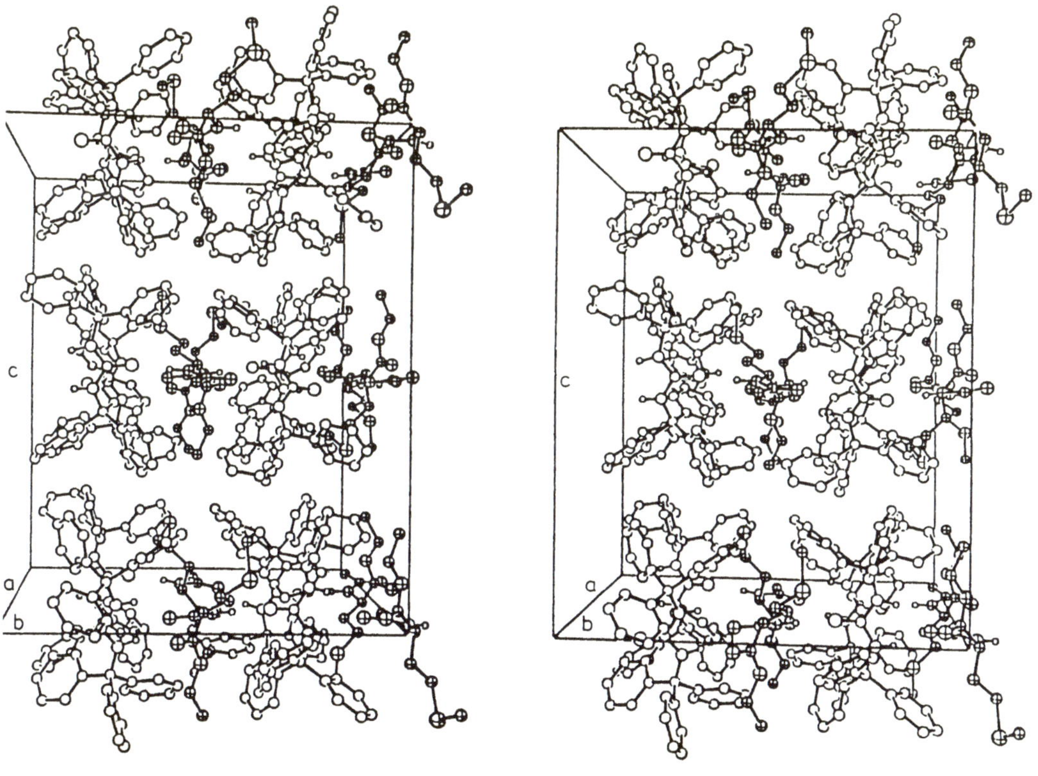

Fig. 27. Stereoview of the crystal structure of 1:1 DTU · ethyl N-acetylmethioninate, with eight entities of the complex in the unit-cell [46]

An identical scheme of the host-guest-host-... linkage has been observed with chiral amide derivatives. The corresponding examples involve the 1:1 complexes of DTU with ethyl N-acetylalaninate (Fig. 26) and with ethyl N-acetylmethioninate (Fig. 27) which formed chiral crystals (space group $P2_12_12_1$) [46]. In the four structures, the characteristic translation between adjacent units along the hydrogen-bonded chain is about 9 Å irrespective of the crystallographic space symmetry; the ester residues of the amino acids are not involved in hydrogen bonds and are more free to move in the crystal lattice. This common pattern of host-guest interaction is illustrated schematically in Fig. 28.

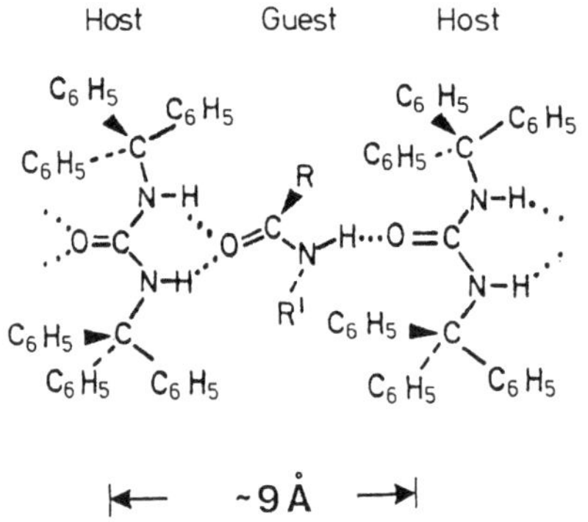

Fig. 28. Schematic illustration of the hydrogen bonding between DTU (*22*) and amide guests (taken from Ref. [24])

Additional observations reflect on the high stability of the DTU host lattice and illustrate its versatile capability to form clathrates with different guests. The crystalline 1:1 complexes of DTU with diethylether (Fig. 29) and diethylamine are isomorphous with the structure of the propanamide adduct (Fig. 24). However, in these structures the hydrogen bonds do not form a continuous pattern but are, rather, confined to

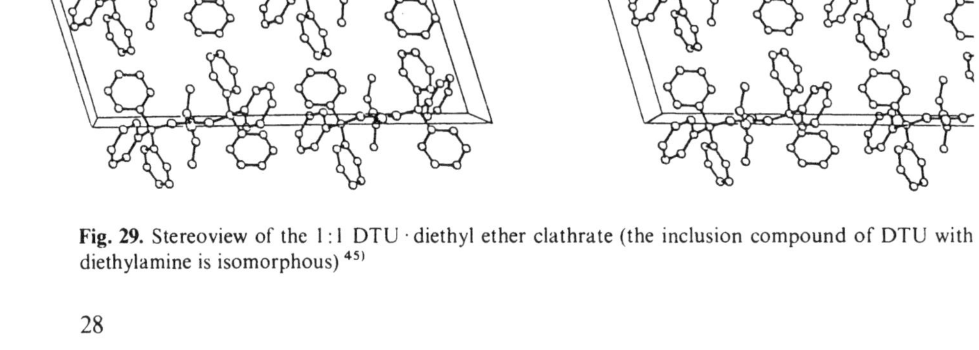

Fig. 29. Stereoview of the 1:1 DTU · diethyl ether clathrate (the inclusion compound of DTU with diethylamine is isomorphous) [45]

Table 5. Crystal data for the 1:1 inclusion compounds of host 22 [23,24,45]

Guests:	(I)[a]	(II)	(III)	(IV)[a]	(V)[b]	(VI)[b]
a (Å)	15.839(2)	16.707(4)	16.926(9)	9.010(3)	9.730(4)	17.108(4)
b	9.088(1)	8.793(2)	8.740(5)	10.800(3)	18.714(7)	19.415(5)
c	24.584(3)	24.376(5)	24.404(13)	19.810(9)	21.362(7)	24.066(5)
α (deg)	90.0	90.0	90.0	105.29(3)	90.0	90.0
β	111.05(1)	106.94(2)	106.99(4)	94.33(3)	90.0	90.0
γ	90.0	90.0	90.0	93.03(2)	90.0	90.0
V (Å^3)	3302.6	3425.6	3452.6	1848.8	3889.7	7993.6
Z	4	4	4	2	4	8
D_c gcm^{-3}	1.242	1.198	1.189	1.239	1.202	1.269
Space group	$C2/c$	$C2/c$	$C2/c$	$P\bar{1}$	$P2_12_12_1$	$P2_12_12_1$

[a] Data at 115 K; [b] data at 128 K

Inclusion compounds:
with propanamide (I); with diethyl ether (II); with diethylamine (III); with ethyl N-acetylglycinate (IV); with ethyl N-acetylalaninate (V); with ethyl N-acetylmethioninate (VI)

individual host-guest entities. The host-to-guest association involves the N—H groups of DTU as proton donors and the O or N sites of the guest species as proton acceptor (see below). Crystal data of the complexes referred to above are given in Table 5. The structural characteristics of other inclusion complexes of DTU, particularly those with 1:2 and 2:1 stoichiometries (e.g., with small alcohols and with N,N,N′,N′-tetramethylsuccinamide, respectively) have not yet been fully analyzed.

Evidently, it is important to have two large substituents on the urea in order to prevent self-association of the host species and leave the N—H and C=O attractive binding sites open for guest molecules with complementary features. This indication emerges from the following study. Crystallization experiments with the mono-substituted N-tritylurea (*23*) revealed that this host prefers to form clathrates with less polar guests [23]. The crystal structure analysis of the 2:1 inclusion complex with N,N-dimethylformamide shows that in this case the host molecules make use of their hydrogen bonding potential by associating with each other (Fig. 30). The host lattice can be best described here as consisting of H-bonded continuous chains of the N-tritylurea moieties with an alternating arrangement of the trityl substituents on both sides of the polar region. The guests are enclosed within hydrophobic lattice cavities formed by the trityl groups of four neighboring hosts. The length of the trans-

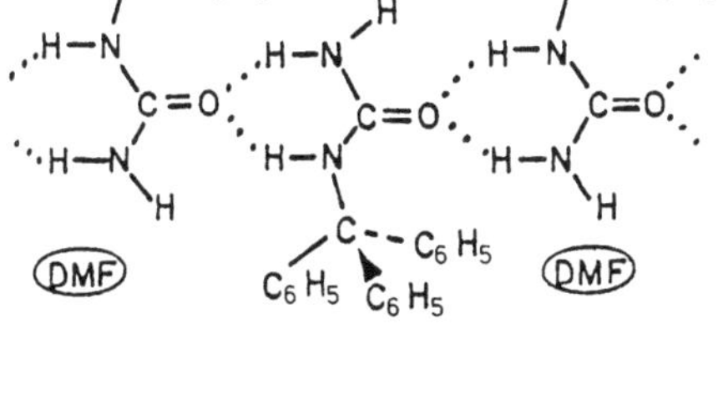

Fig. 30. Schematic illustration of the 2:1 clathrate between N-tritylurea and dimethylformamide, showing the self-association of the host molecules along the *b* axis of the crystal (crystal data: a = 29.614, b = 8.906, c = 16.127 Å, β = 121.04°, space group Cc) [23]

lation vector between identical units is about 9 Å as in the previously described hetero-molecular chains involving DTU.

Selectivity studies with DTU indicated marked discrimination in the clathrate formation [23,45]. As in other types of clathrates, the steric factor is important in differentiation between compounds of similar functionality but different shape. For example, DTU forms crystalline complexes with some alcohols (methanol, ethanol, propanol, 1-butanol) but not with others (2-butanol). It complexes the ethyl esters of N-acetyl derivatives of glycine, alanine, methionine and aspartic acid, but not of proline, serine, phenylalanine and glutamic acid.

Furthermore, a preferential guest inclusion by DTU has been observed from solutions containing a 1:1 guest mixture in ethylacetate [45,47]. Studies of guest binding to the host show the following results (guests ratio found in the complexes after one crystallization step with DTU): 3.5:1.0 for diethyl ether and n-propyl methyl ether, 4.0:1.0 for 2-propanol and 1-propanol, 4.6:1.0 for diethylamine and n-propyl methyl amine, >20:1.0 for acetamide and 2,2-dimethylpropanamide, >20:1.0 for acetamide and N,N-dimethylformamide. The excessive inclusion of guests characterized by a higher symmetry is clearly correlated with the two-fold rotational symmetry of DTU.

Discrimination between guests in complexation with DTU can also be obtained when the factor of functional complementarity is involved [45,47]. Thus, diethyl ether is better complexed than diethyl amine (4.6:1.0) because the N—H(host) ... OR_2(guest) hydrogen bonding is more attractive than N—H(host) ... NHR_2(guest). Similarly, acetamide is better complexed with DTU than diethylamine (>20:1.0) due to its multiple hydrogen bonding function which allows a more effective intermolecular association.

The above selectivities are affected by changes in the molar ratio between the guests and between guest and host in the solution mixture. They show, however, consistent trends. Qualitatively similar results of guest selectivity were obtained with N,N′-bis(tri-*p*-tolylmethyl)urea and a number of related hosts [47]. Additional experiments using ureas substituted with chiral "spacers" are readily envisioned.

4.2 Separation of Structural Isomers by Crystalline Complexation with 1,1-Di(*p*-hydroxyphenyl)cyclohexane

The wheel-and-axle design as source for host-guest compounds was originally proposed by Toda and Hart in 1981 for hosts containing hydroxyl functions [48] (see Ch. 3, Sect. 2.1 of Vol. 140). The 1,1,6,6-tetraphenylhexa-2,4-diyne-1,6-diol (*24*) provides a representative compound. It forms 1:2 crystalline inclusion complexes with a large number of small guest molecules, including a variety of ketones, amines, amides and a sulfoxide [48].

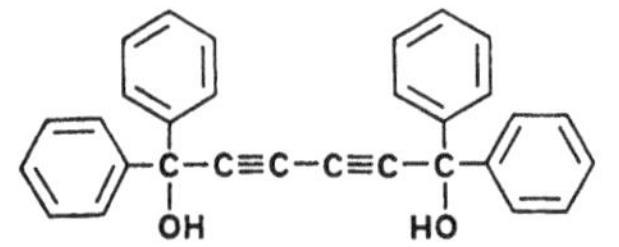

24

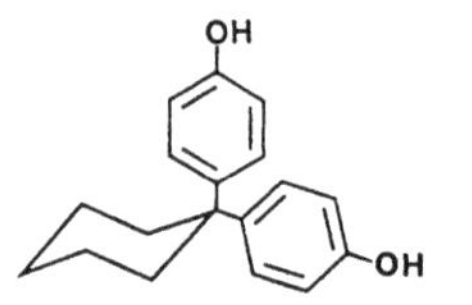

25

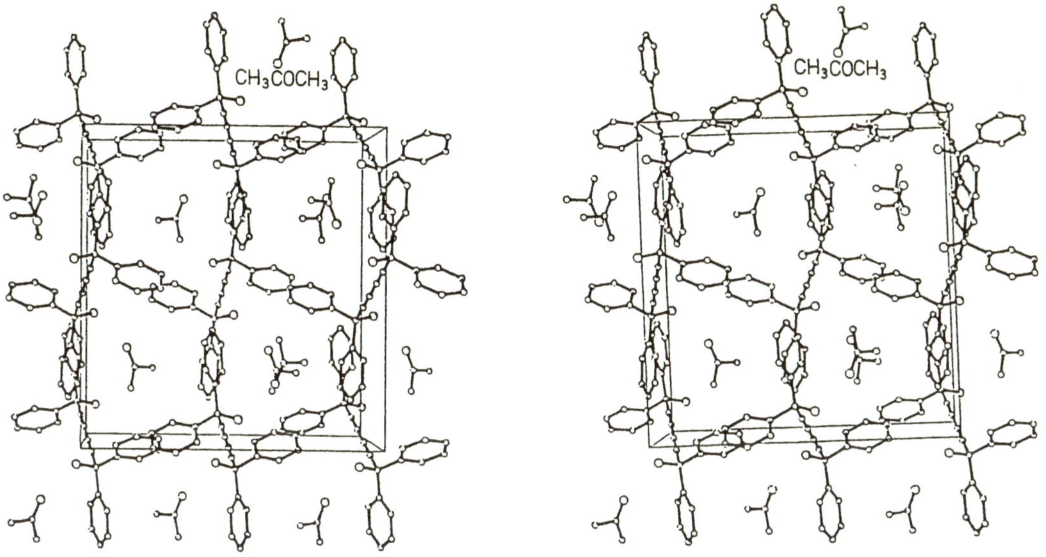

Fig. 31. Stereoview of the *24* · acetone clathrate down the channel axis. Each acetone guest is within a hydrogen bonding distance from one of the hydroxyl sites of the host [48]

Figure 31 illustrates the crystal structure of the 1:2 inclusion complex with acetone [48]. This structure contains open channels roughly parallel to the long axis of the diyne-diol host in which the acetone is hydrogen-bonded to the anti-oriented hydroxyl groups. Reportedly, other acetylenic diols (including chiral acetylenic alcohols) form similar clathrates stabilized by hydrogen bonding through the hydroxyl groups [20].

An efficient separation of close structural isomers by the method of inclusion crystallization may require, however, a more constrained and less open host lattice with strong binding interactions between the host species as well. In this respect it is interesting to examine the problem of separation of phenol and cresols by crystalline complexation with 1,1-di(*p*-hydroxyphenyl)cyclohexane (*25*) [25]. The molecular structure (overall shape and relative disposition of the —OH groups) of this diol host allows a close approach and hydrogen bonding between hosts as well as between host and guest in the crystal lattice. It should be noted that the cresols are characterized by close boiling points (e.g., 202.0 °C for *m*-cresol and 201.8 °C for *p*-cresol), and their separation by conventional distillation is thus very difficult.

The general procedure for the separation of one component from a mixture of two cresols or of phenol and cresol by crystallization is much simpler as follows: Equimolar amounts of two cresols (or phenol and cresol) and *25* are dissolved in ethyl acetate by heating. The resulting solution is kept at room temperature to give a 1 : 1 inclusion complex. The crude complex obtained can be purified by recrystallization from ethyl acetate to give the pure complex of one component which upon heating in vacuum (20 mmHg at 200 °C) gives pure cresol or phenol. The ratio of products obtained by distilling the complex derived from *25* and a 1:1 mixture of two guest components in a single step of crystallization showed which component is more

Table 6. Studies of guest binding to host *25* [25]

Entry	Equimolar guest mixture in ethyl acetate	Guests ratio in the complexes[a]
1	*o*-cresol:phenol	42.7:57.3
2	*o*-cresol:*p*-cresol	35.2:64.8
3	*o*-cresol:*m*-cresol	26.2:73.8
4	Phenol:*p*-cresol	45.7:54.3
5	*p*-cresol:*m*-cresol	34.1:65.9

[a] Determined by gas-chromatography in the crude complexes which were obtained after one step of crystallization

Table 7. Crystal data for the 1:1 inclusion compounds of host *25* [25,50]

Guests:	Phenol	*o*-cresol	*m*-cresol	*p*-cresol	Cyclo-hexanol	Cyclo-hexanone
a (Å)	6.232(3)	6.270(2)	6.250(2)	6.237(4)	6.226(3)	6.326(3)
b	10.849(2)	10.907(4)	10.807(2)	10.856(3)	32.218(6)	30.663(5)
c	14.845(3)	15.446(7)	15.490(2)	15.818(4)	10.810(3)	10.778(5)
α (deg)	95.69(1)	92.87(4)	98.11(1)	99.29(2)	90.0	90.0
β	93.49(3)	93.33(3)	93.41(2)	92.21(3)	103.65(3)	100.38(3)
γ	104.31(3)	102.15(4)	101.11(2)	105.67(3)	90.0	90.0
V (Å^3)	964.0	1028.8	1012.3	1013.9	2107.1	2056.4
Z	2	2	2	2	4	4
D_c gcm^{-3}	1.249	1.215	1.235	1.233	1.162	1.184
Space group	$P\bar{1}$	$P\bar{1}$	$P\bar{1}$	$P\bar{1}$	$P2_1/c$	$P2_1/n$

strongly included in the crystal lattice. Results of the selectivity studies disclosed a sequence of preferential complex formation between *25* and a phenol derivative in the order *m*-cresol > *p*-cresol > phenol > *o*-cresol. The relevant guest ratios, detected by gas chromatography in the material obtained by crystallization from a three-component mixture (two guests + host) in ethyl acetate, are given in Table 6. These observations suggested a possibility of *m*-cresol separation from a cresol mixture (obtained from coal tar), and in fact 98.5% pure *m*-cresol was isolated in 55% yield [25].

Crystallographic analyses of the four inclusion complexes between *25* and one of the phenol derivatives revealed isomorphous structures [25]; the relevant crystal data are given in Table 7. The common crystal structure type can be best described as composed of "layers" of hydrogen-bonded species which lie parallel to the *ab* plane of the crystal (Fig. 32a). Within each such layer there is a characteristic pattern of hydrogen bonds as follows. Molecules of the host related to each other by translation along the *b* axis interact through their terminal hydroxyl groups with one hydroxyl acting as a proton donor to and the other as a proton acceptor from the neighboring moieties. The continuous chains of the host molecules thus formed are linked to one another along the *a* axis through the phenolic guest, utilizing the dual capacity of a hydroxyl group for hydrogen bonding (each guest moiety donates its hydroxyl proton to one chain and provides an acceptor site for a proton from another chain). As illus-

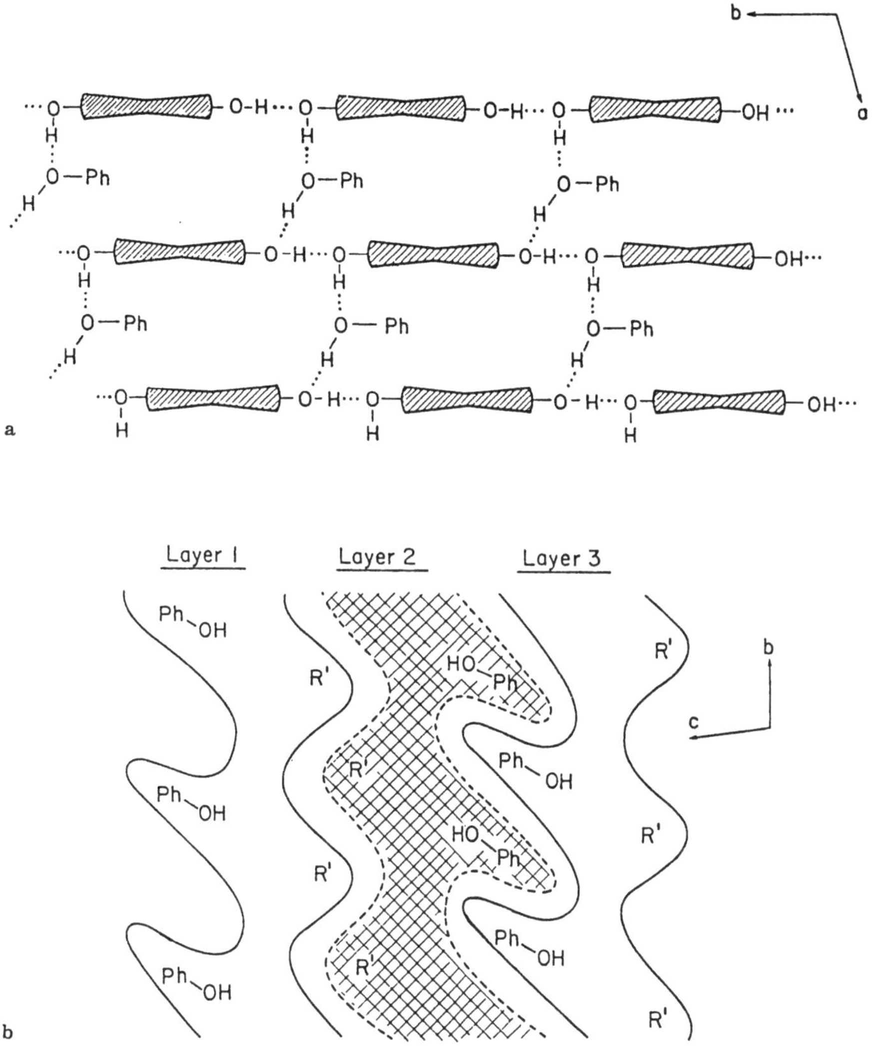

Fig. 32a and b. Schematic illustration of intermolecular arrangements in the crystalline complexes of host *25* (taken from Ref. [25]): **a** the two-dimensional hydrogen bonding pattern parallel to the *ab* plane (the shaded area represents the 1,1-diphenylcyclohexane framework); **b** the van der Waals type packing of the hydrogen bonded layers along the *c* axis (R′ represents the cyclohexyl ends of the host species)

trated in Fig. 32a this results in a relatively unflexible two-dimensional cross-linked array of hydrogen-bonded entities which is characterized by the general pattern host-host-host-... along *b* and guest-(host-host)-guest-(host-host)-... along *a*.

It is considerably easier to induce an expansion of the periodicity along the *c* axis, since along this direction the layered structure is stabilized mainly by weaker dispersion forces. Indeed, the structural variation of the guest species caused a significant increase of the *c* axis from 14.845 Å (for the phenol complex) to 15.818 Å (for the *p*-cresol complex). The interlayer arrangement is shown in Fig. 32b. It is characterized by a convenient steric fit between the convex and the concave surfaces of adjacent layers which are related to each other by crystallographic inversion. The actual crystal structures of the complexes are illustrated in Figs. 33–36.

In the four isomorphous structures the different guest components are located in a similar crystalline environment. Since the species involved are of nearly identical

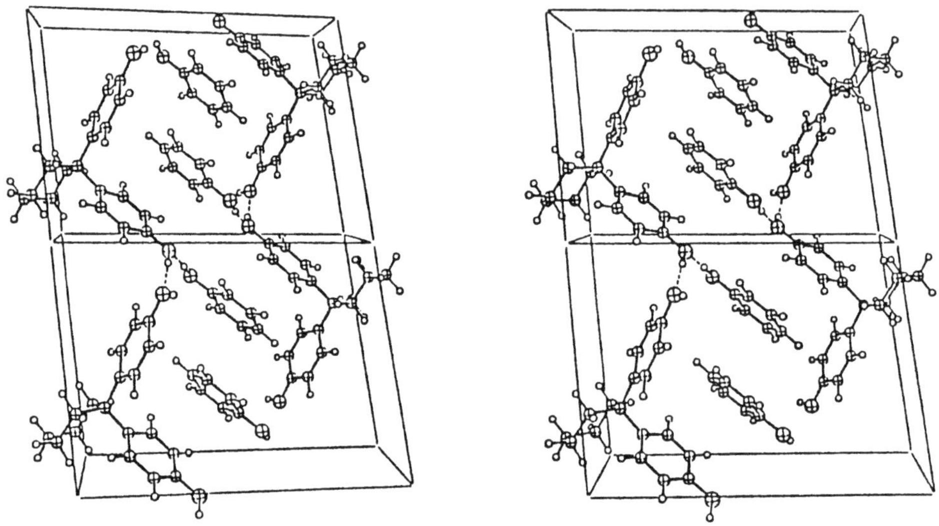

Fig. 33. Stereoview of the 1:1 *25* · phenol inclusion compound, approximately down the *a* axis (*c* is horizontal) (taken from Ref. [25])

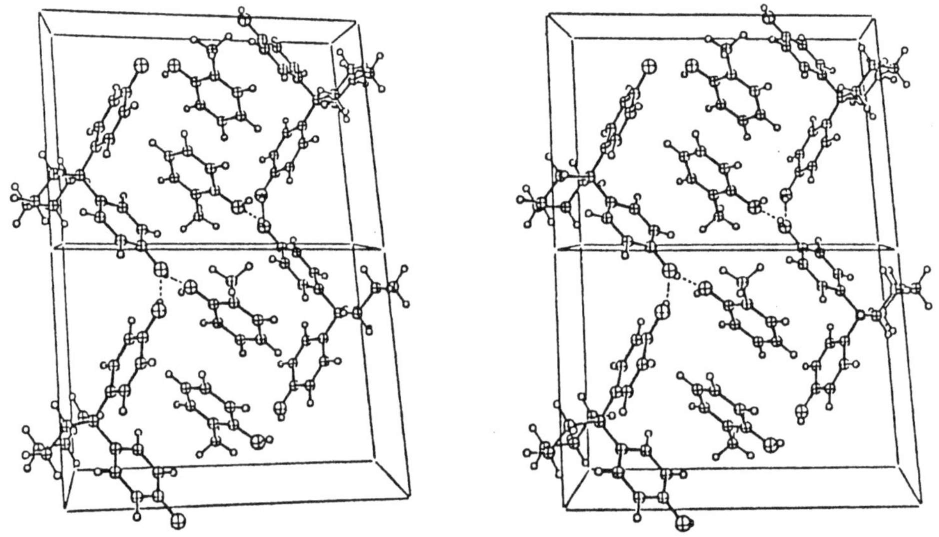

Fig. 34. Stereoview of the 1:1 *25* · o-cresol inclusion compound, approximately down the *a* axis (*c* is horizontal) (taken from Ref. [25])

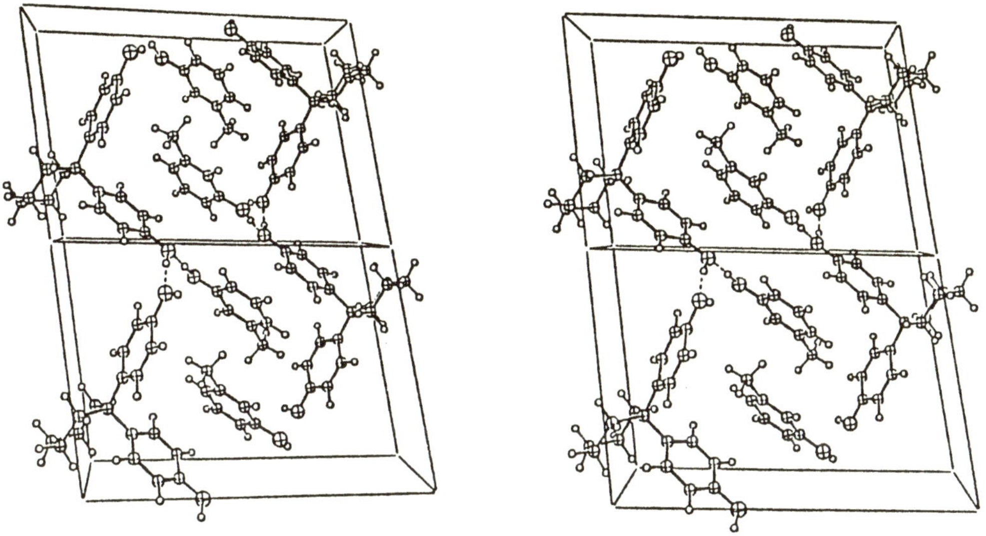

Fig. 35. Stereoview of the 1:1 *25* · m-cresol inclusion compounds, approximately down the *a* axis (*c* is horizontal) (taken from Ref. [25])

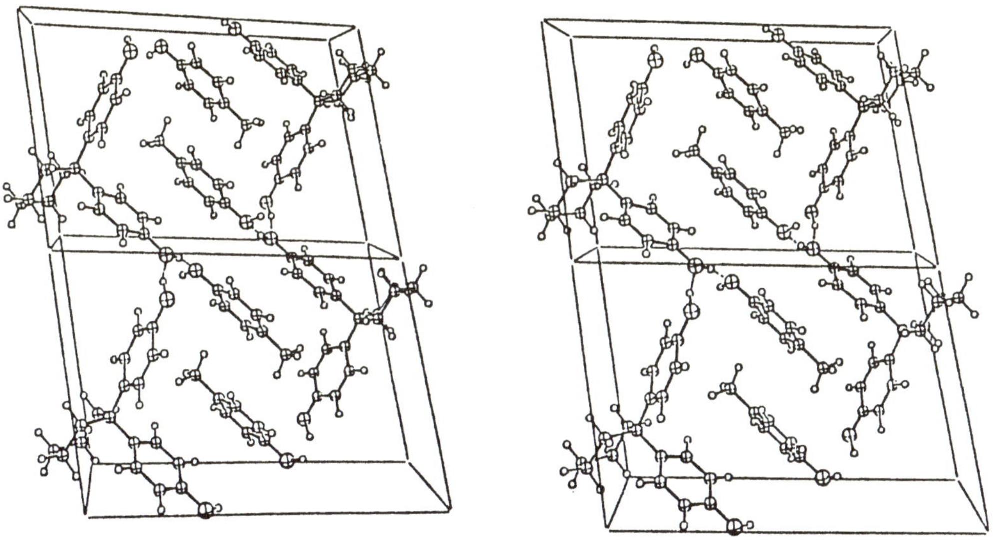

Fig. 36. Stereoview of the 1:1 *25* · p-cresol inclusion compound, approximately down the *a* axis (*c* is horizontal) (taken from Ref. [25])

chemical and physical properties, containing only a small structural variation, the solvation effects before crystallization are expected to be similar as well. In view of the identical hydrogen bonding scheme in these structures (the geometries of the nearly linear bonds indicate strong interaction, the OH ··· O distances ranging from 2.65 to 2.74 Å), preferential complexation of *25* with one guest out of a mixture of two or more structural analogues will be determined mainly by differences in the feature of spatial complementarity between the various constituents. It is possible, in fact, to detail some significant differences between modes of van der Waals interaction in the observed structures by surveying the relevant intermolecular distances. Ordinary nonbonding contacts characterize the structure containing the unsubstituted phenol as guest. However, as can be anticipated from Fig. 32a, the guest phenyls are rather loosely packed along the *a* axis of the crystal, all relevant intermolecular distances are >4.0 Å.

Correspondingly, replacement of the phenol (Fig. 33) by *m*-cresol (with the methyl oriented partially along *a*, Fig. 35) is associated with an increase of only about 0.65 Å in the length of the *c* axis. The methyl substituent is perfectly well accommodated within the expanded lattice, contributing additional interactions of dispersion without an apparent distortion of the previous ones. All intermolecular distances remain well within range of characteristic van der Waals values [49].

A substitution of the methyl group in the *para*-position (Fig. 36) causes a larger expansion of the unit cell along *c*. The crystal packing of the resulting structure is still an efficient one, however the *p*-substituted methyl introduces already some steric hindrance into the structure; the shortest distance of the methyl from its surroundings (adjacent phenyl "walls") is 3.56 Å as compared to the normal van der Waals methyl ··· ··· phenyl distance of about 3.7 Å [49]. This steric misfit could be responsible for the preferential complexation of host *25* with *m*-cresol rather than with *p*-cresol.

Introduction of the Me substituent at the *ortho* position (Fig. 34) has an unfavorable effect on the intermolecular arrangement. The methyl groups of guest molecules interrelated by inversion at 0,0,1/2 essentially collide one into the other. The corresponding Me ··· Me distance is 3.50 Å, about 0.5 Å shorter than the sum of van der Waals radii. Moreover, the $\pi-\pi$ interaction between the partially overlapping guests located around the 1/2,1/2,1/2 center of inversion is considerably reduced in this structure; the interplanar distance has increased from about 3.6 Å in the previous examples to 3.92 Å here. Thus, the inefficient crystal packing of the molecular entities makes the complex with *o*-cresol relatively unstable due to poor interaction between the hydrogen-bonded layers.

A similar procedure was used to separate efficiently between cyclohexanol (b.p. 161 °C) and cyclohexanone (b.p. 155 °C), two structurally similar but chemically different species, by selective crystalline complexation with host *25* [50]. For example, when a solution of this host and a 1:1 mixture of the two guests in ethyl acetate was kept at room temperature for 24 hours, and the colorless crystals thus obtained were subsequently heated in vacuum, the composition of the resulting guest mixture was 94.71% of cyclohexanol and 5.23% of cyclohexanone.

The corresponding crystal structures of the 1:1 complexes between *25* and either one of the guest components are nearly isomorphous (Table 7). In spite of the differences in periodicity and space symmetry, the observed patterns of intermolecular arrangement in these structures are very similar to those of the previously described inclusion complexes of *25* with phenol and cresol derivatives. Thus, the cyclohexanol complex, shown in Fig. 37a, contains identical patterns of hydrogen bonds and van der Waals interactions. Every hydroxyl group in the structure is involved in two hydrogen bonds, acting simultaneously as a proton donor to one site and as a proton acceptor from another site. The arrangement of the hydrogen-bonded layers along

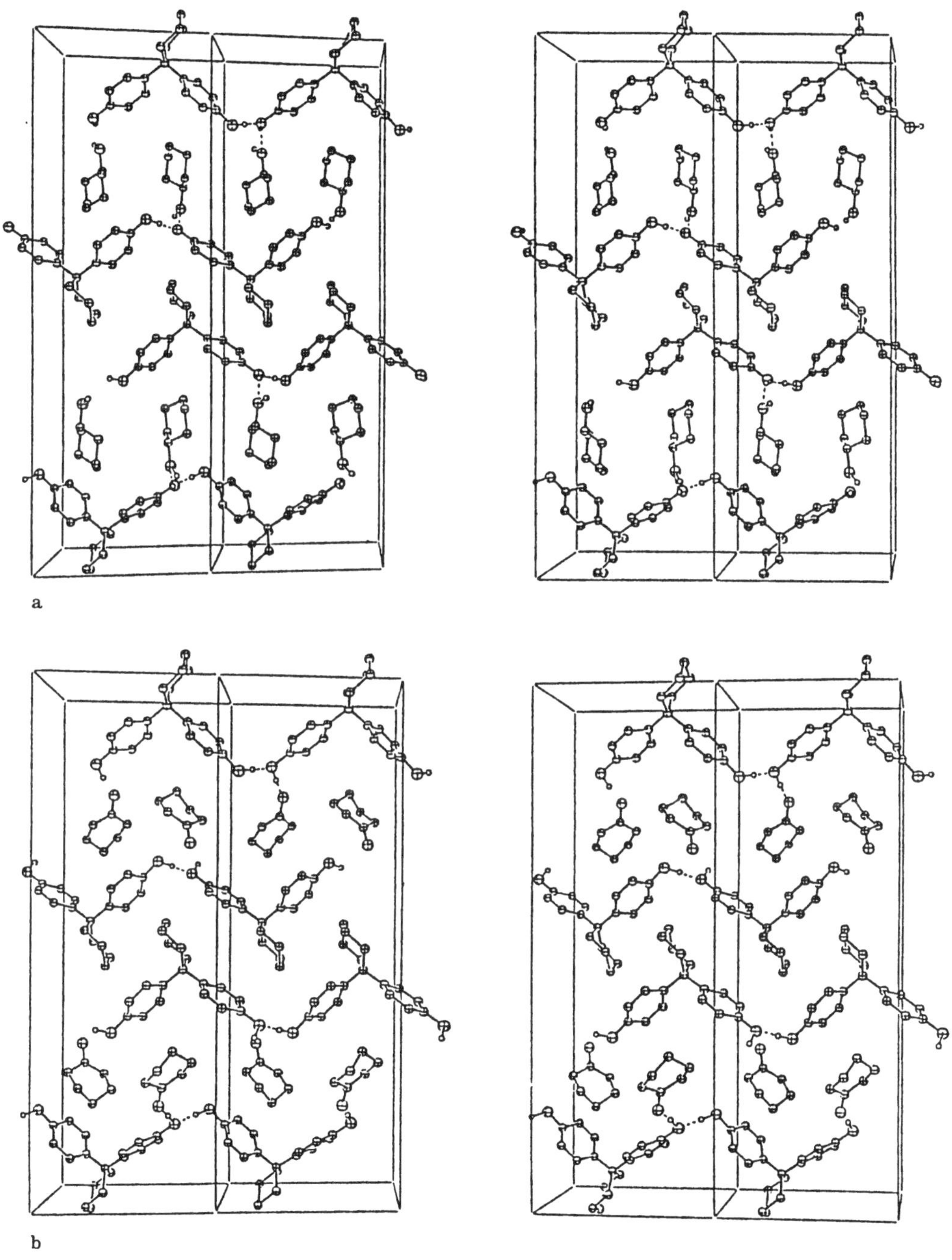

Fig. 37 a and b. Crystal structures of the 1:1 inclusion compounds of *25* with **a** cyclohexanol, and **b** cyclohexanone; stereoviews down the *a* axis, *c* is horizontal (taken from Ref. [50])

the *b* axis is stabilized, on the other hand, by weak dispersion forces. The cyclohexyl fragments (one from the host and one from the guest) located on both sides of each such layer conveniently fit in between the cyclohexyl groups of adjacent layers.

No major differences occur in the overall crystal packing arrangement when cyclohexanol is replaced by cyclohexanone in the structure (Fig. 37b). As before, the host species self-associate to form continuous chains along the *c* axis, but every such OH ··· OH binding site is further liked via an H-bond to one guest only. The less restricted relative disposition of the molecular units with respect to the *a* axis is now modified to optimize other packing interactions. This is associated with a transformation of the space symmetry from $P2_1/c$ in the cyclohexanol complex to $P2_1/n$ in the cyclohexanone complex. It is also reflected in a marked contraction of the *b* axis by 1.56 Å, and an improved steric fit between neighboring chains of molecules. Evidently, the more extensive hydrogen bonding pattern with cyclohexanol makes its complex with *25* energetically more stable, giving rise to the selective complexation.

4.3 Chiral Inclusion Complexes of 2,2′-Dihydroxy-1,1′-Binaphthyl with Phosphinates and Phosphine Oxides

The question that emerges at the climax of this survey relates to the possibility of using crystalline inclusion phenomena for optical resolutions of molecular species. Can this be done effectively with suitably designed host compounds? The definitely positive answer to this question has elegantly been demonstrated by Toda [20] as well as by other investigators (see Ch. 2 of Vol. 140). An optically active host compound will always form a chiral lattice. Therefore, when an inclusion type structure is induced, one enantiomer of the guest moiety should be included selectively within the asymmetric environment.

To this aim, the 2,2′-dihydroxy-1,1′-binaphthyl (*26*) molecule possesses a particularly useful shape. It is chiral and configurationally stable with the two naphthyls oriented nearly at right angles to each other. It contains also functional "sensors" (hydroxyl groups) for an efficient guest binding. Indeed, it has recently been reported that sulfoxides and selenoxides are easily resolved by inclusion complexation with optically active *26* [51,52]. Reversely, *26* was resolved very efficiently by complexation with optically pure guest components. This inclusion crystallization method is very simple and does not require natural products. It can give 100% optically pure compound (which is recovered from the complex by column chromatography) since the

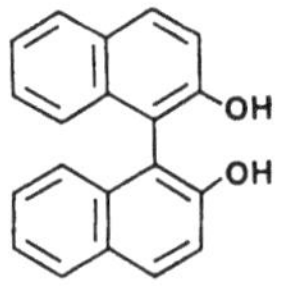

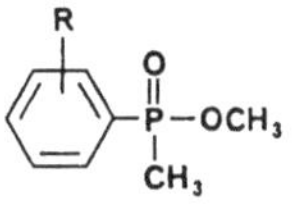

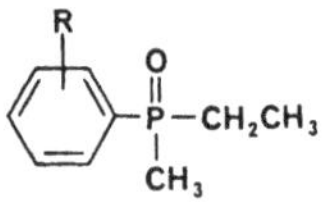

a R=H
b R=o-CH3
c R=m-CH3
d R=p-CH3

26 27 28

Table 8. Optical resolutions of phosphorous compounds *27* and *28* by inclusion crystallization with (—)-*26*

Compound	100% *ee* enantiomers obtained from the:		
	Complex with (—)-*26*		Filtrate[a]
	$[\alpha]_D$ (°)[b]	yield (%)	yield (%)
27a	+46.0	12	—
27b	+45.5	47	14
27c	+46.3	31	50
27d	+47.1	32	14
28a	—23.1	60	30
28b	—[c]	—	—
28c	—24.0	33	16
28d	—[c]	—	—

[a] Isolated by complexation with host (+)-*26*;
[b] measured in THF at the concentration of c 0.2;
[c] no optical resolution occurred

resolution can easily be repeated, or the crude complex first obtained can be purified by recrystallization. This method can also be used for storage of labile optically pure species in the complexed form [52].

A successful resolution of some alkylaryl-substituted phosphinates *27* and phosphine oxides *28* by crystalline inclusion complexation with optically active form of host *26* has also been achieved [53]. The results of these experiments with various guests are summarized in Table 8. The simplicity of the method used is illustrated by referring to the main stages of the reaction between *26* and the phosphoric ester *27a*. Treatment of the racemic phosphinate *27a* with (—)-*26* in benzene led to the formation of a 1:1 complex of (+)-*27a* with (—)-*26*. Two subsequent recrystallizations of the complex from benzene gave a pure compound. Column chromatography of the resulting complex on silica gel gave optically pure (+)-phosphinate *27a*. A similar treatment with (+)-*26* of the filtrate left after the above described separation of the complex gave finally the optically pure (—)-*27a* isomer.

Other phosphorous compounds, as shown in Table 8, were resolved by the same procedure. The three isomeric phosphinates *27b–d* containing a methyl group attached to the aryl substituent could also be resolved, irrespective of the methyl position. From the related phosphine oxides *28a–d*, however, only those with R=H (*28a*) and $R=m\text{-}CH_3$ (*28c*) could be well resolved; no satisfactory resolution could be obtained for the other isomers of *28*. The efficiency of the optical resolution of alkyl-aryl-substituted sulfoxides and selenoxides was found to depend similarly on the type of substitution on the aryl ring.

Two enantiomeric 1:1 inclusion complexes of a representative phosphoric ester (+)-*27a* with (+)-*26* and (—)-*26* were studied crystallographically in order to characterize the geometric pattern of intermolecular interactions [53]. The crystal structures of the two compounds (monoclinic space group $P2_1$) are characterized by similar features (Fig. 38). They consist of continuous chains of hydrogen-bonded species which are aligned along the polar axis of the crystal. Along the chains, adjacent

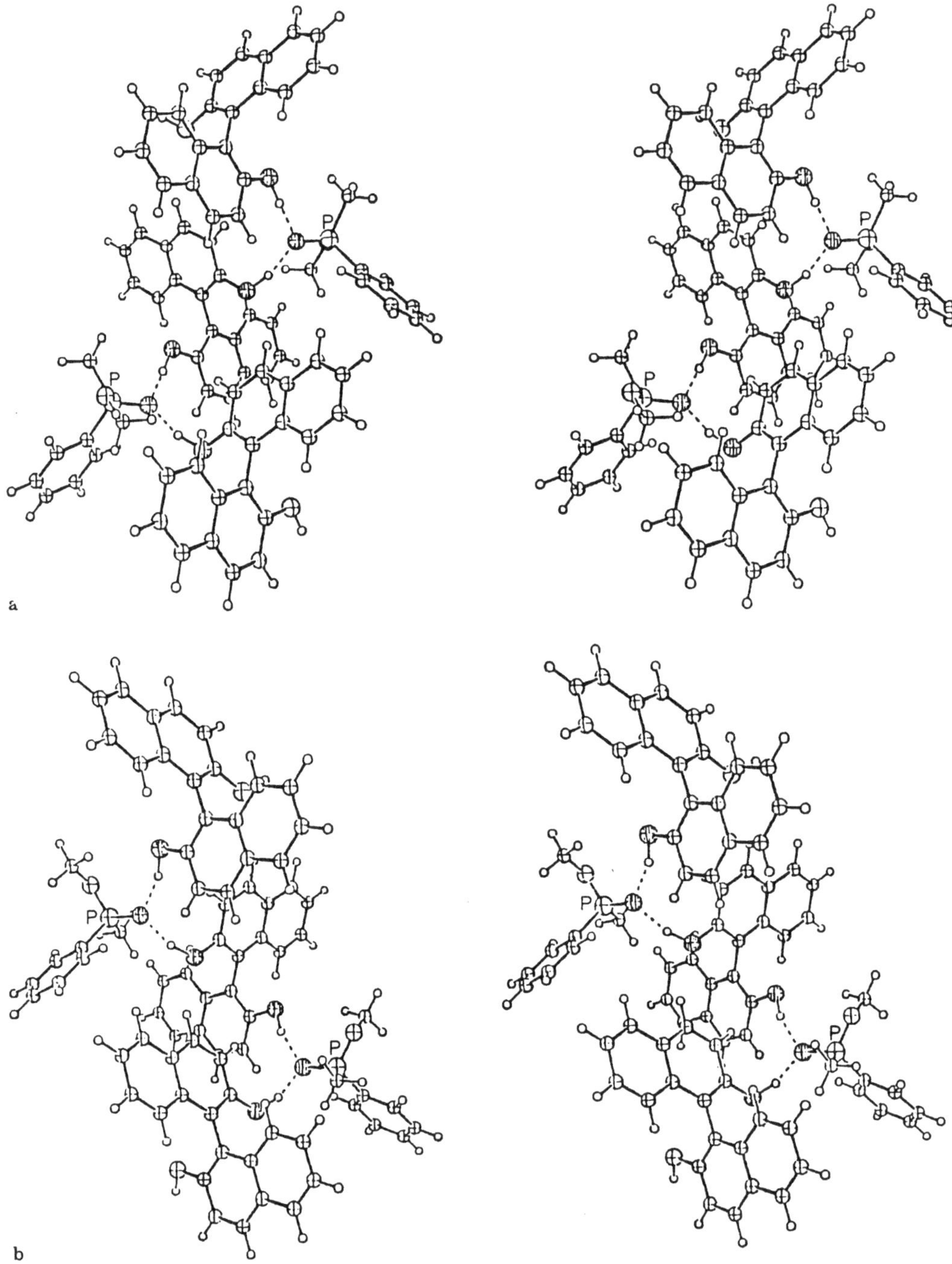

Fig. 38a and b. Stereoviews of the hydrogen bonding interactions between host and guest in the 1:1 inclusion complexes (taken from Ref. [53]) of **a** (−)-*26* with (+)-*27a* (crystal data: $a = 9.132$, $b = 11.809$, $c = 11.708$ Å, $\beta = 112.50°$, space group $P2_1$); **b** (+)-*26* with (+)-*27a* (crystal data: $a = 9.384$, $b = 11.809$, $c = 11.462$ Å, $\beta = 111.44°$, space group $P2_1$)

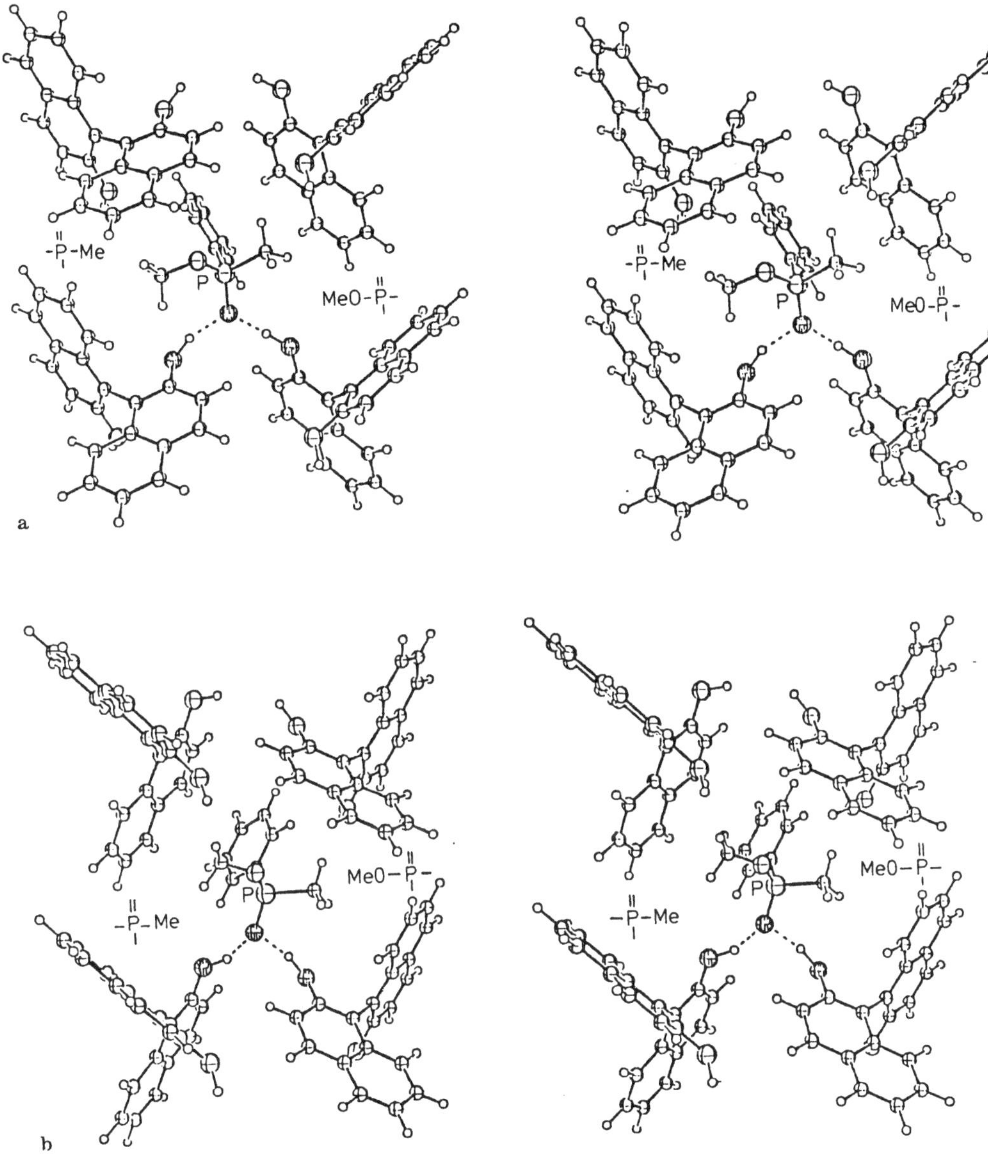

Fig 39a and b. Stereoviews of the crystalline environment around the phosphoric ester guest (+)-*27a* in its complexes with **a** (—)-*26*, and **b** (+)-*26*. For clarity, the positions of the two adjacent guest moieties are shown schematically (taken from Ref. [53])

binaphthol hosts related to each other by the 2_1 screw symmetry are bridged through their hydroxyl groups by the phosphoric ester guest. The polar P=O nucleophile of the latter acts as a proton acceptor from two different hosts. Correspondingly, each one of the hosts donates its two hydroxyl protons to different guests located on opposite sides of the chain. The two structures are thus dominated by a fixed

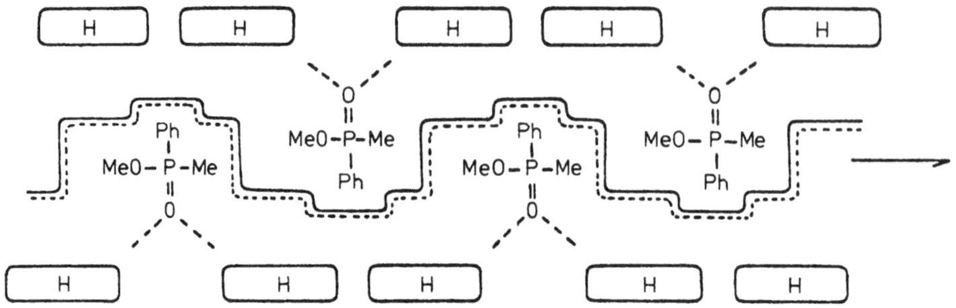

Fig. 40. Schematic illustration of the interface between two neighboring chains of the hydrogen-bonded entities in the enantiomeric complexes of host *26* with guest *27a*. The host molecule is denoted by "H" [53]

pattern of strong hydrogen bonds around the twofold screw axes which is associated with an identical length of the polar *b* axis (11.809 Å). Side packing of the hydrogen-bonded chains along the *a* and *c* directions is stabilized by weaker van der Waals forces (hydrophobic interactions between host and guest, partial overlap of the aromatic fragments), thus being more sensitive to structural variations.

Figure 39 illustrates the crystalline environment of the phosphinate guest species (+)-*27a* in the two enantiomerically related host lattices of *26*; each phosphinate is surrounded by and in contact with two adjacent guests and four binaphthol hosts. An illustration of the interface between two neighboring chains of the H-bonded entities (Fig. 40) indicates that, in addition to the hydrogen bonds, the most significant nonbonding interaction of each guest with its immediate environment in the crystal involves the —OMe and —Me substituents on phosphorus. Inspection of the structural details reveals that the configurational relationships between the constituents in the crystals of the complex between (+)-*26* and (+)-*27a* (cf. Figs. 38b and 39b) are less complementary than in the complex between (—)-*26* and (+)-*27a* (cf. Figs. 38a and 39a). In the former, the larger methoxy group is turned outward with respect to the hydrogen bonding site, "pushing" away the neighboring chains and effecting a strained arrangement (which is reflected in a large dihedral angle) of the binaphthyl unit. In the latter, the methoxy substituent is turned towards the binding site, the smaller —Me group turning outward. The observed Me ··· MeO nonbonding distances between guest species located in adjacent chains are short (3.65 Å) in both structures. This provides a possible explanation of the observation that phosphorous guests *27* and *28* with larger three-atom alkyl substituents on P (such as $—C_3H_7$ or $—O—C_2H_5$) did not form similar inclusion complexes with *26*, and could not be resolved.

The possibility to resolve the two enantiomers of *27a* (or *26*) by crystalline complexation with optically active *26* (or *27a*) is mainly due to differences in topological complementarity between the H-bonded chains of host and guest molecules. In this respect, the spatial relationships which affect optical resolution in the above described *coordination-assisted clathrates* are similar to those characterizing some optically resolved *molecular complexes* [54]. This should encourage additional applications of the lattice inclusion phenomena to problems of chiral recognition.

Furthermore, all the examples shown above in Sects. 4.1–4.3 emphasize the significance of both steric and functional features in selective crystallizations. The latter are needed not only for binding between the individual host and guest constituents, but also for effecting a continuous and relatively unflexible pattern of intermolecular arrangement in the crystal lattice. This observation appears to be a very useful one in the systematic design of novel clathrate-based synthetic receptors.

5 Acknowledgements

The author is most grateful to Profs. Harold Hart, Fumio Toda, and Edwin Weber for sharing their original ideas with him, for providing the compounds for structural interpretations, and for their encouragement and collaboration. The cooperation of Dr. Stephen Czerkez in the early stages of the study reviewed here is also appreciated. Special thanks are due to Mrs. Zafra Stein for her invaluable assistance.

6 References

1. Wöhler, F.: Liebigs Ann. Chem. *69*, 297 (1849); Clemm, A.: Liebigs Ann. Chem. *110*, 375 (1859)
2. Powell, H. M.: J. Chem. Soc. *1948*, 61 and *1954*, 2658
3. Hagan, M. (ed.): Clathrate Inclusion Compounds, New York, Reinhold 1962
4. Atwood, J. L., Davies, J. E. D., MacNicol, D. D. (eds.): Inclusion Compounds, Vol. 1–3, London, Academic Press 1984
5. Pedersen, C. J.: J. Am. Chem. Soc. *89*, 7017 (1967)
6. Cram, D. J., Trueblood, K. N.: Top. Curr. Chem. *98*, 43 (1981)
7. Lehn, J.-M.: Pure Appl. Chem. *51*, 979 (1979) and *52*, 2303, 2441 (1980)
8. Stoddart, J. F. in: Progress in Macrocyclic Chemistry, Izatt, R. M., Christensen, J. J. (eds.), New York, Wiley 1981, Vol. 2, pp. 173
9. Vögtle, F., Müller, W. M., Watson, W. H.: Top. Curr. Chem. *125*, 131 (1984)
10. Goldberg, I. in: Vol. 2 of ref. 4, pp. 261
11. Weber, E.: in Synthesis of Macrocycles: The Design of Selective Complexing Agents (Progress in Macrocyclic Chemistry, Vol. 3), Izatt, R. M., Christensen, J. J. (eds.), New York, Wiley 1987, pp. 337
12. (a) Cram, D. J.: Science *219*, 1177 (1983);
    (b) Cram, D. J., Stewart, K. D., Goldberg, I., Trueblood, K. N.: J. Am. Chem. Soc. *107*, 2574 (1985)
13. Tabushi, I., Yamamura, K., Nonguchi, H., Hirotsu, K., Higuchi, T.: J. Am. Chem. Soc. *106*, 2621 (1984)
14. Canceill, J., Cesario, M., Collet, A., Guilhem, J., Pascard, C.: J. Chem. Soc., Chem. Commun. *1985*, 361
15. (a) MacNicol, D. D. in: Vol. 2 of ref. 4, pp. 1;
    (b) Farina, M. in: Vol. 2 of ref 4, pp. 69;
    (c) Collet, A. in: Vol. 2 of ref. 4, pp. 97;
    (d) Davies, J. E. D., Finocchiaro, P., Herbstein, F. H. in: Vol. 2 of ref. 4, pp. 407;
    (e) MacNicol, D. D. in: Vol. 2 of ref. 4, pp. 123;
    (f) Gerdil, R.: Top. Curr. Chem. *140*, 71 (1987);
    (g) Giglio, E. in: Vol. 2 of ref. 4, pp. 207
16. Arad-Yellin, R., Brunie, S., Green, B. S., Knossow, M., Tsoucaris, G.: J. Am. Chem. Soc. *101*, 7529 (1979)
17. Popovitz-Biro, R., Chang, H., Tang, C. P., Shochet, N. R., Lahav, M., Leiserowitz, L.: J. Am. Chem. Soc. *100*, 2542 (1978)
18. Iwamoto, T.: Isr. J. Chem. *18*, 240 (1979)

19. Tsoucaris, G., Knossow, M., Green, B. S., Arad-Yellin, R.: Mol. Cryst. Liq. Cryst. *96*, 181 (1983)
20. Toda, F.: Top. Curr. Chem. *140*, 43 (1987)
21. Hart, H., Lin, L.-T. W., Ward, D. L.: J. Am. Chem. Soc. *106*, 4043 (1984)
22. Weber, E.: Private communication, 1987
23. Goldberg, I., Lin, L.-T. W., Hart, H.: J. Incl. Phenom. *2*, 377 (1984)
24. Hart, H., Lin, L.-T. W., Goldberg, I.: Mol. Cryst. Liq. Cryst. *137*, 277 (1986)
25. Goldberg, I., Stein, Z., Tanaka, K., Toda, F.: J. Incl. Phenom. *6* (1988), in press
26. Weber, E., Seichter, W., Goldberg, I.: J. Chem. Soc., Chem. Commun. *1987*, 1426
27. Goldberg, I.: J. Incl. Phenom. *4*, 191 (1986)
28. Goldberg, I., Doxsee, K. M.: J. Incl. Phenom. *4*, 303 (1986)
29. Rudi, A., Reichman, D., Goldberg, I., Kashman, Y.: Tetrahedron *39*, 3965 (1983)
30. Goldberg, I.: J. Am. Chem. Soc. *102*, 4106 (1980)
31. (a) O'Brien, J. P., Allen, R. W., Allcock, H. R.: Inorg. Chem. *18*, 2230 (1979);
(b) Lavin, K. D., Riding, G. H., Parvez, M., Allcock, H. R.: J. Chem. Soc., Chem. Commun. *1986*, 117
32. Allcock, H. R.: Acc. Chem. Res. *11*, 81 (1978)
33. For a new proposal for the classification and nomenclature of host-guest-type compounds see: Weber, E., Josel, H.-P.: J. Incl. Phenom. *1*, 79 (1983)
34. El Murr, N., Lahana, R., Labarre, J.-F., Declercq, J.-P.: J. Mol. Struct. *117*, 73 (1984)
35. Goldberg, I.: J. Incl. Phenom. *1*, 349 (1984)
36. Goldberg, I., Stein, Z.: Acta Crystallogr. *C40*, 666 (1984)
37. Goldberg, I.: J. Am. Chem. Soc. *104*, 7077 (1982)
38. Cameron, T. S., Labarre, J.-F., Graffeuil, M.: Acta Crystallogr. *B38*, 168 (1982)
39. Galy, J., Enjalbert, R., Labarre, J.-F.: Acta Crystallogr. *B36*, 392 (1980)
40. Goldberg, I., Stein, Z., Lin, L.-T. W., Hart, H.: Acta Crystallogr. *C41*, 1539 (1985)
41. Goldberg, I., Lin, L.-T. W., Hart, H.: Acta Crystallogr. *C41*, 1535 (1985)
42. Allcock, H. R. in: Vol. 1 of ref. 4, pp. 351
43. Weber, E., Csőregh, I., Stensland, B., Czugler, M.: J. Am. Chem. Soc. *106*, 3297 (1984)
44. Weber, E.: 4th International Symposium on Molecular Inclusion Phenomena, Lancaster, England (1986)
45. Hart, H., Lin, L.-T. W., Ward, D. L.: J. Chem. Soc., Chem. Commun. *1985*, 293
46. Goldberg, I., Hart, H.: Unpublished results.
47. Hart, H., Lin, L.-T. W., Ng, K.-K. D., Ward, D. L., Goldberg, I., Toda, F.: 4th International Symposium on Molecular Inclusion Phenomena, Lancaster, England (1986)
48. Toda, F., Ward, D. L., Hart, H.: Tetrahedron Lett. *22*, 3865 (1981)
49. Pauling, L. in: The Nature of the Chemical Bond, New York, Cornell University Press 1960
50. Goldberg, I., Stein, Z., Kahi, A., Toda, F.: Chem. Lett. *1987*, 1617
51. Toda, F., Tanaka, T., Nagamatsu, S.: Tetrahedron Lett. *25*, 4929 (1984)
52. Toda, F., Mori, K.: J. Chem. Soc., Chem. Commun. *1986*, 1357
53. Toda, F., Mori, K., Stein, Z., Goldberg, I.: J. Org. Chem. *53*, 308 (1988)
54. Goldberg, I.: J. Am. Chem. Soc. *99*, 6049 (1977)

Functional Group Assisted Clathrate Formation — Scissor-Like and Roof-Shaped Host Molecules

Edwin Weber[1] **and Mátyás Czugler**[2]

1 Institut für Organische Chemie und Biochemie der Universität Bonn, Gerhard-Domagk-Straße 1, D-5300 Bonn-1, FRG
2 Central Research Institute of Chemistry Hungarian Academy of Sciences, Pusztaszeri út 59–67 H-1025 Budapest, Hungary

Table of Contents

A strategy has been developed for the design, synthesis, and testing of new clathrate hosts that possess relationship complementarity to specific guest-compounds.

The new approach starts from particular host geometries (related to scissors or a roof) and makes extensive use of functional group interactions between host and guest molecules allowing planned inclusion properties. Functional sensor groups are characterized as H-bond donors and/or acceptors of different strength. The crystalline supramolecular systems formed in this way are members of the new type of "coordinatoclathrates" (coordinative group assisted clathrates) which usually are more stable than the conventional clathrates. They form highly selectively, and are predictable within certain limits. Also, they provide insight into the elementary interactions of functional groups on which molecular recognition is generally based. For comparative studies, the corresponding apolar host analogues typical of van der Waals interactions are covered as well.

The article is devided into sections which put the emphasis of discussion either on chemical (Sects. 1–3) or on crystallographic aspects (Sect. 4). Section 5 shows points of contact between coordinatoclathrate formation and biochemical problems.

1 The Starting Observation

The outset was an observation by chance concerning 1,1′-binaphthyl-2,2′-dicarboxylic acid (*1*). This compound, when obtained by the common procedure[1], gave an amorphous powder, but when crystallized from ethanol, resulted in colorless transparent crystals [2]. The crystals contain solvent which is retained very strongly in the lattice and resists drying conditions, e.g. vacuum (15 Torr) at room temperature, without appreciable decomposition [3]. Forced drying conditions (0.1 Torr, raised temperature) are required to decompose the adduct, while at ambient conditions it is storable and stable for nearly an unlimited time. From elemental analysis, it follows that two moles of ethanol correspond exactly to one mole of the acid [2]. These observations suggest the presence of a clathrate [4–6] (crystal lattice inclusion, cf. Chapter 1 in Vol. 140 of this series, 'Molecular Inclusion and Molecular Recognition — Clathrates I') where the characteristic groups (carboxyl and hydroxyl) are

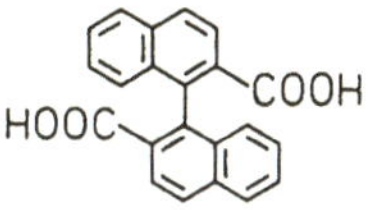

1

important, e.g. assisting the host-guest binding coordinatively; a stimulus to reflect on it.

2 Functional Group Assisted Clathrate Formation

2.1 An Old Matter or a Missing Event?

Polar and coordinatively active functional groups are structural elements frequently found in the constitution of crystal inclusion hosts, mainly including conventional host molecules [7]. Typical examples are urea (*2*), thiourea (*3*), hydroquinone (*4*), Dianin's compound (*5*), deoxycholic acid (*6*) or simply water (Fig. 1). This was the reason to assume that functional groups play an important part in the construction of crystal inclusion compounds.

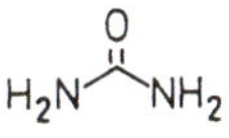

2

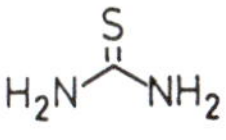

3

4

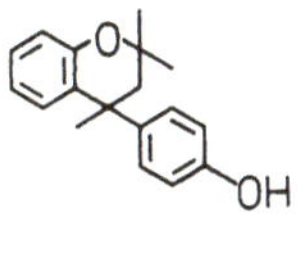

5

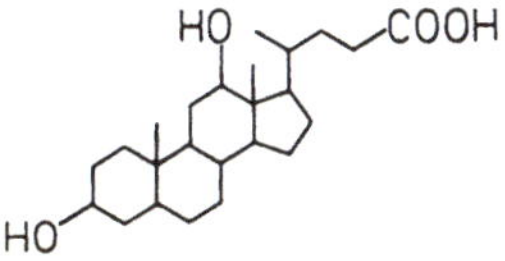

6

Fig. 1. Constitutions of well-known clathrate (lattice inclusion) hosts equipped with polar (H-bond-active) functional groups

Ordinary tetragonal urea (*2*), e.g. on crystallization from appropriate solvents forms a hexagonal non-close-packed crystal lattice which shows long "infinitely" extended channels. They are apt to accommodate solvent molecules or other organic species matching the channel dimensions (e.g. unbranched hydrocarbons)[9–11]. Actually this particular inclusion behavior is an unexpected fact for a molecule with such a low molecular weight (M = 60.0). Looking at the inclusion structures[11], the reason becomes quickly obvious. Figure 2 shows for a n-alkane inclusion compound that the urea molecules form a specific H-bridge network, at which each oxygen is bound to four nitrogen atoms, and each nitrogen to two oxygen atoms of adjacent urea molecules[12]. H-bonds are also responsible of the helical grouping of the urea molecules in the channel wall and thus for the helicity of the respective inclusion lattice[6]. Contribution to a direct binding of guest molecules in the channel interior via H-bonds occurs, however, only in very exceptional cases. There are van der Waals forces well to the fore.

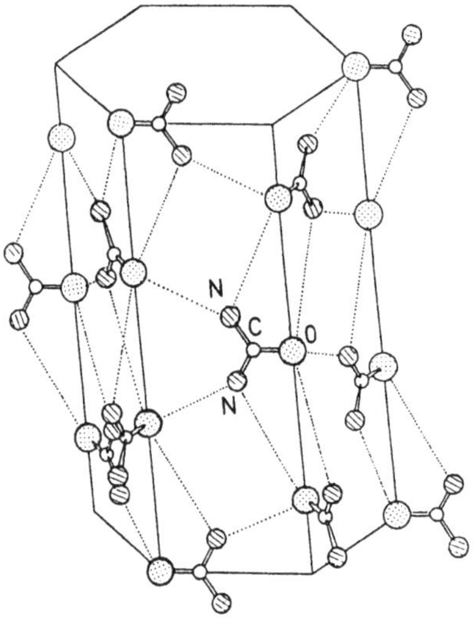

Fig. 2. The H-bonded hexagonal channel network of urea (*2*) typical for a *n*-hydrocarbon inclusion compound (H-bonds as dotted lines) (Adapted from Ref. 12)

A similar behavior is found for thiourea (*3*), except for the channel diameter which is expanded (from 5.25 for urea to 6.1 Å for thiourea); consequently there is enough space available to accommodate more voluminous guest compounds, e.g. branched hydrocarbons[11].

Also deoxycholic acid (*6*) crystallizes in an inclusion lattice with channel-shaped cavities[13]. Figure 3 shows that they are formed by facing molecules of deoxycholic acid[14]. This characteristic structural unit is a double layer of head-to-tail linked deoxycholic acid molecules at which specific H-bridges between hydroxy and carboxy groups are the decisive fact. The channels as such (e.g. in case of the orthorhombic crystal, see Fig. 3) are lined with lipophilic groups. Thus only van der Waals contacts are kept between the included guest molecules (also for polar molecules like acetone, Fig. 3) and the molecules of the channel wall.

In the crystal inclusion compounds of the phenols, e.g. of hydroquinone (*4*) or of

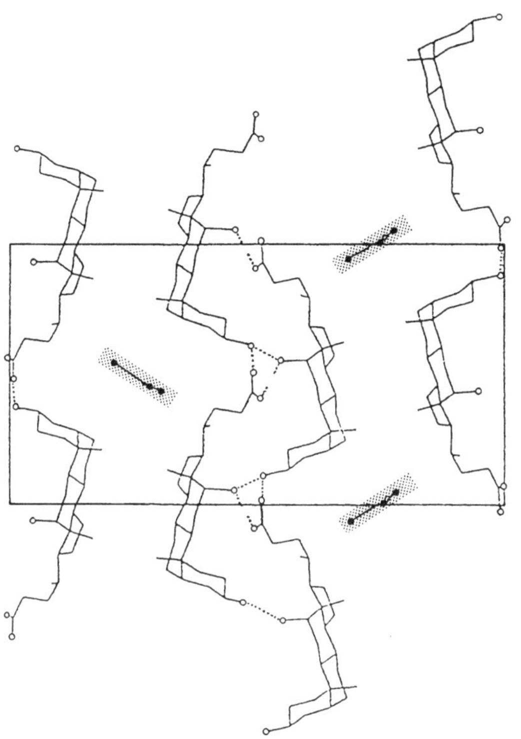

Fig. 3. Channel inclusion compound of deoxycholic acid (*6*) with acetone. The crystal packing is affected by head-to-tail H-bond-mediated double layers of host molecules (H-bonds as dotted lines, guest molecules shaded) (Adapted from Ref. 13)

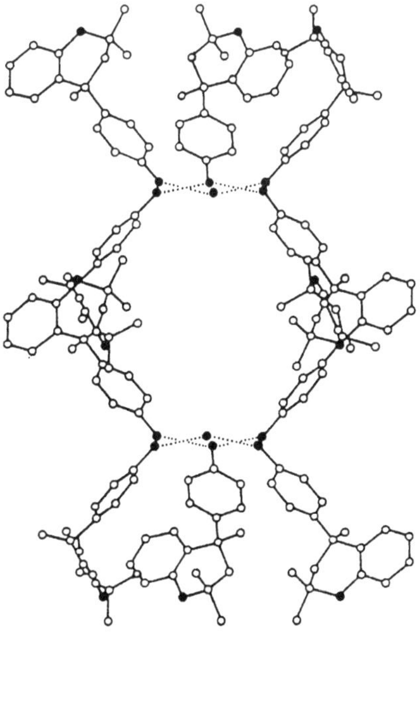

Fig. 4. Inclusion cage of Dianin's compound (*5*). The matrix is constructed via a cyclic H-bonded hexagonal system of host molecules (on top and on bottom of the macrocage; O atoms as bold dots, H-bonds as dotted lines); bulky parts of the host molecules interlock (equatorial of the cage). The cage can be filled with molecules of fitting size (e.g. one molecule of chloroform) (Adapted from Ref. 16)

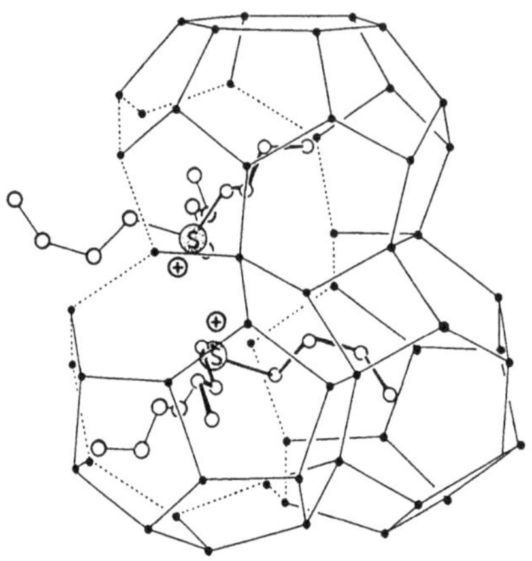

Fig. 5. Constitution of a typical hydrate clathrate (tri-n-butylsulfonium ion as guest). The guests are accommodated within H-bond-mediated water polyhedrons (apexes are equivalent to water oxygens) (Adapted from Ref. 6)

the Dianin's compound (*5*), the hydroxy groups are used to construct hexagonal cyclic H-bridge systems [15,16]. In case of the Dianin's compound (Fig. 4), cages in the crystal lattice are created, having a diameter of approx. 6.2 Å, where spatially fitting molecules from different classes of compounds can be included [16]. The host lattice acts as a sterical barrier.

The same applies to the historic gas-hydrates (hydrate clathrates, Fig. 5) [17,18]. However, on principle, only such molecules are suited for inclusion into the complicated H-bridge networks of gas-hydrates which do not interfere with the H-bridges of water, but have a hydrophobic nature. More recent hosts related to this inclusion principle are given in Chapter 3 of this book.

The examples might have illustrated that functional groups (e.g. OH, COOH, NH_2), as they are a component of classical crystal inclusion compounds[4-7], are usually used for construction, cross-linking, and stabilization of the host lattice (Fig. 6a), and are not used, as could have been, for direct binding of guest molecules, e.g. via coordination or H-bonding (Fig. 6b). To speak with a newly developed classification system on inclusion compounds [19] (see Chapter 1 of Vol. 140), those are "true" clathrates and not "coordinatoclathrates" (cf. Fig. 6, for a more detailed specification see Fig. 15 in Chapter 1 of Vol. 140). As in the case of urea and thiourea, a rather stable, but nearly invariable host lattice with rigidly

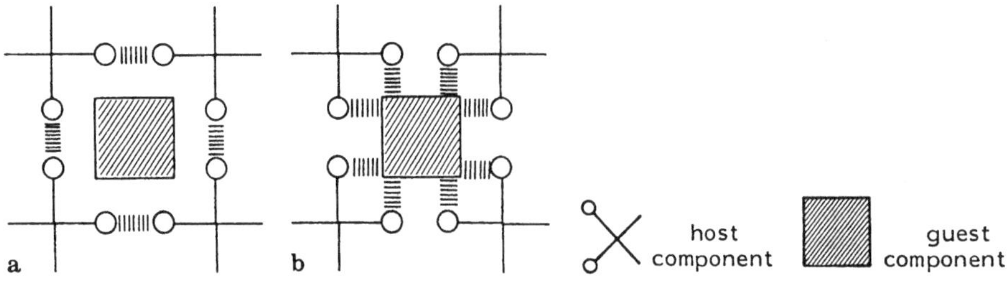

Fig. 6. Diagrammatic (two-dimensional) representation of different modes of lattice inclusions involving coordinative (H-bond) interactions (indicated by broken lines): **(a)** cross-linked matrix type of inclusion (host-host interaction, "true" clathrate); **(b)** coordinatoclathrate type of inclusion (coordinative host-guest interaction, coordination-assisted clathrate)

definite cavity dimensions results. That is why the chemical nature of inclusion partners in case of such crystal compounds (with exception of gas hydrates) generally play only a minor part. Whether a molecule is included or not, depends at first approximation on its size and shape (guest selection by molecular size and shape). Examples, deviating from this observation, are rarely found in the literature before 1984 or show no general principle [20]. To define such a principle is subject of the following section.

2.2 Nature of the Concept (Coordinatoclathrate Concept) and Basic Host Design

It is commonly accepted [21] but unwritten law: bulkiness and crystal inclusion are closely related. In Figs. 2–5 (Sect. 2.1) this principle is confirmed by bulky cross-linking of actually non-bulky host molecules, e.g. using polar groups ("bulky crystal network"). By way of contrast, we aim at transferring the element of bulkiness into the host skeleton ("molecule inherent bulkiness") and use the polar groups differently, e.g. for direct guest binding. This is the basic tenor of the new strategy and serves as a general definition regarding former reflections (cf. hydroquinone). A more detailed description is given below.

The origin of the strategy of "*coordinatoclathrate inclusion*" was the desire to get hold of more, and above all, more highly effective tools for specific host/guest adjustment [22]. As mentioned above, conventional clathrate inclusion formers (see Fig. 1), and in addition many host-compounds of recent date, are mainly qualified to select guest molecules according to size and shape. Imagine a mixture of molecules with comparable dimensions, but belonging to different classes of substances (e.g. acetone, isopropanol, 2-chloropropane), they should possibly be selected by chemical aspects, too.

Beside a crystal cavity of suitable size, additional information is required, e.g. designed polarity gradients in the cavity or carefully located and specific binding sites, respectively. At best, the crystal inclusion will enjoy an ideal "lock and key" relation [23] between host and guest components from a chemical and spatial point of view.

The latter could be obtained, if particular donor substituents (e.g. specific functional groups) facing corresponding acceptor groups of the guest molecule are added to the host and vice versa, thus using specific host-guest interactions of polar

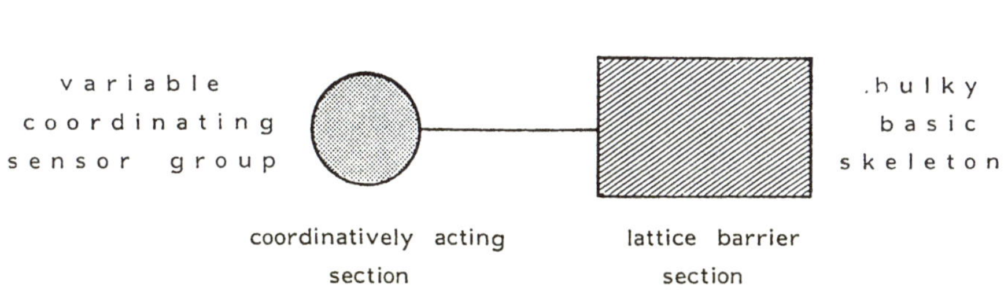

Fig. 7. Abstracted structure of a host molecule characteristic of coordinatoclathrate formation [2]

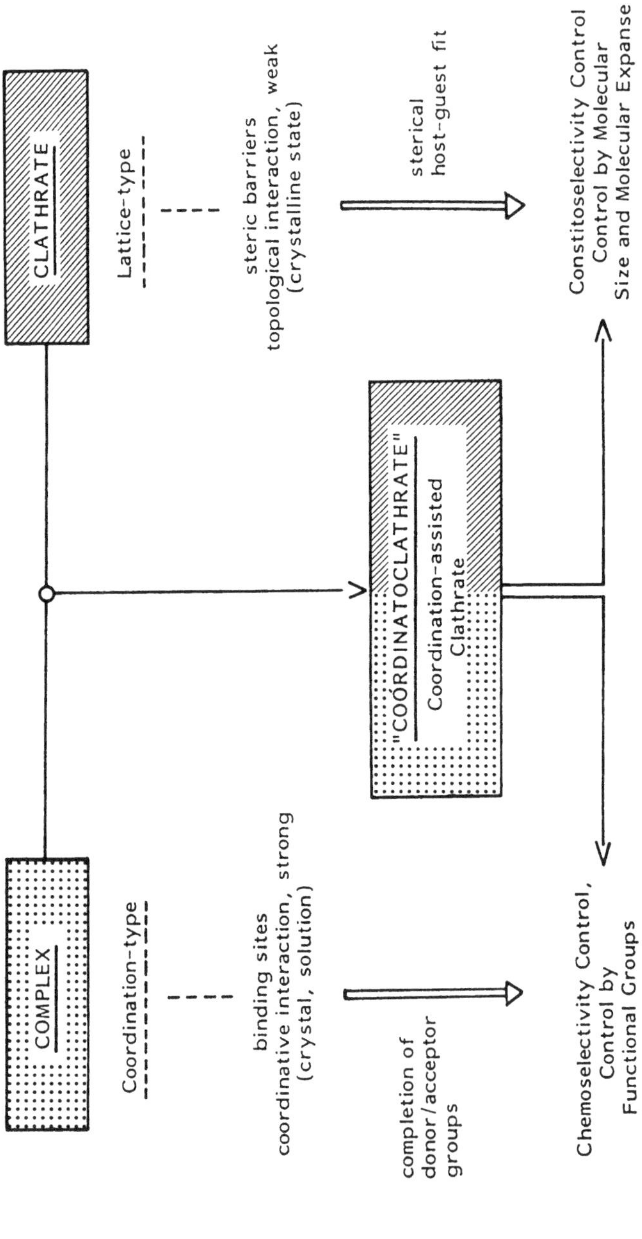

Fig. 8. Coordinatoclathrate concept: definitions, relations, and functions of control [2]

nature such as dipol-dipol attraction or H-bonds at guest selection [24] (see Fig. 6). Basic problems are to prevent or to disguise interactions with molecules of the same kind, i.e. host with host or guest with guest. This is not trivial, since many polar groups (e.g. C—OH, COOH etc.) are simultaneously donors and acceptors. Consequently high inclusion selectivities depend on a very precise selection and balance of polar and functional groups in the host-guest unit. Here, bulky parts of the molecular framework are important. They are required to construct cavities, i.e. non-close-packed host lattices (see above), but also to prevent the polar groups of the individual host molecules from contacting one another undisturbed, i.e. satisfying themselves.

Accordingly the most important feature of a host compound designed for coordinatoclathrate formation [2] is a bisection in the molecule, as schematically shown in Fig. 7, providing:

1) a *bulky basic skeleton* which makes the clathrate-typical lattice cavities available,
2) *appended functional groups (sensor groups)* which manage the coordination to the included guest substrate.

Combining features typical of both complexes and clathrates (coordinatoclathrate) should provide new possibilities of host-guest control [2]. They are indicated by the relations specified in Fig. 8, e.g. chemoselectivity or selectivity for functional groups on the one hand, caused by the complex part, and on the other hand constitutional selectivity or selectivity for molecular size and expanse due to the clathrate branche of the diagrammatic family tree of a coordinatoclathrate shown in Fig. 8.

Another advantage of using oriented polar and coordinative bonds between the host and the guest molecule, as is intended by the coordinatoclathrate principle, is facilitation of lattice design. Moreover, it is assumed that a possible coordinatoclathrate inclusion, because of the relatively strong binding forces (dipole-dipole attraction, H-bridges) [24] acting in the host-guest unit, involves little disorder and higher stability than conventional-type clathrates for which weak van der Waals interactions in the crystal aggregates are more typical [25] (cf. Sect. 2.1). It is important to give the hereby shown strategy a background based on molecules.

3 Suggested Coordinatoclathrate Hosts and Chemical Proving of the Concept

3.1 General Considerations

The "coordinatoclathrate principle" is intended for host molecules endowed with both a bulky basic skeleton and appended sensor groups. This design of a general structure requires some differentiations.

Molecules aim at being packed as closely as possible in the crystal [26–28]. *Bulkiness* of a molecule, however, is an impediment. It presupposes a molecular constitution rather unbalanced in its spatial dimensions and a certain extent of conformational rigidity. Here we encounter a parallel to an everyday occurrence: trunks can easily be arranged to a compact ordered stack, unlike bulky root stocks which cannot. The latter, by contrast, yield a relatively disordered, labile stack interspersed with voids. The same holds for molecules.

Here, however, it is possible to obtain stabilization of the low-dense lattice build-up of bulky molecules via intermolecular adhesion and orientation forces. Molecules with planar structural elements are advantageous in this respect since they are apt to support the lattice aggregate, and at the same time they are able to partition off cavities effectively. It is very convenient to use aromatic units.

Figure 9 goes more deeply into these considerations. The geometric figure (A) which is obtained by cross-connection of two plane subunits is representative of a bulky molecule. There are several possibilities of arranging these molecules in a two-dimensional pattern (top view), e.g. two terminals of the cross-shaped molecules in each case are in contact giving a molecular packing of particularly low density (a) (free space shaded), or (b), (c), and others where the free space is further subdivided and modified. How the host molecules are arranged (a–c) depends on the polarity conditions, e.g. whether hydrophilic or hydrophobic terminals or regions of the molecules touch.

Figure 10 where some facts are assumed in respect to molecular polarity and the type of arranging the molecules explains that the lattice cavities are not of uniform chemical nature, but are subdivided in polar (broken circles) and apolar regions (continuous circles). On possible inclusion of guest molecules, the guest will be oriented according to the polarity gradient, or interact coordinatively with the host lattice ("coordinatoclathrate principle"). Considered in three dimensions, a decisive factor is, whether a newly added layer of molecules arranges in a congruent or a shifted position. In the former case channels, in the latter case cavity-type voids are created in the respective crystal lattice.

Evidently, *symmetry attributes* of the host (and guest) molecules play also a role in these considerations [26,29]. We prefer to design host molecules by using a two-fold symmetry element [30]. A direct connection between host symmetry and the coordinatoclathrate principle, however, does not necessarily emerge from the given statement, though, of course, symmetry conformity of host and guest molecules are general parameters of selection.

Concerning the *nature of the sensor groups* at the host molecule, one should aim at binding contacts to the corresponding guest molecule as selective and strong as possible. Functional groups qualified as H-bond mediators meet the requirements

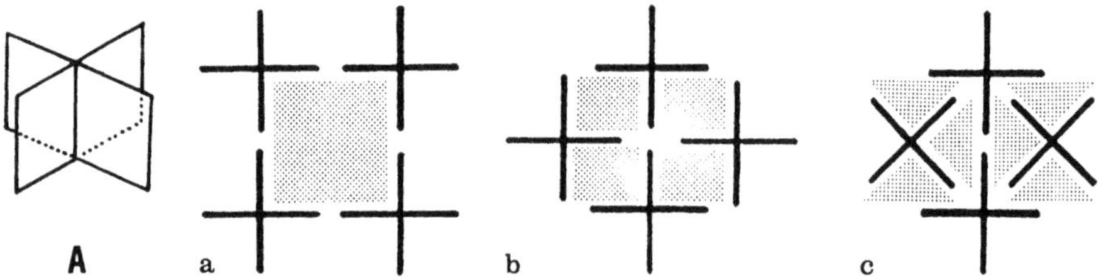

Fig. 9. Diagrammatic representation of a bulky host constitution (A) and **(a)**—**(c)** of crystal lattice-analogous arrangements of A (two-dimensional versions; shaded areas represent the lattice voids)

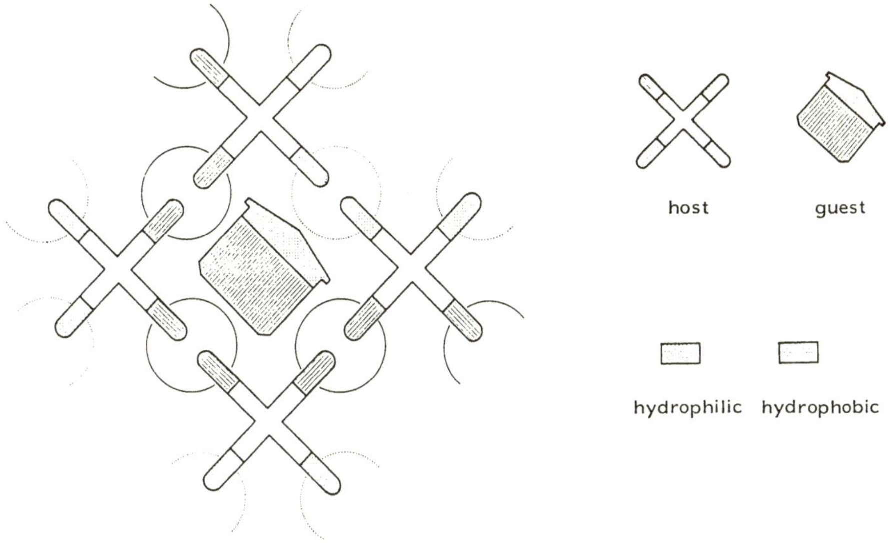

Fig. 10. Relation between amphiphilic nature and orientation of crystal lattice and guest. Hydrophilic and hydrophobic interactions are indicated by circles in dotted and non-interrupted style of line representation, respectively

best. Thus, different OH-, NH-, CO-groups, and other functionalities containing heteroatoms are suitable behaving either like H-donors or H-acceptors, or both at the same time.

Within the limits of this article, we attached importance to keeping the range of the used sensor groups easy to survey and to limiting variation in respect to the bulky basic skeleton to only a few selected structures which are closely connected with the geometric figure shown in Fig. 9, i.e. molecules shaped like scissors or roofs.

3.2 1,1′-Binaphthyl-2,2′-dicarboxylic Acid (*1*) — The Initial Touchstone

3.2.1 Structural Relation to a Pair of Scissors

A common pair of tailor's scissors (Fig. 11a) has the twofold symmetry (C_2) we prefer. On opening the edges, the scissors gain bulkiness. One can say a pair of scissors is "polar", too. One part serves for cutting, the other for handling.

It is possible to translate the symbolism of the scissors and its basic structure into the field of molecules (cf. Fig. 11b). Indeed, the scissor-shaped bulky binaphthyl compound equipped with two appending carboxy groups *1* (1,1′-binaphthyl-2,2′-dicarboxylic acid, Fig. 11c), as mentioned at the beginning (Sect. 1), strictly meets the general structure of an assumed coordinatoclathrate host (cf. Fig. 7). Also, the compound is in keeping with the considerations on an expectedly favorable lattice build-up (see Sect. 3.1). For checking, the crystal inclusion properties of *1* were studied in detail [2].

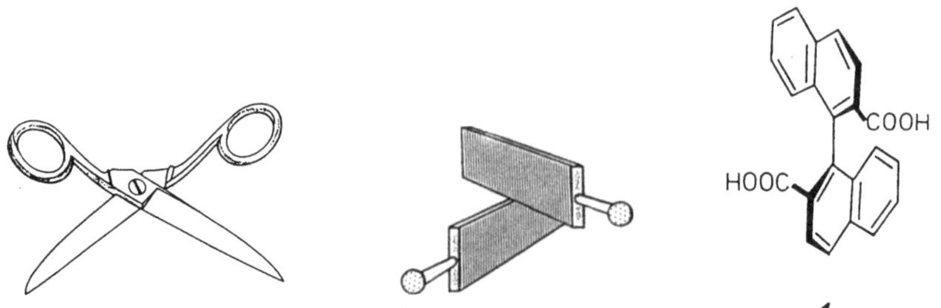

a b c *1*

Fig. 11. Graphic development **(a)–(c)** of a molecular constitution (coordinatoclathrate host *1*) compared to a pair of scissors

3.2.2 Formation of Inclusion Compounds, Host-Guest Stoichiometries

The question arises, whether and to what extent the dicarboxylic acid *1* is capable of binding other solvents besides ethanol (starting observation, cf. Sect. 1) in the crystal lattice. For this purpose, to begin with, crystallization experiments using further alcohols (straight-chain, branched, univalent and polyvalent) were carried out. It was found that *1* is apt to form crystal inclusions on a large scale, i.e. with alcohols of various constitutions. A list of different examples is given in Table 1 (Entries 1–16).

It is important to note that, depending on the spatial requirements of the used alcohol and on the number of hydroxyl groups existing in the alcohol, different stoichiometries (*1*: alcohol) predominate in the crystal lattice. For instance, methanol (Entry 1) and ethanol (Entry 2) are accommodated in the crystal of *1* with 1:2 stoichiometry, whereas the isomeric butanols (Entries 5–8) give rise to 1:1 stoichiometry. The same holds for the inclusions of benzyl alcohol (Entry 13) and of trichloroethanol (Entry 14) as bulky guest substitutes and for those of the bifunctional representatives ethylene glycol (Entry 15) and propylene glycol (Entry 16).

Since H-bonding as a supporting factor for inclusion formation seemed very likely, one passed over to solvents having a very high potency for proton donorship. In this context it was found that carboxylic acids have also an affinity to intercalate into the lattice of *1*. Examples belonging to this sort of host-guest inclusion were obtained for acetic, propionic, and lactic acids (Entries 17–19); the stoichiometries vary from 2:3 to 2:1. But also in the area of carboxylic acid derivates free of hydroxyl, by preference amides, many fewer esters which are suitable partners for inclusion formation are found. For instance formamide, N-methylformamide, and N,N-dimethylformamide (Entries 20–22) all led to inclusions with 1:2 (host: guest) stoichiometry, whereas the host-guest aggregates formed with esters, e.g. dimethyl carbonate or diethyl carbonate (Entries 23 and 24), gave no reproducible stoichiometry ratio after drying under usual conditions (cf. Table 1) because of their low stability.

On the other hand, CH-acidic solvents such as acetylacetone, acetonitrile, nitromethane, and dimethyl sulfoxide (Entries 25–28) yield stable crystal inclusions, each having a strict 1:1 stoichiometry. Finally, respective crystallization experiments using solvents of even less polarity or ability to form H-bonds have been carried out. The

Table 1. Clathrate inclusion compounds of *1*: stoichiometries and thermal stability characterization

Entry	Guest compound	Host:guest mol ratio[a]	Thermal dec [°C][b]
1	methanol	1:2	146 (+82)
2	ethanol	1:2	88 (+10)
3	1-propanol	2:1	72 (−25)
4	2-propanol (isopropanol)	1:2	86 (+ 4)
5	1-butanol	1:1	72 (−46)
6	2-butanol (sec-butanol)	1:1	92 (− 7)
7	2-methyl-1-propanol (isobutanol)	1:1	71 (−37)
8	2-methyl-2-propanol (t-butanol)	1:1	141 (+58)
9	1-pentanol	1:2	123 (−15)
10	2-methyl-1-butanol	2:1	135 (+ 7)
11	2-methyl-2-butanol	1:2	164 (+62)
12	4-methyl-1-pentanol	1:1	154 (− 9)
13	benzyl alcohol	1:1	120 (−85)
14	trichloroethanol	1:1	117 (−34)
15	ethylene glycol	1:1	165 (−32)
16	propylene glycol	1:1	149 (−66)
17	acetic acid	2:3 (1:1)	115 (− 3)
18	propionic acid	2:1	139 (− 2)
19	lactic acid	1:1	140 (−)[e]
20	formamide	1:2	136 (−74)
21	*N*-methylformamide	1:2	108 (−75)
22	*N*,*N*-dimethylformamide (DMF)	1:2	117 (−35)
23	dimethyl carbonate	2:1[c]	<25
24	diethyl carbonate	[d]	<25
25	acetylacetone (2,4-pentanedione)	1:1	64 (−70)
26	acetonitrile	1:1	119 (+38)
27	nitromethane	1:1	126 (+25)
28	dimethyl sulfoxide (DMSO)	1:1	155 (−34)
29	diethyl ether	[d]	<25
30	bromobenzene	1:1	116 (−40)

[a] Determined by NMR integration after a drying period of 12 h at 0.5 torr for each compound. [b] Value indicates the beginning of the clathrate decomposition (either onset of opacity or release of the gaseous component). Specification in parentheses gives the relative thermal stability (difference between the decomposition point of the clathrate and the boiling point of the respective neat guest solvent at atmospheric pressure). [c] Decomposes under vacuum drying at ambient temperature. [d] Unstable at atmospheric pressure. [e] No description of the boiling point for the neat guest solvent in the literature.

results show that attractive powers emanating from such solvent components are, as a rule, insufficient to form a stable inclusion aggregate. We were just able to isolate a 1:1 stoichiometric inclusion compound of the relatively non-polar, but difficultly vaporizable bromobenzene (Entry 30), whereas diethyl ether as a guest component seems too volatile to be retained in the host lattice under ambient conditions [2].

Amines are expected to form a salt in contact with *1*. For the present, compounds like amines have generally been regarded as less suitable guest components. Nevertheless, highly interesting conditions were found in case of the aggregate where *1*, imidazole, and water are combined in a single crystal lattice. The special features of this inclusion aggregate are reported separately (Sect. 5.1).

3.2.3 Stabilities and Decomposition Properties of the Crystal Inclusions

With some exceptions, the isolated crystal inclusion compounds are almost indefinitely stable under ambient conditions (cf. Table 1) and allow storage in air over a long period with no appreciable decomposition [2]. Also under vacuum conditions (0.5 Torr, 12 h, 25 °C) the majority of the inclusion compounds, being discussed here, exhibit a remarkably strong guest fixation (cf. Sect. 1). Only on heating (under reduced pressure, 0.5 Torr), a decomposition with growing opacity of the formerly transparent crystals or with spontaneous bursting of the crystal and evolution of a gas is observed within a specific temperature range for each compound (Table 1).

In the case of the methanol, ethanol, t-butanol, 2-methyl-2-butanol, acetonitrile, and nitromethane inclusion compounds (Entries 1, 2, 8, 11, 26, and 27 in Table 1), these temperature ranges lie above the boiling point of the corresponding guest solvent (cf. values given in parentheses in Table 1). It is shown most remarkably by methanol, for which the boiling point and the point for the beginning of the decomposition differ by 70 °C. This indicates a particularly strong clathration (e.g. comparable to the corresponding tri-o-thymotide solvent inclusion which is representative of high thermal stability and strong clathration [31]. For the bulk of the isolated inclusion compounds, however, the decomposition point falls rather into the range of the boiling point of the corresponding pure guest component or just below. It may illustrate a way to distinguish between thermally stable and less stable inclusion compounds.

IR spectra (KBr) of the crystalline inclusion compounds highlight bands which document a coordinative host-guest interaction in accordance with our concept. The values for the OH-stretching modes of the lattice-included alcohols, e.g. indicate strong H-bridge bonding as they are very close to those found in the pure liquid alcohols rather than of the alcohols in the gaseous state (e.g. 3374 in the lattice-included species versus 3676 cm^{-1} for gaseous ethanol) [32]. The absorptions in the 1700 cm^{-1} region arising from the carbonyl groups of the host molecule, as expected, suggest stronger interactions in the alcohol cases than in the others. Results of solid-state NMR measurements (^{13}C—CP-MAS spectra) are not relevant to the problem, at least for the time being.

3.2.4 Guest Selectivity Behavior, Separation of Solvent Mixtures

If mixtures of two or more potential guest molecules are offered, the host lattice of *1* allows the selective accommodation of solvent molecules [2]. In many cases, a pratically 100% discrimination of one guest species (>95% by NMR integration) is achieved by a single crystallization process using *1*, e.g. from an equimolar two-component solvent mixture. Table 2 summarizes important results (Entries 1–15).

Selection at inclusion formation (see relative guest excess) is derived from steric as well as from chemical points of view. Hence, high discrimination is found for a combination of potential solvents differing in the functional group characteristic, e.g. if belonging to different classes of substances (see Entries 8–10). But also within the same class of substance, so far as a series of homologues and different substituted or branched compounds are concerned, discrimination is effected in up to a 90% ratio [33].

Table 2. Preference of guest binding of host *1* from a two-component solvent system

Entry	Recrystalln Solvent compd mixture (equimol ratio)	Relative guest excess, % g.e.
1	methanol/*ethanol*[a]	46[b]
2	methanol/*t-butanol*	91
3	*methanol*/toluene	14
4	*ethanol*/2-propanol	79
5	*ethanol*/t-butanol	92
6	ethanol/*benzyl alcohol*	20
7	ethanol/*ethylene glycol*	>95
8	*ethanol*/acetic acid	>95
9	ethanol/*dimethylformamide*	>95
10	*ethanol*/acetonitrile	>95
11	1-propanol/*2-propanol*	29
12	1-butanol/*t-butanol*	74
13	isobutanol/*t-butanol*	78
14	acetonitrile/*dimethyl sulfoxide*	>95
15	*acetonitrile*/toluene	42

[a] Solvents printed in italics refer to those preferentially enclathrated.
[b] Determined by NMR integration of the isolated crystals after a drying period of 12 h at 0.5 torr.

For instance, the crystallization of *1* from an equimolar mixture of methanol/ethanol gave a distinct discrimination of methanol and yielded the corresponding ethanol inclusion compound in a relative guest excess (g.e.) of 46% (Entry 1). The discrimination coming from a 1:1 mixture of methanol/t-butanol was found to favor clearly t-butanol (g.e. 91%, Entry 2). This finding again shows the preferred binding of the higher homologues of methanol by the host lattice. However, in the case of the solvent mixtures ethanol/2-propanol and ethanol/t-butanol, the lower homologue, that is ethanol, is the preferred inclusion component (Entries 4 and 5).

A remarkable occurrence with the high preference for ethanol (g.e. 95%) takes place with the solvent pair ethanol/acetic acid (Entry 8) since the removal of acetic acid from any polar media is observed to be difficult. Preference for ethanol (g.e. >95%) is also observed in the presence of acetonitrile (Entry 10), but not in the presence of dimethylformamide (Entry 9). Here the ratio is reversed (>95% g.e. of dimethylformamide). In respective mixtures with ethylene glycol or benzyl alcohol, ethanol is not the favored guest component either (Entry 6 and 7). The high preference for ethylene glycol with respect to ethanol may proceed from a more extensive H-bonding pattern in the inclusion lattice in case of the glycol because of its bivalency. Benzyl alcohol, in the main, seems structurally well adapted to the aromatic moieties of the host molecule. The results of Entries 11–13 are emphasized by the fact that the higher branched analogues of propanol and butanol are favored for inclusion into *1*.

The practically complete discrimination of acetonitrile in favor of dimethyl sulfoxide (Entry 14) is also remarkable since both solvents are of the same category (dipolar aprotic) and, in addition, they have comparable polarities [34]. These facts are retained even when acetonitrile is of tenfold excess in a respective mixture with dimethyl sulfoxide. In comparision, the preferred uptake of acetonitrile in a mixture with toluene (Entry 15) is a relatively trivial differentiation of molecules.

Selectivity at formation of a respective inclusion compound and its thermal stability behavior might differ (cf. Tables 1 and 2), since for both representations different processes should be taken into consideration. Formation of a crystal inclusion compound is normally controlled by kinetics, whereas the thermal stability (decomposition property) is a result of thermodynamics. Thus, we speak of "formation selectivity", on the one hand, and of "binding selectivity", on the other.

3.3 Scissor-like Host Molecules Having Altered Building Blocks

3.3.1 Examples of Selected Compounds

To probe the range of application of the new inclusion strategy (coordination-assisted clathrate formation) in different ways, directed structural modifications were undertaken starting from the basic constitution *1*, for instance as to the molecular skeleton (basic structure) and/or the sensor section (functional groups). The formulae *7–24* show different possibilities of such variations.

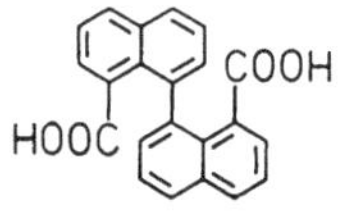

7

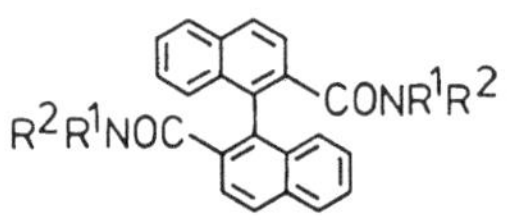

9 R^1 =H, R^2=H
10 R^1 =H, R^2=CH_3
11 R^1 =CH_3, R^2= CH_3

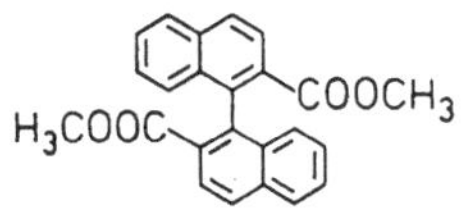

8

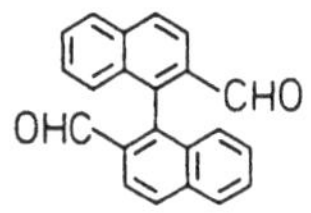

12

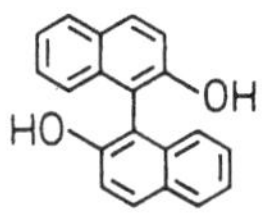

13

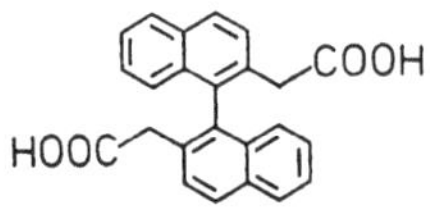

14

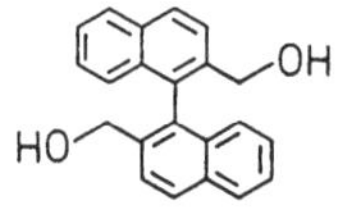

15

The fundamental components (basic skeleton and functional groups), of which *7* [35] is made up, are the same as in *1*, however, they are linked in a different way (8,8'- instead of 2,2'-position). On the other hand, compounds *8–13* (*8* [1], *9* [36], *10–12* [37], *13* [38]) have a skeleton identical to *1* and the same positioning of substituents, but contain functional groups differing from *1*, e.g. $COOCH_3$, $CONH_2$, $CONHCH_3$, $CON(CH_3)_2$, CHO, OH, respectively. At *14* [37] and *15* [1] (cf. *1* and *13*, respectively), modification is associated with a relation of homology (insertion of one CH_2 unit, each, between skeleton and functional group).

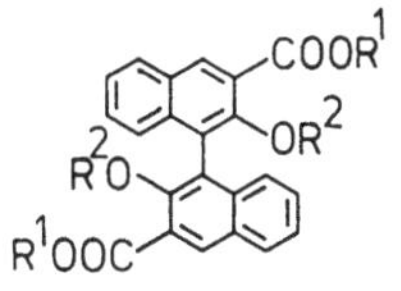

16 R^1 =H, R^2=H
17 R^1 =H, R^2=CH_3
18 R^1 =CH_3, R^2=CH_3

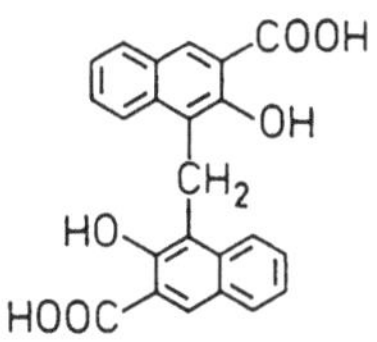

19

Enlargement to four functional groups in all (two hydroxyl and two carboxyl groups) is characteristic of *16* [39]; *17* [37,40] and *18* [37,40] are examples of comparison. Compound *19* [41] has an additional CH_2-group inserted into the rotation axis of the molecular hinge compared with *16*. Strictly speaking, this is no longer a molecule shaped like a pair of scissors. Enlargements including the basic skeleton, but without touching the scissor-like structure, are displayed in the bianthrylcarboxylic acids *20* [37] and *21* [42].

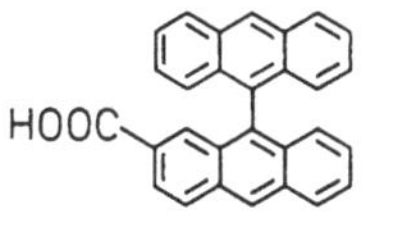

20

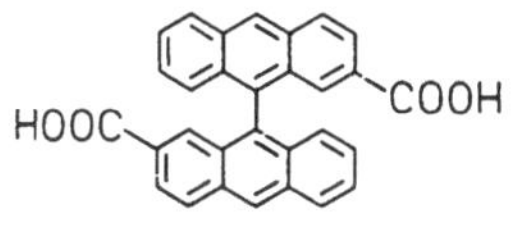

21

A completely different access to molecules with a scissor-like shape is opened up via spiro-linkage. Typical examples following this building principle are the spiro-bifluorenes *22* [43] and *23* [43] which differ also in the functional groups. In case of the spirocompounds, the flexible hinge is not applicable and the edges of the molecular scissors are fixed at an angle of 90°.

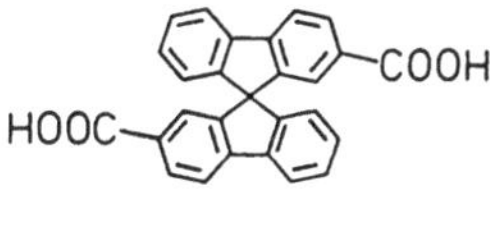

22

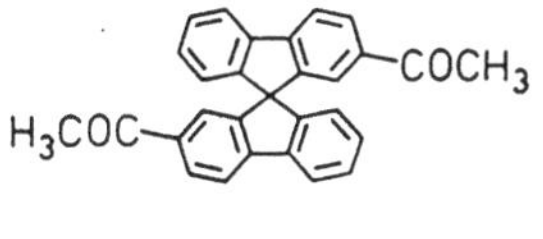

23

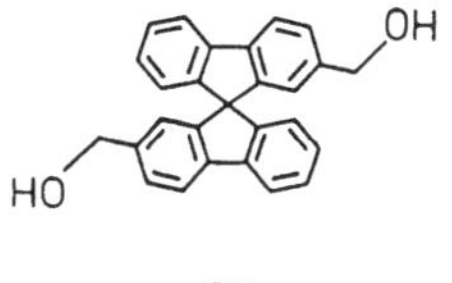

24

3.3.2 Inclusion Properties

Because of the superior inclusion properties exhibited by *1* [2], one may infer that the positional isomer with reference to the carboxylic group, 7, also behaves as a good inclusion host. However, this is not true. So far, we have not succeeded in isolating any inclusion compound of an uncharged molecule using *7*, except those of a salt-like nature (see Sect. 4.2.2.). Obviously the functional groups of *7* are located in the molecule in a way that works against the net bulkiness of the skeleton (connected with the crystal build-up); in *1* they cooperate.

1,1′-Binaphthyls incorporating functional groups other than carboxyl (e.g. carbonamide, *N*-methylcarbonamide, formyl, hydroxyl, but not methoxycarbonyl and *N*,*N*-dimethylcarbonamide), however in the privileged 2,2′-position, again show crystal inclusion properties (Table 3) [37]. For instance, the biscarbonamide *9* forms stable inclusion compounds with dimethylformamide, acetic acid, propionic acid, and dioxane. By contrast, the *N*,*N*-dimethyl-substituted analogue *10* yields only an inclusion with acetic acid which is probably a result of the reduced H-bond

Table 3. Clathrate inclusion compounds of scissor-like hosts other than *1*

Host no	Guest compound	Host: guest mol ratio[a]	Host no	Guest compound	Host: guest mol ratio[a]
9:	acetic acid	1:1	*10*:	acetic acid	1:1
	propionic acid	1:1			
	dimethylformamide	1:1	*12*:	nitromethane	2:1
	dioxane	1:1		pyridine	2:1
				dioxane	2:1
13:	methanol	1:2			
	cyclopentanol	1:2	*16*:	dimethylformamide	1:2
	ethylene glycol	2:3		1-propanol,	[b]
	lactic acid	1:2		t-butanol	
	cyclopentylamine	1:1		dimethyl sulfoxide	[b]
	diisopropylamine	1:1	*17*:	bromobenzene	2:1
	di-t-butylamine	2:3		toluene	2:1
	dicyclohexylamine	1:1		acetone	[b]
	2,5-diamino-2,5-dimethylhexane	1:1		dimethyl sulfoxide	[b]
	4-chlorobenzylamine	1:1			
	4-hydroxybenzylamine	1:1	*19*:	dimethylformamide	1:2
	3-methylaniline	1:1			
	3,5-dimethylaniline	1:1	*20*:	dimethyl sulfoxide	1:1
	2,6-dimethylaniline	1:1			
	3-hydroxyaniline	1:2	*22*:	ethanol	2:1
	2-amino-6-methylpyridine	1:1		2-propanol	1:1
	imidazole	1:2		dimethylformamide	1:2
	piperidine	1:1		dioxane	1:1
	dimethylformamide	2:3		benzene	1:1
	dimethyl sulfoxide	1:2			
	acetone	1:1	*23*:	tetrahydrofuran	2:1
	acetylacetone	1:2		benzene	1:1
	dioxane	2:3		1-bromopentane	2:3
	tetrahydrofuran	2:3			

[a] Determined by NMR integration as specified in Table 1. [b] No clear stoichiometry.

mediatorship and *11*, lacking any mobile H, is completely ineffective. For the dialdehyde *12* mainly H-acceptor properties were assumed, hence protic guest molecules should be favored. Surprisingly, up to now, no crystal inclusions of *12* either with alcohols or with acids as guest molecules have been found, but only those with dipolar-aprotic solvents (Table 3). This behavior indicates, that the formyl substituent is a rather poor sensor group, e.g. in respect to hydrogen bonding.

Compound *13*, having two phenolic hydroxyls as potential sites instead, is rather different in its behavior [37]. No doubt, this compound is inferior to *1* in the variety of forming crystal inclusions as far as alcohols and acids as guests are concerned. However, it is an advantage that different amines (primary, secondary) or pyridines are included in the lattice to form inclusion compounds other than being derived from simple salt formation. Amines are even the favored inclusion partners of *13* as shown by the number and stability of the isolated inclusion compounds (Table 3). A prospective interpretation of this exceptional position is hydrogen bonding which seems particularly effective in the present host-guest combination. Recently Toda and Goldberg [44] reported the successful (chiroselective) inclusion of sulfoxides and phosphine oxides into the host lattice of *13* (see Chapter 3 in volume 140 of this series and Chapter 1 of this book).

Considering the competitive experiments, e.g. from two-component solvent mixtures (see Table 4), it is not surprising that the amines are always favored on inclusion formation (Entries 1—5). A remarkable point is also the ability of *13*, properly speaking its crystal lattice, to accommodate relatively voluminous guest molecules, among them many ring compounds (Tables 3 and 4). Hence, *1* and *13* are complementary in their inclusion behavior to some extent.

Hoping to achieve selectivity with regard to spatially more demanding guest species as in the case of *1*, the acetic acid-analogous compound *14* has been synthesized. Evidently the additional methylene units neighboring the functional groups, however, allow *14* too much conformational flexibility. Accordingly no lattice inclusion properties are observed from our tests [37]. The same holds for *15* (an analogue of *13* extended by a methylene group at each substituent).

From the beginning, compound *16* was expected to show only moderate "coordinatoclathrate behavior" since it should have a propensity for *intra*molecular H-

Table 4. Selective guest inclusion of host compound *13*

Entry	Recrystalln solvent compd mixture (I/II)[a]	Host:I:II mol ratio[b]
1	dimethylamine/acetone	1:1:0
2	diisopropylamine/acetylacetone	1:1:0
3	dicyclohexylamine/methanol	1:1:0
4	dicyclohexylamine/cyclopentanol	1:1:0
5	dicyclohexylamine/diphenylamine	1:1:0
6	morpholine/dioxane	1:x:y[c]
7	acetone/acetylacetone	1:x:y

[a] Equimolar ratio. [b] Determined by NMR integration as specified in Table 1.
[c] No clear discrimination in favor of I or II.

bonding rather than *inter*molecular H-bridge formation with guest molecules as binding partners regarding the salicylic acid subunits. The experimental finding [37] confirms this presumption. The only inclusion compound having a defined stoichiometry which has been isolated is the 1:2 crystal inclusion with dimethylformamide (Table 3); besides 1-propanol, t-butanol, and dimethyl sulfoxide are accommodated with no clear stoichiometry.

The methoxy analogous compound *17* is not expected to have the same high potential of forming intramolecular hydrogen bonds opposing a possible host-guest binding. Nevertheless, we succeeded in the isolation of crystal inclusion compounds with typical aprotic (dipolar and apolar) guest molecules like dimethyl sulfoxide, acetone, bromobenzene [45], and toluene (Table 3). Moreover, the respective inclusion stoichiometries cause some difficulties in reproduction. The most unexpected result however is that *17* does not form a crystal inclusion with dimethylformamide; the tetramethyl compound *18* gave no crystal inclusion at all [37].

On the other hand, since the angular derivative *19*, whose constitution is characterized by two quasi-isolated hydroxynaphthalenecarboxylic acid subunits and whose structural analogy to a pair of scissors is removed for the most part, also yields an inclusion compound with dimethylformamide with strict stoichiometry of 1:2 [37], and only and exclusively this one (Table 3), it is obvious that the free salicylic acid unit might be the decisive factor for the preferred binding of dimethylformamide of this class of compounds.

The unequivocal proof is furnished by the crystal inclusion behavior of simple 2-hydroxy-3-naphthalenecarboxylic acid *25a* [46], and its 1-chloro derivative *25b* [37], since both allow the formation of a crystalline adduct ("clathratocomplex" [19]) with dimethylformamide with the expected 1:1 stoichiometric ratio [37]. Thus, the salicylic acid function (2-hydroxycarboxylic acid group) is shown to be an excellent sensor, or a good complementary site for the dimethylformamide molecule in solid state inclusion.

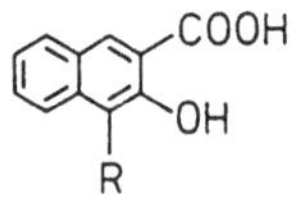

25a R=H
25b R=Cl

In case of the bianthrylmonocarboxylic acid *20*, one may predict the formation of a lattice inclusion at least with dimethylformamide, but we did not succeed in obtaining it [37]. Instead a stable 1:1 stoichiometric inclusion compound of *20* is readily obtainable (Table 3). The bianthryldicarboxylic acid *21*, which is a direct analogue of *1*, is not available in sufficient quantity to be tested in respect to its lattice inclusion properties.

Changing of the flexible scissor-like element, as in *1*, to an orthogonal and rigid version of this element, as in *22*, reduces the activity of inclusion formation to a certain degree. Nevertheless very different guest molecules are readily accommodated in the crystal lattice of *22*, they are proton donors (ethanol, 2-propanol) [47], H-bridge acceptors (dimethylformamide, dioxane), or benzene as an unpolar solvent [48]

(Table 3). The crystal inclusion compounds of the protic solvents are distinguished by relatively high points of thermal decomposition and, as before, the inclusion compound with dimethylformamide exhibits the highest tendency of formation. This is the result of competitive experiments.

According to the coordinatoclathrate predict, the spiro compound *23* will not allow the formation of inclusion compounds with dimethylformamide and other polar solvents, but with benzene, tetrahydrofuran, and 1-bromopentane (Table 3). Due to the limited number of guest inclusions, a lattice cavity of rather restricted dimensions is suggested for *23*; e.g. toluene, cyclohexane or dioxane are not suitable guest partners for *23*, whereas lower homologues (cf. benzene, tetrahydrofuran) are readily included [37]. The behavior of a reduced analogue of *23*, the hydroxymethyl — substituted spiro compound *24*, is in some way comparable since an inclusion compound with benzene is the only one known; interestingly it is formed exclusively with optically resolved but not with racemic *24* [49].

3.4 Host Molecules Related to a Roof

There is no need for much fantasy to see the relation between a roof-shaped (Fig. 12) and a scissor-like basic structure (cf. Fig. 11). Nevertheless, the following point should be noted: Scissor-like molecules (cf. *1*) provide considerable (inherent) bulkiness merely due to the basic skeleton. On the other hand, as far as roof-shaped molecules are concerned [cf. *26* (Fig. 12) and *27–42*], the presence of polar substituents, suitably positioned, preferably at the top ridge of the molecular roof, are also instrumental in developing molecular bulkiness.

26

Fig. 12. Graphic design of a roof-shaped clathrate host (parent compound *26*)

Special features which characterize the prototypical roof-shaped compound *26* are a rigid molecular skeleton and the type of sensor groups used. They render *26* a potential coordinatoclathrand [50]. The capability of *26* in forming inclusion compounds is evident from Table 5. Here, nearly thirty different crystal inclusions of *26* are specified including as guest molecules various alcohols, acids, aprotic dipolar, and rather apolar species.

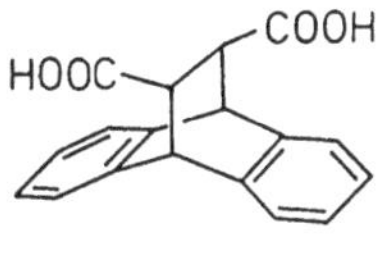

26

Considering this variety of crystalline inclusion compounds, *26* is close to *1* (cf. Table 1) and like *1*, the stoichiometries (host: guest) found for the different aggregates of *26* largely correspond to the expected ratios. Thermal stabilities in most cases are relatively high.

In contrast to *1*, however, inclusions of low-volume substrate species are hardly formed with regard to some of the series of compounds mentioned above. This is

Table 5. Clathrate inclusion compounds of roof-shaped hosts

Host no	Guest compound	Host : guest mol ratio[a]	Host no	Guest compound	Host : guest mol ratio[a]
26:	1-propanol	1:1	*35*:	dioxane	1:1
	1-butanol	1:1			
	t-butanol	1:1	*36*:	dioxane	2:1
	1-pentanol	1:1			
	1-octanol	2:1	*37*:	methanol	2:1
	ethylene glycol	1:2		t-butanol	1:1
	2-methoxyethanol	(1:1)[b]		ethylene glycol	1:1
	formic acid	1:2		epichlorohydrin	2:1
	acetic acid	1:1		diacetone alcohol	2:1
	propionic acid	1:1		dimethylformamide	3:2
	2-chloropropionic acid	1:1		dimethyl sulfoxide	1:1
	valeric acid	1:1			
	lactic acid	(1:1)[b]	*38*:	benzyl cyanide	1:2
	tartaric acid	2:1		dimethyl sulfoxide	1:2
	thioacetic acid	(2:1)[b]		dioxane	2:1
	mercaptoacetic acid	1:1			
	propionic aldehyde	1:1	*39*:	dimethylformamide	2:1
	acetone	1:1		dioxane	1:1
	dimethylformamide	1:1			
	acetonitrile	(1:1)[b]	*40*:	methanol	1:1
	benzyl cyanide	1:1		t-butanol	1:1
	dimethyl sulfoxide	1:1		cyclohexanol	1:1
	tetrahydrofuran	1:2		acetic acid	1:1
	dioxane	2:1		2-chloropropionic acid	1:1
	o-dichlorobenzene	1:1		lactic acid	1:1
	2,6-dimethylnitrobenzene	1:1		propionic aldehyde	2:1
	2-nitrophenol	1:2		acetone	2:1
				dimethylformamide	2:1
28:	ethylene glycol	1:1		dimethyl sulfoxide	1:1
	acetic acid	1:1		tetrahydrofuran	2:1
	propionic acid	2:3		dioxane	2:1
				morpholine	1:1
31:	methanol	1:1		piperidine	1:1
	formic acid	1:3		pyridine	2:1
	acetic acid	1:1		nitrobenzene	1:1
	propionic acid	1:1			
			41:	t-butanol	1:1
33:	tetrahydrofuran	2:1		dimethyl sulfoxide	1:2
	dioxane	2:1		dioxane	1:1

[a] Determined by NMR integration as specified in Table 1. [b] Unstoichiometric or low stability of the compound at atmospheric conditions.

quite clearly seen in the series of the unbranched alcohols: inclusion formation of *26* occurs neither with methanol nor with ethanol, but only beginning with 1-propanol to higher homologues up to and including 1-octanol (Table 5). Correspondingly, also for the carbonyl compounds and nitriles, the lower molecular mass representatives of these substance series are either not or only rather weakly accommodated into the host lattice of *26*. Referring to the carboxylic acids, however, differences in the size of molecules primarily influence the inclusion stoichiometries (e.g. formic acid 1:2, acetic acid and propionic acid 1:1, Table 5). This is perhaps a result of their high potency of forming H-bonded inter-substrate dimers.

The coordinatoclathrate relation appears from the examples as given in the following [50]: bivalent ethylene glycol forms a stable inclusion compound with *26*; gradual conversion into corresponding methyl ethers (2-methoxyethanol, monoglyme, respectively) leads to a loss of the inclusion formation (Table 5). On the other hand, the cyclic ethers tetrahydrofuran and dioxane allow isolation of thermally relatively stable crystal inclusions (range of decomposition 110–120 and 140–145 °C, respectively). This suggests that the functional groups of *26* are not completely available for guest binding, or can be used otherwise, e.g. for host-host interaction in order to stabilize an inclusion matrix. The stoichiometries (host: guest) differing in some cases from the expected ratios, e.g. 1:1 at the inclusion compound with dimethylformamide (half an equivalent of dimethylformamide per carboxylic group), but 1:2 in case of *1* (one dimethylformamide per carboxylic group), also point to the same fact.

Nevertheless, the inclusion of dimethylformamide in the lattice of *26* is found to be so much favored that this inclusion compound is pratically always obtained from competitive experiments [50] (Table 6), even in the presence of a larger excess of the second component. The results of other solvent combinations are less clear, showing that steric and electronic affects between host and guest superimpose in a way difficult to separate from one another. For instance, out of a mixture of acetic acid/1-butanol, the alcohol is selectively included by *26*, but from a mixture of acetic acid/1-octanol, it is the acid which is preferred; t-butanol/1-octanol yields the inclusion compound with 1-octanol. A mixture of 1-butanol/t-butanol leads to the formation of both kinds of inclusion compounds in about equal amounts. The t-butanol, however, is much more weakly bound in the lattice of *26* and thus evaporates completely from the crystal in a few days of storage in air, whereas the stoichiometric ratio of the respective 1-butanol inclusion compound remains unchanged for weeks. Hence 1-butanol compared with t-butanol is able to form the thermodynamically more stable inclusion compound with *26*, but from a kinetic point of view, inclusion formation with both butanol isomers is lacking in selectivity. Regarding a two-component solvent mixture of carboxylic acids, the spatially less demanding component is preferentially included by *26*, e.g. formic acid > acetic acid or propionic acid. But keep in mind, this fact contradicts the behavior of alcohols (Table 6).

Probing the effect of functional group modification was achieved by using the compounds *27–34* [37,51] for the recrystallization experiments. These compounds are expected to show functional complementarity different to *26*. Table 5 summarizes the results. Inclusion compounds with protic and aprotic guest species are formed of *28*, *31*, and *33*, respectively. All the other potential hosts are ineffective. Hence it is demonstrated that the COPh groups of *33* are not suitable for coordinative

Table 6. Selective guest inclusion of some roof-shaped hosts from two-component solvent systems

Host no	Recrystalln solvent compd mixture (I/II)[a]	Host : I : II mol ratio[b]	Host no	Recrystalln solvent compd mixture (I/II)[a]	Host : I : II mol ratio[b]
26:	1-propanol/1-butanol	1:0:1	*37*:	methanol/ethanol	2:1:0
	1-propanol/t-butanol	1:x:y[c]		1-butanol/t-butanol	2:0:1
	1-butanol/t-butanol	1:x:y		acetone/diacetone alcohol	2:0:1
	1-butanol/acetic acid	1:1:0			
	1-butanol/DMF	1:0:1			
	t-butanol/1-octanol	1:0:1	*40*:	methanol/ethanol	1:1:0
	t-butanol/ethylene glycol	1:0:1		methanol/cyclohexanol	1:0:1
	formic acid/acetic acid	1:1:0		cyclohexanol/ethanol	1:1:0
	formic acid/propionic acid	1:1:0		cyclohexanol/1-butanol	1:1:0
	formic acid/acetamide	1:1:0		cyclohexanol/t-butanol	1:0:1
	acetic acid/ethylene glycol	1:0:1		acetic acid/acetone	1:1:0
	acetic acid/1-octanol	1:1:0		propionic acid/acetone	1:0:1
	acetic acid/acetamide	1:1:0		propionic acid/propionic aldehyde	1:0:1
	acetic acid/2-chloropropionic acid	1:x:y		2-chloropropionic acid/propionic acid	1:1:0
	acetic acid/lactic acid	1:1:0		2-chloropropionic acid/lactic acid	1:1:0
	propionic acid/lactic acid	1:1:0		acetone/acetylacetone	1:1:0
	propionic acid/2-chloropropionic acid	1:x:y		acetone/acetonitrile	1:1:0
	DMF/piperidine	1:1:0		benzyl cyanide/nitrobenzene	1:0:1
	DMF/dioxane	1:1:0		4-cyanobenzaldehyde/4-chlorobenzylamine	1:0:1
	DMF/bromobenzene	1:1:0		morpholine/dioxane	1:1:0
	DMF/toluene	1:1:0		pyridine/benzene	1:1:0

[a] Equimolar ratio. [b] Determined by NMR integration as specified in Table 1 . [c] No clear discrimination in favor of I or II.

binding with protic guests, but use their bulk to form "true" clathrates with THF and dioxane. By way of contrast, the amide functions of *28* and *31* only combine with highly protic guests such as acids. Nevertheless steric effects apply also to the amide functional groups, since *29* and *30* which are the lower bulky analogues of *31* have unfavorable solubilities and failed in inclusion formation. On the other hand, the unsubstituted amide *28* may contribute four instead of only two hydrogens for guest binding and lattice build-up.

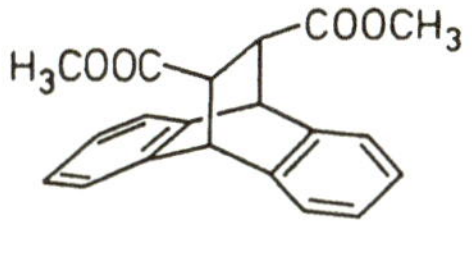

27

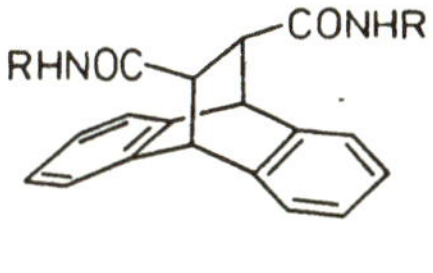

28 R=H
29 R=n-C_3H_7
30 R=i-C_3H_7
31 R=t-C_4H_9

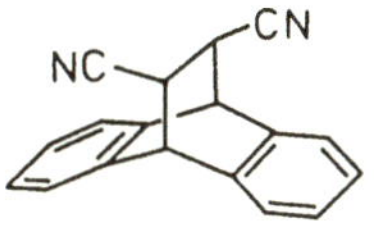

32

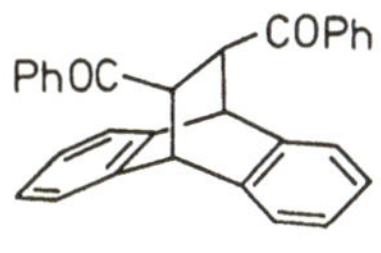

33

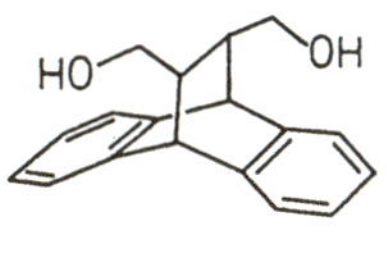

34

Retaining one of the COOH groups of *26* and modification or omission of the other one yields unsymmetric host compounds *35*–*37* [52], respectively. The interesting thing about it is that *35* and *36* behave much the same as *33*, i.e. formation of "true" clathrates with dioxane, whereas *37* forms inclusion compounds with guests typical of keeping up H-bonds to the carboxylic group [2,48]. Thus the inclusion behavior of carboxylic acids *35* and *36* is comparable to functional group free species (cf. Sect. 3.5) with no direct host-guest contact; *37* however is near to *26* with supposed strong host-guest interactions [37]. Considering an explanation of this remarkable behavior would necessarily lead into an analysis of the crystal packings and would require crystal structures (see Sect. 4.2).

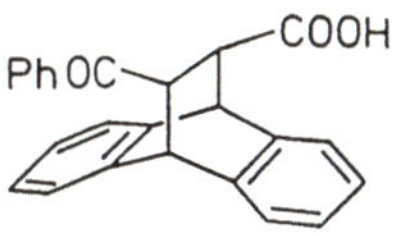

35

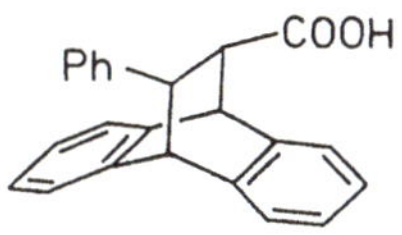

36

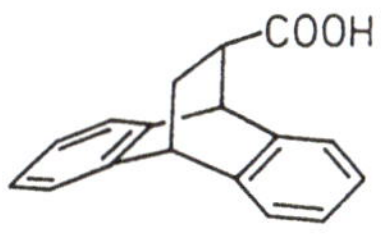

37

Another remarkable fact is that methanol acts as a suitable guest for *37* whereas long-chain alcohols are ineffective. The opposite is true for *26*. Since ethanol which is the next higher homologue of methanol does not lead to a corresponding crystal inclusion with *37* (the same applies for *31*), crystallization of *37* (or *31*) from respective solvent mixtures provides an easy way to separate methanol from ethanol, or other higher alcohols [37]. The same holds for a solvent mixture 1-BuOH/t-BuOH since 1-BuOH is clearly discriminated against by *37*. Of pratical interest is also the clear discrimination against acetone in a mixture with diacetone alcohol (e.g. for the separation of aldol condensation products).

The next point in question refers to the effect of sterically different positioning of the functional groups at the roof-shaped skeleton. As is clearly shown from Table 5, the inclusion properties of the host compound *38* [53] having a *syn* instead of *anti* position of the carboxylic groups (cf. *26*) are restricted, especially in the uptake of hydroxylic guests (inclusion compounds are only formed with dimethyl sulfoxide, benzyl cyanide, and dioxane). This suggests that intramolecular H-bonding of the adjacent functional groups may have occured, neglecting the interaction with protonizable guests.

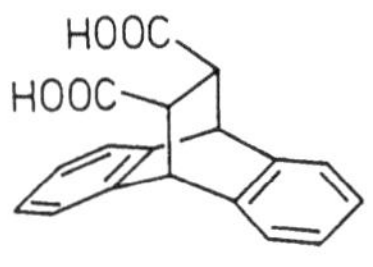

38

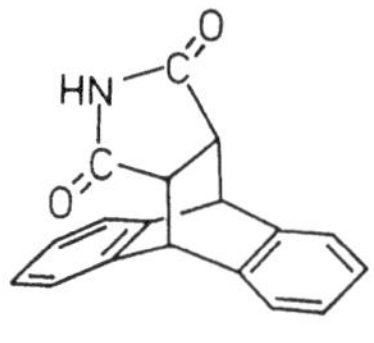

39

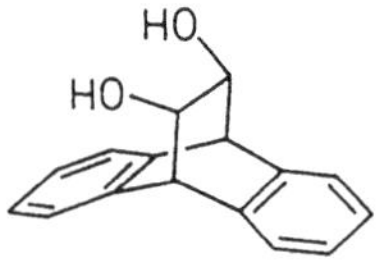

40

Modification of the carboxylic groups in *38* was effected by introducing functions as in *39* and *40* [54]. Both compounds proved to be inclusion hosts (Table 5). Although the two hosts are potential proton donors and acceptors, they show very different inclusion behavior [37]. The imide *39* forms inclusions with typical aprotic hosts (DMF, dioxane) and is thus in contrast with the amides *28* and *31* (see Table 5). The diol *40* provides broad inclusion properties which come near to *26*. Naturally there are some differences. The most striking one is that unbranched alcohols higher than methanol are not accommodated in the lattice of *40*. The stoichiometry ratios (host:guest) observed for the inclusion compounds of *40*, either 1:1 or 2:1 (Table 5), allow a possible distinction between protic and aprotic guest solvents, respectively. This may be interpreted as a result of the presence or the lack of coordinative binding participation as defined by the "coordinatoclathrate conception" (Sect. 2.2.).

The selective inclusion properties of *40* (Table 6) offer several possibilities of compound separation which are of interest in analytics and for preparation purposes [37]. The separation of methanol from a mixture with ethanol, or of propionic aldehyde from propionic acid, or of 2-chloropropionic acid from propionic acid or lactic acid, etc., are a few examples.

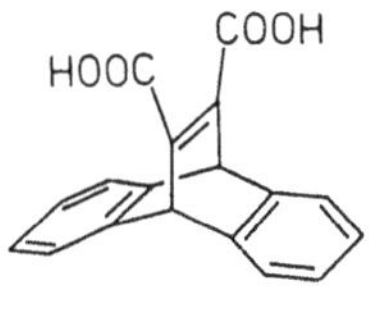

41

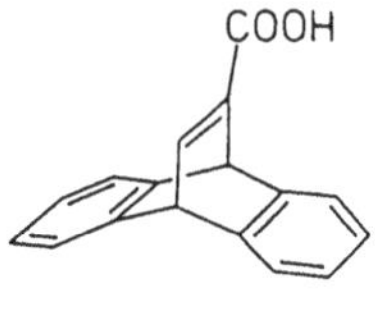

42

The unsaturated structures *41* and *42* [55] provide a third possibility of arranging functional sensor groups at the same roof-shaped skeleton in a given geometry. Here the sensors project in a perpendicular position with respect to the top ridge of the molecule instead of being inclined to one (cf. *37*, *38*) or both sides (cf. *26*) of the roof. In *41*, the carboxylic groups may interact intramolecularily, comparable to *38*. Consequently *41* and *38* display rather similar (and poor) inclusion properties [37] (Table 5), with one important difference however: unlike *38*, *41* allows the formation of an inclusion compound with t-butanol and, in addition, the stoichiometric ratios of the respective dioxane clathrates differ (1:1 in case of *41*, but 2:1 for *38*). In contrast to saturated monocarboxylic acid *37*, the unsaturated analogue *42* fails completely in inclusion formation [37].

3.5 Analogues Lacking Functional Groups and Related Compounds

Incorporation of coordinatively active complementary (functional) groups (in the host *and* the guest molecule) is an essential part of the "coordinatoclathrate idea" (see Sect. 2.2). In other words, in the absence of functional groups (Fig. 13), a so-called "coordinatoclathrate" is unimaginable. Also, under these circumstances coordinative linkage of the host molecules with each other to form a recticular matrix is not applicable (see. Sect. 2.1).

However, if Fig. 15a in Chapter 1 of Vol. 140 of this series ("Molecular Inclusion and Molecular Recognition — Clathrates I") applies, the absence of functional groups does not necessarily imply the failure of crystal inclusion formation, for instance when the molecular skeleton, as a result of inherent bulkiness (i.e. without coordinative binding assistance), provides favorable preconditions to form a clathrate (cf. Sect. 3.1). However, the selective binding character of a host owing to the so-called "sensor groups" (see Figs. 7 and 8) is of course lacking. In order to prove these assumptions, a series of accordingly modified compounds *43–55* have been synthesized. They display molecular skeletons related to the hosts described in the previous sections (3.2–3.4), but lack the typical functional groups of a coordinatoclathrate former.

The compounds *43* [56] and *44* [57] correspond to the compounds *1* and *7–18* (see Sects. 3.2 and 3.3), intended as coordinatoclathrate hosts, in the bulky binaphthyl hinge. Compound *44* provides methyl groups in a position (2,2′) which was previously substituted by functional groups (cf. *1*), whereas *43* features the plain

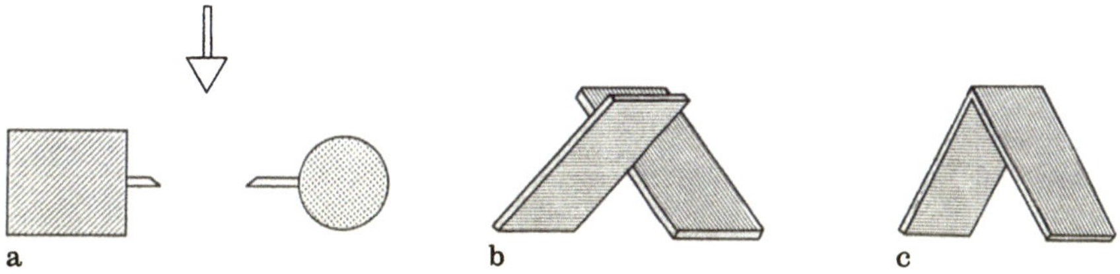

Fig. 13. Dismissal of functional groups (sensor groups) (**a**) effected at scissor-type and roof-shaped hosts (diagrammatic representation) (**b** and **c**)

1,1'-binaphthyl base. The hydrocarbon *45* [58] possesses a partially hydrated binaphthyl skeleton which modifies the original geometry of scissors to some extent.

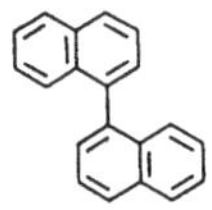

43

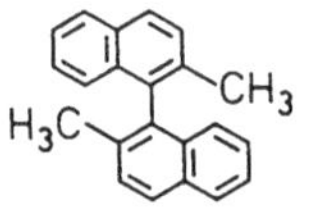

44

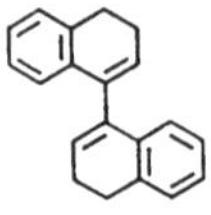

45

Considering the current knowledge (see above), one may assume that *43*, as well as *44* or *45*, are capable of forming crystal inclusions only with low-voluminous hydrocarbons using weak van der Waals forces. Contrary to all expectations, polar

Table 7. Clathrate inclusion compounds of functional group-free binaphthyl and bianthryl hosts

Host no	Guest compound	Host:guest mol ratio[a]	Host no	Guest compound[b]	Host:guest mol ratio[a]
43:	methanol	1:1	*47*:	2,3-dimethyl-2-butene (7.3)	1:1
	ethanol	1:1		cyclopentane (6.0)	2:1
	1-propanol	1:1		cyclohexane (7.0)	2:1
	1-butanol	2:1		cycloheptane (7.2)	[c]
	propionic acid	2:1		cyclooctane (7.8)	[d]
	acetonitrile	2:1		cyclopentene (6.0)	[d]
	3-aminopinane	1:1		cyclohexene (6.9)	2:1
	isoquinoline	1:1		cycloheptene (7.2)	[c]
	tetrahydrofuran	2:1		cyclooctene (7.7)	[c]
	dioxane	2:1		1,3-cyclohexadiene (6.9)	1:1
	cyclohexene	1:1		1,4-cyclohexadiene (6.9)	1:1
	2,3-dimethyl-2-butene	1:1		benzene (6.9)	1:1
	t-butyl chloride	2:1		cycloheptatriene (7.2)	1:1
				toluene (7.8)	[c]
				o-xylene (8.2)	1:1
				m-xylene (8.7)	1:1
				p-xylene (8.8)	1:1
44:	methanol	2:1		cyclopentanone (6.6)	[c]
	cyclohexane	(3:1)[c]		cyclohexanone (7.8)	1:1
				cyclohexene oxide (7.1)	1:1
46:	dioxane	(3:1)[c]		tetrahydrofuran (6.0)	1:1
	furan	(2:1)[c]		dioxane (6.8)	1:1
	pyridine	2:1		morpholine (6.8)	1:1
	cyclohexane	(3:1)[c]		piperidine (6.9)	1:1
	benzene	2:1		pyridine (6.9)	1:1
	toluene	2:1			
	m-xylene	(2:1)[c]			
	mesitylene	1:1			
	o-dichlorobenzene	2:1			

[a] Determined by NMR integration as specified in Table 1. [b] Numerical data in parentheses give the largest extension (Å) of guest molecules (based on van der Waals radii from space filling models, taking into consideration in most probable conformations, Ref. 65). [c] Unstoichiometric or low stability of the compound at atmospheric conditions. [d] Traces of solvent included.

and even protic solvents are the guest species accommodated into the host lattice of *43* [37] (Table 7). So far, there is no reasonable explanation at hand for this particular behavior. However, recently indications appeared in outlines that weak intermolecular H bridges can exist between OH groups and aromatic π-systems (the OH group is assumed to be positioned perpendicular to a π-bond) [59]. One may also imagine a cluster formation of two or more guest molecules which mutually mask off their hydrophilic sites (cf. dimer formation of carboxylic acids in the clathrates of *26*; see Sect. 4.2.1), so as to be compatible with the apolar hydrocarbon-type host lattice. In favor of the latter assumption and against a possible polar interaction are the rather low thermal stabilities observed for these inclusion compounds. On the other hand, it is well known from the hydrate clathrates that molecules very different in chemical nature, e.g. water and a noble gas, are successfully assembled in a crystalline aggregate (cf. Sect. 2.1). But in these cases (and others) the apportionment of polarities on host and guest are reversed. Hence, the nature of the present crystal inclusions remains a critical point. Racemic binaphthyl *43*, in that respect, is polymorphous [60] and undergoes a solid-state transformation at moderate temperature rise (spontaneous separation into enantiomers) [61]. It is not too far from reality to assume that this background also plays a role for the unusual inclusion behavior of *43*.

The methyl-analogous compound *44* shows poor a tendency to crystallization which is possibly the reason why we succeeded in obtaining only a very limited number of inclusion compounds [37] (guest species methanol and cyclohexane; Table 7). The ability of a hydrocarbon host to combine with a polar guest molecule is retained, though. The hydrocarbon *45* with a partially saturated skeleton gave no respective inclusion compounds, although a "solvate" with acetic acid of the corresponding 7,7′-dimethyl derivative is reported [62].

Whereas the hydrocarbons with binaphthyl constitution *43* and *44* show rather unusual clathrate behavior, the crystal inclusion properties of the bianthryls *46* [42] and *47* [63] are in keeping with predictions made for potential hosts without polar groups. Accordingly, compounds *46* and *47* reject protic guests (like alcohols, carboxylic acids, etc.), without exception, but allow the formation of a great many crystal inclusions with *hydrocarbon* and a few with *dipolar-aprotic* guest molecules, among them ketones and heterocycles [64] (Table 7). The hydrocarbon host molecule *48* [43], for which the spiro compounds *22* and *23* are the underlying models, behaves accordingly [64] (Table 8).

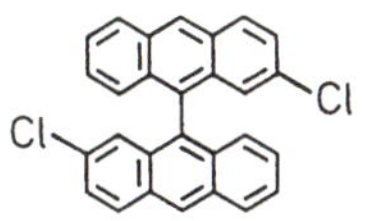

46

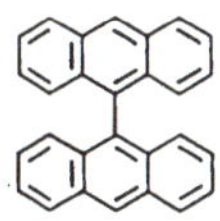

47

A careful examination of the results given in Tables 7 and 8 reveal that with the exception of 2,3-dimethyl-2-butene (in the case of *47*) only *cyclic* guest molecules are taken up into the lattices of the host compounds *46*–*48*, but not the respective open-chain analogues. Saturated 2,3-dimethylbutane, as a compound for comparison, is not accommodated either, either by *46* or by *47*. Moreover, only cycles with distinct ring sizes (*five-* to *eight-membered* rings) are effective, indicating the presence

Table 8. Clathrate inclusion compounds of functional group-free spiro-type hosts

Host no	Guest compound	Host:guest mol ratio[a]	Host no	Guest compound	Host:guest mol ratio[a]
48:	cyclopentane	1:1	*49*:	benzene	3:2
	cyclohexane	1:1			
	cycloheptane	[b]			
	cyclopentene	1:1	*51*:	cyclohexane	3:1
	cyclohexene	1:1		benzene	1:1
	cycloheptene	2:1		tetrahydrofuran	2:1
	1,3-cyclohexadiene	1:1		dioxane	1:1
	1,4-cyclohexadiene	1:1		morpholine	1:1
	benzene	1:1		piperidine	2:1
	cycloheptatriene	1:1		pyridine	1:1
	p-xylene	2:1			
	cyclopentanone	1:1			
	cyclohexanone	1:1	*52*:	benzene	1:1
	cyclohexene oxide	1:1		pyridine	1:1
	tetrahydrofuran	1:1			
	dioxane	1:1			
	morpholine	1:1	*53*:	benzene	1:1
	piperidine	1:1		dioxane	1:1
	pyridine	1:1		pyridine	1:1

[a] Determined by NMR integration as specified in Table 1. [b] Traces of solvent included.

of crystal cavities with a defined geometry and similar dimensions (guest dimensions in Table 7). Referring to the alicyclic guests, cycloheptane and cyclopentane are the maximum and minimum molecular sizes, respectively, to be tolerated by *46* and *47*. Step-by-step introduction of double bonds into the larger ring homologues (e.g. as in cycloheptene, cycloheptatriene, or cyclooctene) causes a defined flattening of the rings. This obviously improves the spatial conditions and leads to the formation of crystal inclusions with *47* and *48* also in the range of seven- and eight-membered ring compounds (host:guest stoichiometries 2:1 or 1:1). Consequently it is also possible to distinguish in the field of cyclic guest compounds between *saturated* and *unsaturated* or *aromatic* molecules (e.g. cyclopentane from cyclopentene or toluene from methylcyclohexane).

Likewise it is possible to differentiate between *substituted* and *unsubstituted* alicycles using inclusion formation with *47* and *48*; only the unbranched hydrocarbons are accommodated into the crystal lattices of *47* and *48* (e.g. separation of cyclohexane from methylcyclohexane, or of cyclopentane from methylcyclopentane). This holds also for cycloalkenes (cf. cyclohexene/methylcyclohexene), but not for benzene and its derivatives. Yet, in the latter case no arbitrary number of substituents (methyl groups) and nor any position of the attached substituents at the aromatic nucleus is tolerated on inclusion formation with *46*, *47*, and *48*, dependent on the host molecule (Tables 7 and 8). This opens interesting separation procedures for analytical purposes, for instance the distinction between benzene and toluene or in the field of the isomeric xylenes.

Further important results of compound separation (two-component solvent mixtures) using hosts *47* and *48* are taken from Table 9 and are as follows [64]: *47* allows

Table 9. Selective guest inclusions of hosts *47* and *48* from two-component solvent systems

Host no	Recrystalln solvent compd mixture (I/II)[a]	Host:I:II mol ratio[b]	Host no	Recrystalln solvent compd mixture (I/II)[a]	Host:I:II mol ratio[b]
47:	toluene/benzene	1:x:y[c]	*48*:	benzene/toluene	1:1:0
	toluene/o-,m-xylene	1:x:y		benzene/o-,m-xylene	1:1:0
	toluene/pyridine	1:0:1		benzene/p-xylene	2:0:1
	toluene/dioxane	1:0:1		benzene/cyclohexane	1:x:y
	toluene/cyclohexane	1:0:1		benzene/cyclohexene	1:x:y
	cyclohexane/o-,m-,p-xylene	1:1:0		cyclohexane/cyclohexene	1:x:y
	cyclohexane/pyridine	1:1:0		cyclohexane/o-,m-xylene	1:1:0
	pyridine/tetrahydrofuran	1:1:0		cyclohexane/p-xylene	1:1:0
	pyridine/n-heptane	1:1:0		dioxane/THF	1:x:y
	pyridine/chloroform	1:1:0		dioxane/morpholine	1:x:y
	pyridine/dimethylformamide	1:1:0		morpholine/piperidine	1:x:y

[a] Equimolar ratio. [b] Determined by NMR integration as specified in Table 1. [c] No clear discrimination in favor of I or II.

separation of cyclohexane from toluene, from xylene, and from pyridine, or of pyridine from toluene and from tetrahydrofuran, or of dioxane from toluene; *48* allows separation of cyclohexane from p-xylene, of p-xylene from benzene, etc. Differentiation between cyclohexane and benzene, or between cyclohexane and cyclohexene, however, is not complete with *47* and *48*.

The formation of crystal inclusion of *47* and *48* with cyclic ketones of suitable ring size (cyclopentanone, cyclohexanone) and with cyclohexene oxide are also important facts. Corresponding inclusion compounds with alcohols or amines could not be obtained. With reference to the heterocyclic guest molecules, the suitability of the ring size is likely to be the decisive factor for guest inclusion.

Comparison (Tables 7–9) shows that *47* and *48* are similar in their host properties, but they are not equivalent in hehavior. Thus, host compound *48* is more qualified to select according to spatial aspects (see benzene derivatives) and, as a rule, it also forms the thermally more stable inclusions. This may be attributed to the rigid molecular geometry of the spirane *48*, whereas the biaryl *47* allows sterical adaptation to different guests via the flexible hinge to a certain degree.

48

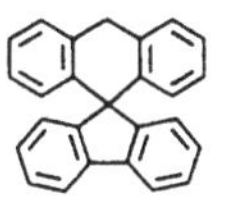

49

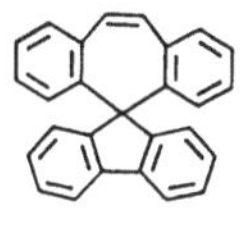

50

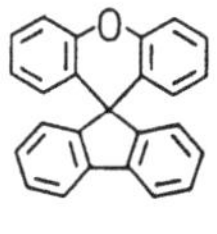

51

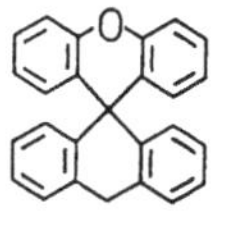

52

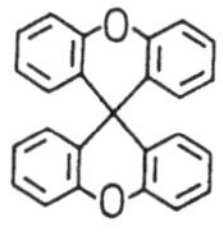

53

The series of spiranes *49–53* [37,66], having a gradually altered constitution, demonstrate that in the range of the scissor-like molecules broad structural variabilities are possible. Compounds *49–53* differ in their molecular geometries from *48* in that the extra C and/or O atoms cause a more or less distinct bend at the "edges" of the molecular scissors. The general inclusion properties, however, are not fundamentally (except for *50* which gave no inclusions), but only gradually affected by this operation (Table 8), e.g. expressed in a reduced number of inclusion compounds [37,64] compared to *48*. The inclusion stoichiometries are also different in some cases. Obviously, the mode of constitutional modifications as in *49–53* (restrictive of *50*) has only a secondary effect on the over-all lattice build-up and on the net geometry of the formed lattice voids.

The hydrocarbons *54* and *55* [51] which are functional group-free analogues of *26* and *38* display no activities of inclusion formation, either with polar or with apolar solvents [37]. This result is another proof that mostly for the roof-shaped type of compounds, functional groups play a fundamental role in the construction of a low-density packed crystal lattice.

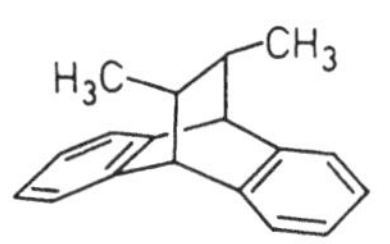

54

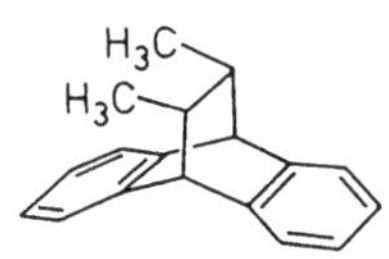

55

Additional clathrate inclusions of this particular type of functional group free hosts have been studied by Toda and coworkers (see Chapter 3 in Vol. 140 of this series, "Molecular Inclusion and Molecular Recognition — Clathrates I").

The question arises about the crystal parameters which could be relevant for the clathrate properties and may reflect the observed inclusion selectivites.

4 Proving the Concept by X-Ray Crystallography

Crystal structure studies supply essential information about the solid state build-up of inclusion compounds, in principle. Due to the complexity of the problem, a full quantitative description of interactions between host and guest species in inclusion compounds can not be given by theoretical treatment as yet. Not only the most prominent primary interactions should be dealt with, but also effects due to the neighboring unit cells must be taken into account, since they also exert an influence on the entity to be considered. Such an *ab initio* approach has not been adopted until now. For this reason, the following discussion tends to describe some major effects in a somewhat qualitative manner. Because the argumentation relies on crystallographic data, directional parameters associated with the intra-aggregate attraction and repulsion of groups and segments are well to the fore. They involve H-bonding, van der Waals forces, and polarization.

At the beginning of these studies, the question was asked, whether such kinds of molecular associates may be really regarded as inclusion compounds, or as something else which is sometimes called a "solvate". Now the associates considered here are well-defined inclusion compounds. Their solid-state structures demonstrate the particular way guest molecules are held in the crystal lattices. Another supporting argument is that many of the hosts described here provide a series of inclusions, whereas the use of the term "solvate" implies that they are individual cases. We are convinced that a study of the so called "solvates" in the structural literature may eventually lead to the discovery of other classes of inclusion compounds unknown as yet. For example, literature search in the Cambridge Crystallographic Database reveals 3471 entries named as organic "solvates" (state January 1986, Table 10.). It can be proven that many of these are in fact inclusions [67].

Accordingly, systematic studies offered by the designed host compounds are of importance precisely in the case of such classification problems. Therefore, the following discussion of the results of X-ray structure investigations is grouped to reveal that structural systematics exist for related host molecules ("coordinatoclathrate concept", see Sect. 2.2). This is best achieved by sorting inclusion compounds with the same *guest* molecules into one group. Classification of these solid associates will lead to the formulation of certain *recognition* principles for given guest molecules [67]. These may be understood as a characteristic pattern of their interaction with the host matrix in the crystal. Such an interaction pattern is mutually determined

Table 10. Output excerpt from a compound name search in the Cambridge Crystallographic Database showing independent occurrences of the search string "solvent" as a result of logical operations cross-referenced with different possible combinations of other denominations. (The final figure seems to be a pessimistic estimate)

INFORMATION FOR REFERENCES IN FILE 19 PRODUCED WHEN THE FOLLOWING TEMPORARY FILES WERE AVAILABLE

1	255	WORD	CLATHRATE	
2	2	WORD	INCLUSION	
3	32	WORD	ADDUCT	
4	563	WORD	COMPLEX	
5	366	WORD	HYDRATE	
6	541	WORD	AQUA	
8	354	NOT	5	6
9	895	MERGE	6	8
10	881	NOT	9	1
13	876	NOT	10	4
14	3538	WORD	SOLVATE	
15	3495	NOT	14	13
16	3487	NOT	15	1
17	3487	NOT	16	2
18	3484	NOT	17	3
19	*3471*	NOT	18	4

by accessible functional groups of both the host and guest molecules and of their respective shapes (*complementary* principle, see Fig. 8). Thus, alterations which do not grossly deformate structures of the individual host molecules will yield, somewhat trivially, to homology in their patterns of interactions with the same types of guest molecules in their crystals as well. Such systematics lay the basis of a kind of *supramolecular chemistry* which relies upon interactions between neutral associates of molecules.

4.1 Structures of Free Host Molecules

Usually the structures of inclusion compounds are more numerous. However, crystallographic studies involving free host molecules are no less important. At least they serve their purpose as a reference thus giving a broader basis for the understanding of the corresponding inclusions, too (cf. Refs. 12 and 16). Unfortunately, the study of free host structures is not always possible owing simply to the lack of proper single crystals. Such an unlucky case is compound *1*. Actually host compounds are intended to result in a loosely packed arrangement (cf. Sect. 3.1) and this means correspondingly an energetic instability of the would-be crystal. Hence we are caught on the horns of a dilemma. The outstanding inclusion ability of this host may be also due to the fact that a structure formed solely from the molecules of *1* is so unstable that virtually in all instances, inclusions are formed with quite a variety of solvents (cf. Table 1.).

In contrast to *1*, the related pure host *7* may be obtained in crystalline form [68]. The crystal structure of *7* is built *via* helical chains of alternating intra- and intermolecular H-bonding through the carboxyl functions. This structure supplies the information that the carboxyl groups are therefore already positioned in an appropriate way to facilitate analogous H-bonding in the known "inclusions" of *7*. As discussed later (Sect. 4.2.2), these are exclusively salt-type associates and as such, intimately interact with the carboxyl groups. Hence one may infer that displacement of the carboxyl functions from position 2 in *1* to position 8 in *7* reduces the ability of inclusion formation. Similar reasons such as the solid-solubility differences observed in the classical naphthalene/chloronaphthalene systems (alpha- *vs.* beta-substituted derivatives, cf. Ref. 28 may also be applied here.

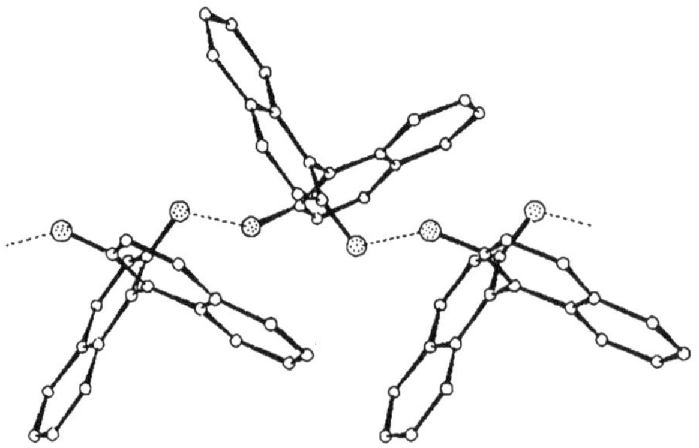

Fig. 14. H-bonding detail in the crystal structure of *13* (H-bonds as broken lines; O atoms dotted; H atom positions are not given in the reference) [69]

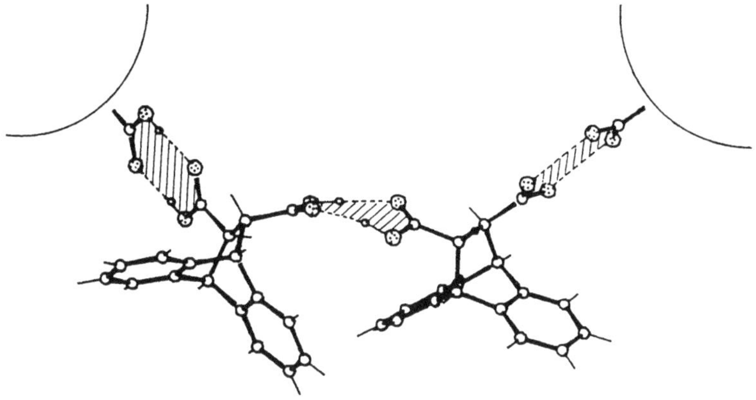

Fig. 15. Packing excerpt from the crystal structure of *26* [71] showing a central dimeric entity forming the asymmetric unit. The characteristic H-bond rings are indicated by hatching (H-bonds as broken lines; O atoms dotted; H atoms connected with non-heteroatoms are shown as sticks only)

The crystal structure of *13* was studied by Struchkov and his coworkers [69]. It is essentially characterized by its constituent molecules being placed in infinite H-bond helices (Fig. 14).

Compound *15* has an ill-defined structure [70] due to poor crystal quality. It is, however, interesting to note that *15* became resolved (space group $P2_12_12_1$, $Z = 8$) in an attempted cocrystallization experiment in the presence of optically pure *1*. The two molecules of *15* in the asymmetric unit may be seen as a sort of a "self inclusion", where molecules are again linked into infinite H-bond chains analogous to *13*.

The structure of *26* (Fig. 15) throws light upon the inclusion systematics of this roof-shaped host. A basic motif, apparent from the structure of the asymmetric unit

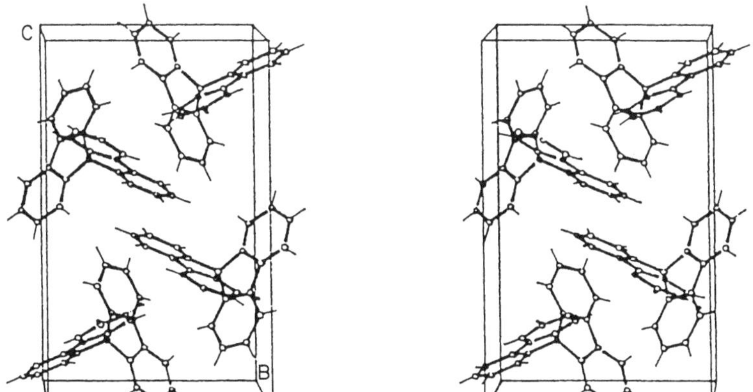

Fig. 16. Stereo drawing of the packing in the crystal structure of free host compound *48* [74] (H atoms are shown as sticks only)

of the pure host, is the formation of H-bonded dimers[71]. This pattern then returns in all known inclusions of *26* (see below), explaining the most common 1:1 stoichiometry. In spite of being a dicarboxylic acid, like *1*, *26* behaves (on inclusion) like a monocarboxylic acid with only one of its carboxylic groups free for binding to guest molecules, thus giving 1:1 host:guest ratios. The only exception where a 2:1 host:guest relation occurs is the aggregate of *26* with formic acid (Sect. 4.2.1). Nevertheless, this case neither contradicts the basic rule concerning the way of self-dimerization of the host *26* nor its consequences.

The structure of the parent compound *43* of the 1,1′-binaphthyl host family has been determined many times in independent laboratories and in different (racemic and resolved) crystal forms [60,61,72,73]. A common feature of both types of structure is that molecules of *43* form helices. Lateral contacts between such helices play an important role in the respective crystals. These spatial arrangements also emphasize the importance of the question concerning the presence of alpha or beta substituents' positioning as mentioned earlier [28].

Crystals of another pure hydrocarbon host *48* are formed in dimethyl sulfoxide [74]. Constituents of this structure are also arranged into helical arrays (Fig. 16).

4.2 Function of Sensor Groups in Binding Guest Molecules

Clathrate formation is determined by and large by an interplay of many different factors stemming from the shape and electronic properties of the host and guest assemblies. To some extent the role of the functional groups is transparent. They perform important control over the contact features by virtue of e.g. H-bonds to an eventual second (or even third) component of the aggregate. Their function parallels those of the recognition (sensoric) functions of higher organisms, hence the term "sensor group" seems appropriate. The great majority of the structures under discussion entails —COOH groups. In the following paragraphs, how this particular group (and in a few instances other groups as well) recognizes different classes of solvent guest molecules will be scrutinized.

4.2.1 Inclusion Compounds Involving Readily Deprotonizable (Proton-Donating) Guest Molecules: Alcohols and Acids as Guest Species

Alcohols as Guest Species

The high tendency of host *1* to form crystal inclusions was first demonstrated by its aggregates with alcohols [2]. At the same time, the nature of host-guest binding was also established by X-ray crystallography. The crystal lattices of these inclusions of *1* show similarities in their lattice parameters and symmetries (Table 11). The structures of the inclusion compounds between *1* and MeOH, EtOH, 2-PrOH, 2-BuOH, t-BuOH, 1-PrOH and ethylene glycol also illustrate some relations as shown by some selected characteristic examples (Figs. 17 and 18). Close resemblance in the binding mode of the guest entities is to be seen here. The first thing we learn from these structures refers to the —COOH groups of host *1*, as they perform their task, as expected, by building H-bonds to their opposing functions in the guest (cf. Table 12). The systematics of H-bonding patterns found in these crystals is also remarkable.

Table 11. Crystal data and packing coefficients for the alcohol inclusions of *1*

Compound[a]	*1a*	*1b*	*1c*	*1d*	*1e*	*1f*	*1g*
Space group	$P2_1/n$	$C2/c$	$C2/c$	$P\bar{1}$	$P2_1/n$	$P2_1/n$	$P2_1/n$
Unit cell							
a(Å)	15.642	11.737	12.051	10.160	12.009	10.603	14.276
b (Å)	14.532	14.522	14.776	14.050	12.747	14.377	9.533
c (Å)	9.292	13.769	14.362	15.167	14.982	15.664	15.556
α (deg)	90	90	90	100.37	90	90	90
β (deg)	95.14	101.50	102.53	104.4	105.52	104.2	109.19
γ (deg)	90	90	90	94.8	90	90	90
V (Å^3)	2104	2300	2496	2044	2210	2315	1999
Z	4	4	4	2	4	4	4
C_k[b]	0.75	0.75	0.74	0.71	0.75	0.71	0.77
D_c (gcm^{-3})	1.283	1.254	1.230	1.210	1.252	1.195	1.343

[a] Designation: *1a* = *1* · MeOH (1:2), *1b* = *1* · EtOH (1:2), *1c* = *1* · 2-PrOH (1:2), *1d* = *1* · 1-PrOH (2:1), *1e* = *1* · 2-BuOH (1:1), *1f* = *1* · t-BuOH (1:1), *1g* = *1* · ethylene glycol (1:1). [b] Packing coefficients calculated from the volume increments in Ref. 106.

Table 12. H-bonding data in the alcohol inclusion of *1* (cf. Table 11, and Figs. 17 and 18). E.s.d's are given in parentheses for the parameters involving non-hydrogen atoms only. Mean values and r.m.s.d.'s for the three groups (see footnote)[a] are 2.61(4), 2.66(4), and 2.74(7) Å from 10, 20, and 10 data of the unmarked, starred, and double-starred entries, respectively, of this table and of the corresponding entries of Table 13

Compound[b]	D ... A (Å)	H ... A (Å)	D-H (Å)	D-H ... A (deg)
1a	2.633(4)	1.66	0.98	171
	2.588(4)	1.54	1.05	172
	2.788(4)*	1.83	1.05	150
	2.734(4)*	1.79	0.97	161
1b	2.62(1)	[c]		
	2.66(1)*	[c]		
1c	2.676(3)	[c]		
	2.694(3)*	1.73	1.01	158
1d	2.589(3)	1.61	0.99	169
	2.727(3)*	1.74	1.03	160
	2.626(3)**	1.75	0.88	173
	2.595(3)**	1.64	0.97	166
	2.618(3)**	1.66	0.97	169
1e	2.565(3)	1.51	1.06	171
	2.693(3)*	1.75	0.95	173
	2.640(3)**	1.55	1.13	161
1f	2.547(8)	[c]		
	2.677(8)*	1.59	1.10	169
	2.611(9)**	[c]		
1g	2.615(3)	1.65	1.00	161
	2.654(2)	1.74	0.92	176
	2.917(2)*	1.99	0.97	157
	2.739(2)*	1.81	0.93	176

[a] Parameters marked with * derive from an alcohol donor and an acid (C=O) acceptor. Parameters marked with ** derive from a carboxyl(-OH) donor and carboxyl (C=O) acceptor (intermolecular). Entries not marked represent inverted relations. [b] For compound designation see Table 11. [c] No hydrogen atomic positions were determined for these sites.

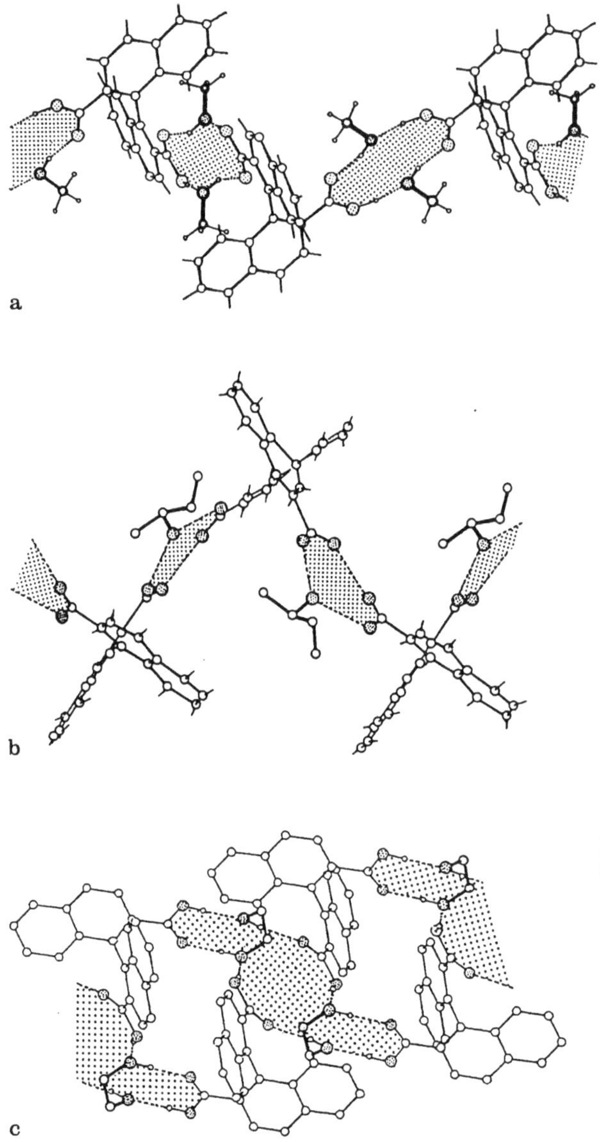

Fig. 17. Characteristic closed H-bond loops (shaded regions) in the alcohol inclusions of *1* [2]. Packing excerpts show the rings formed in the coordinatoclathrates **(a)** *1* · MeOH (1:2), **(b)** *1* · 2-BuOH (1:1), **(c)** *1* · ethylene glycol (1:1) (H-bonds as broken lines; O atoms dotted; H atoms of the host connected with non-hetero-atoms are shown as sticks only, or are omitted completely)

Hydrogen bonding is a phenomenon of considerable interest [75]. It is responsible for holding together many organic crystals. Also, H-bonding plays a major role in determining the conformations of nucleic acids, proteins and polysaccharides. Thus, H-bonds are involved in the formation of tertiary structures and in biomolecular recognition. Consequently, thorough understanding of their properties should assist us in modelling such complicated systems by artificial mimics of molecular association and recognition using simpler aggregates like the clathrates discussed here and the many other systems presented in different chapters of this volume. Unlike other kinds of weak interactions, H-bonding can be made more simply describable, in a similar

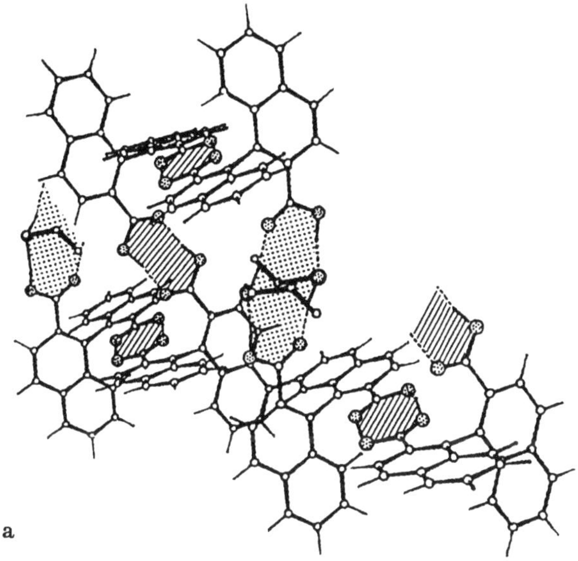

a

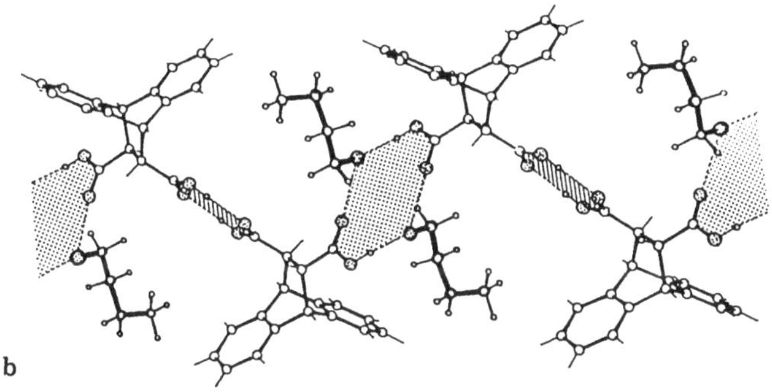

b

Fig. 18. Closed H-bond loops (dotted and hatched regions indicating host-guest and host-host interactions, respectively) in alcohol inclusions of *1* and *26*. Packing excerpts show the rings formed in the coordinatoclathrates: **(a)** *1* · 1-PrOH (2:1)[77], **(b)** *26* · 1-BuOH (1:1)[71] (H-atoms of the hosts are shown as sticks only)

way, as covalent bonds, i.e. in terms of distances and angles, which provides another advantage. Other kinds of interactions are less well describable and hence understood — they lack the accessibility of easy description typical of H-bonds. The interaction is not restricted to pairs of molecules within an asymmetric unit but, as revealed by the respective packing diagrams (abstracted in Figs. 17–19), involves a multitude of neighboring molecules.

The example specified as IIa in Fig. 19 reveals the pattern found invariably in the methanol, ethanol, and 2-propanol inclusions of *1*. It is characterized by a loop of H-bonds which always involves two guest molecules opposing each other through a center of symmetry and two carboxyl groups of two symmetry-related molecules of *1* thus having adverse chirality (Fig. 17a). The loop of H-bonds seems to be formed with

Type	Specification	Example
I 10 $-D_G A_H D_H A_H D_H A_G-$	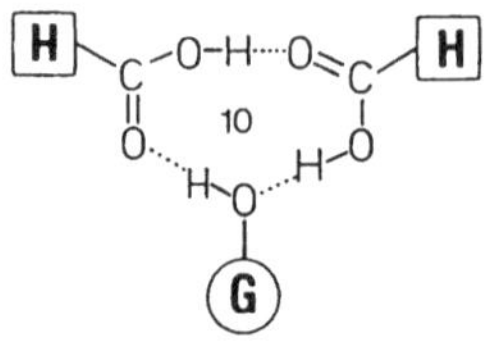	*1*·2-BuOH (1:1) (see Fig. 17b) *1*·t-BuOH (1:1)
IIa 12 $-(D_G A_H D_H A_G)_2-$	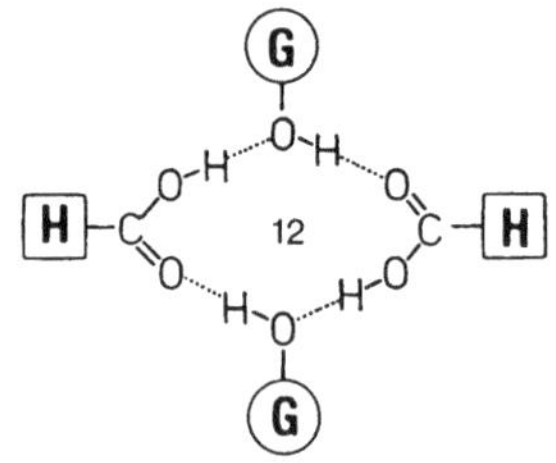	*1*·MeOH (1:2) (see Fig. 17a) *1*·EtOH (1:2) *1*·2-PrOH (1:2)
IIb 12 $-(D_G A_H D_H A_G)_2-$ 8 $-(D_H A_H)_2-$	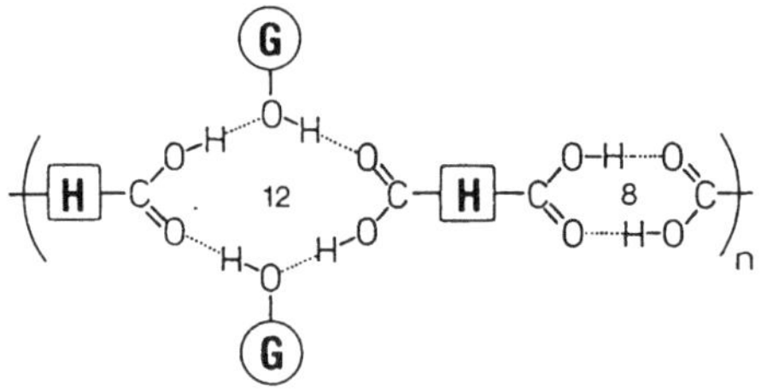	*1*·1-PrOH (2:1) (see Fig. 18a) *26*·1-BuOH (1:1) (see Fig. 18b)
III 24 $-(D_G A_H D_H A_G)_4-$	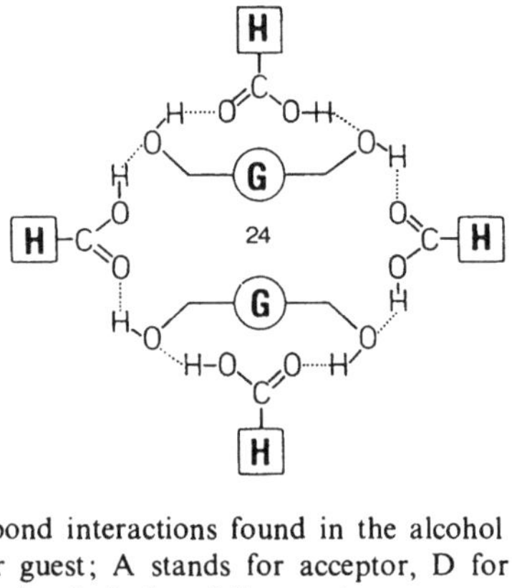	*1*·$HOCH_2CH_2OH$ (1:1) (see Fig. 17c)

Fig. 19. Systematics of the H-bond interactions found in the alcohol inclusions of *1* and *26* (the bold H stands for host, G for guest; A stands for acceptor, D for donor; the number in the center of the rings indicates the size, inclusive of H atoms)

directed donor-acceptor interactions corresponding to a homodromic arrangement [76]. This observation also applies to other clathrates of *1* with alcohols. Arrangements such as above are thought to possess slightly enhanced stability [76].

Examination of the steric relations in these complexes (cf. Fig. 30) suggests that the more voluminous branched alcohols cannot follow the same principle. Indeed, in the 2-butanol and also in the t-butanol inclusion compound, a different ring system is built (Fig. 17b and type I in Fig. 19). While the short-chain alcohols form twelve-membered H-bond loops, the branched butyl alcohols are embedded into a ten-membered asymmetric loop. The stoichiometry of the asymmetric unit also changes from 1:2 (host:guest) ratio to 1:1. The so-built ring system of homodromic H-bonds still contains a mirror-related pair of hosts *1*, but comprises only one guest molecule.

The ethylene glycol clathrate of *1* displays 1:1 stoichiometry, too. However, this guest has two —OH functions appended to a relatively small molecule. For that reason, it is able to form a rather huge (24-membered) centrosymmetric ring system (Fig. 17c and also type III in Fig. 19). It involves four carboxyl and hydroxyl groups, each coming from four host and two guest molecules, respectively. Yet, the building principle remains the same, involving host molecules of *1* of opposite chirality and H-bonds with homodromic direction.

A more deviating stoichiometry is found in the case of the inclusion compound of *1* with *1*-propanol [77]. Here the assistance of two independent host molecules is required and results in a 2:1 stoichiometry. Nevertheless, even this unusual host:guest ratio gives rise to a similar H-bond pattern (Fig. 18a and *type IIb* in Fig. 19) as found for the inclusions of *1* with simpler alcohols (cf. Fig. 17a), namely the 12-membered ring system. Now, another interesting fact arises, signalling the flexibility of host *1* in its inclusion behavior. This is the formation of host dimers through H-bonds to ensure clathration.

Precisely, this behavior is found for the host *26* (see Sect. 4.1), another properly tailored carboxylic acid (cf. Sect. 4.5). The crystal structure of the 1-butanol associate of *26* (Fig. 18b) shows the same 12-membered H-bond pattern around a center of symmetry as found for the inclusions of *1* with MeOH, EtOH, and 2-PrOH and exactly the same building principle (dimeric host and 12-ring formation) as in the 1-PrOH aggregate of *1*. Thus, they both belong to the same *type IIb* of building blocks (Fig. 19).

From these observations, we have noticed the similarity of the simple lattice inclusions to the more sophisticated assemblies of molecules (e.g. cyclodextrins [76]) and proteins [78]) where the formation of H-bonded loops was first detected and described. Conclusively the motive for the formation of simple inclusion crystals and of more complex associates between high and low molecular weight compounds, such as enzyme-substrate complexes, can be traced back to the same source.

The results clearly show the capacity of these hosts to act in coordinatoclathrate formation with hydroxylic group-containing guests which means a mutual coordinative relationship between the —COOH group of the host and of the —OH group of the guest is demonstrated. Another way of expressing the factual findings is to speak of mutual recognition between carboxylic hosts and hydroxylic guests originating from a particular association of the functional groups in the first place. Although the individual mode of the association shows some variation depending on the fine

Type and specification	Example

I

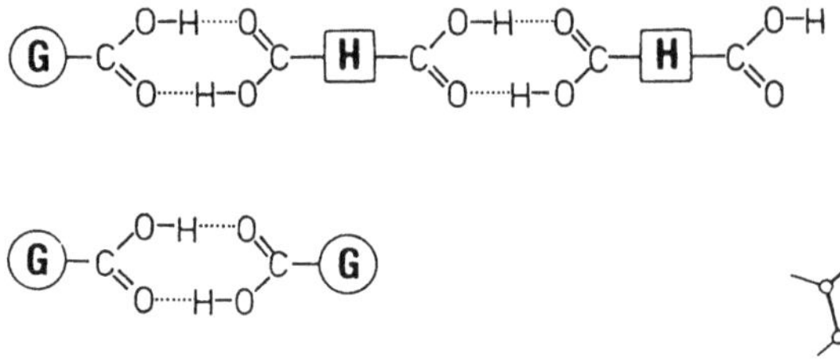

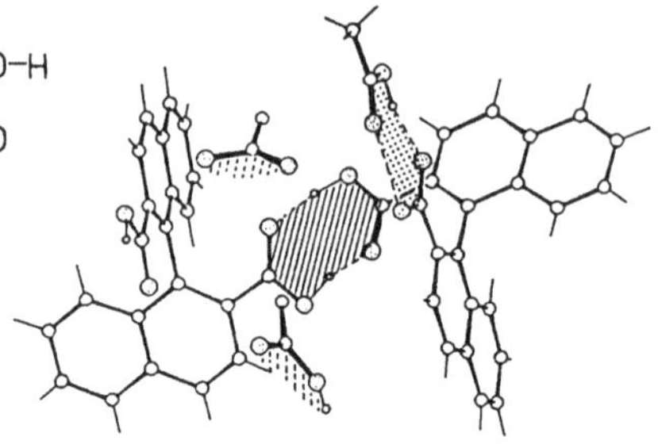

1 · H_3CCOOH (2:3)

II

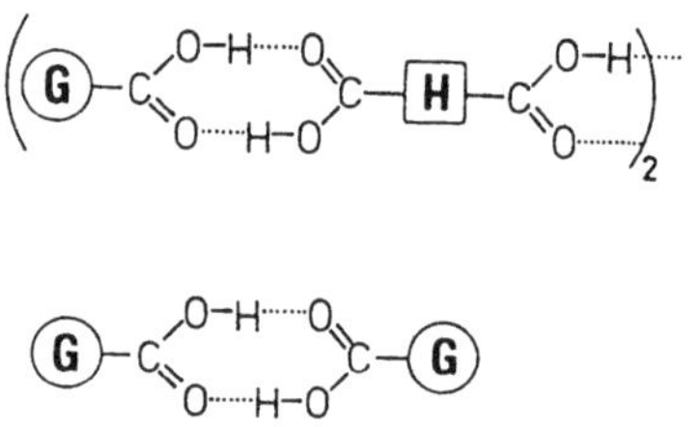

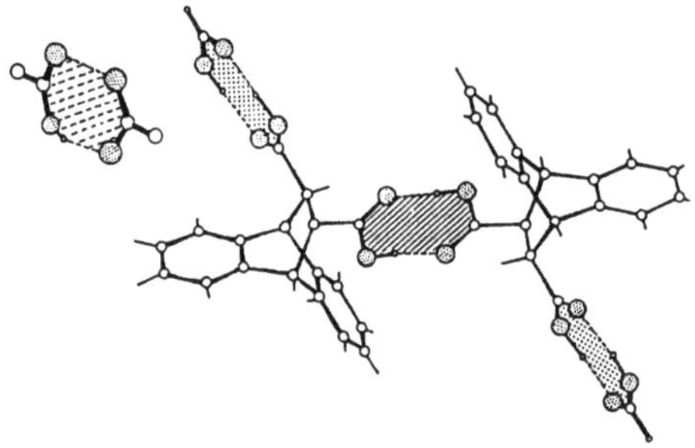

26 · HCOOH (1:2)

III

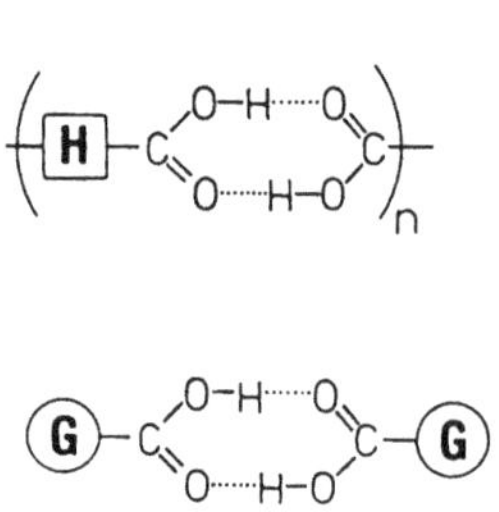

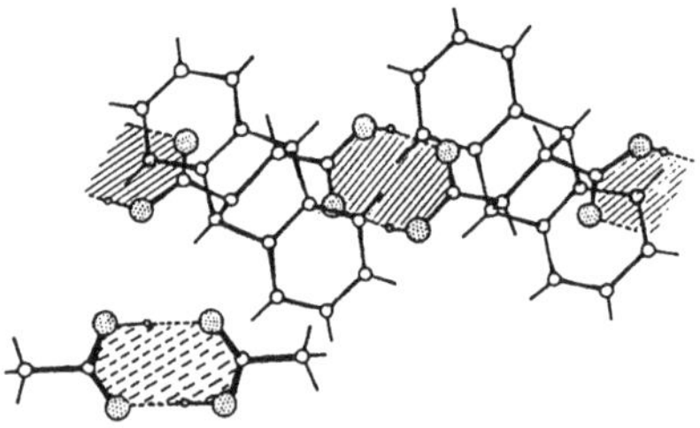

26 · H_3CCOOH (1:1)

26 · H_3CCH_2COOH (1:1)

Fig. 20. Schematic and packing excerpt illustrations of the principal H-bond interactions found in the carboxylic acid clathrates of *1* [79] and *26* [50,71]. The bold H stands for host, G for guest. H-bond rings coming from either host-host, host-guest, or from guest-guest dimers are indicated by

tuning of the given host-guest combination, as e.g. expressed also in the number of the contributing partners, the principle of forming the associates is more general. It involves
— closed loops of homodromic H-bonds including —COOH and —OH;
— both enantiomers of the racemic acids are incorporated into this block (at least this applies for racemic alcohols);
— the full H-donor/acceptor capability of the attending groups is exploited.

A further general comment relates to the geometry of these H-bonds (Table 12) indicating strong interactions even when considering the reliability of H-atom positions. An interesting tendency is also seen in the mean values of the O ... O distances between different categories of H-donor/acceptor moieties (Table 12). The data seem to reflect the common difference in the chemical behavior of an alcoholic —OH (as a better H-acceptor) and of a carboxyl —OH (a stronger Brønsted-acid). For a verification, however, more accurate (H-atom position by e.g. neutron diffraction) and a larger number of data are required.

Acids as Guest Species

Inclusions of acids by *1* and *26* give risè to interesting findings. They surprise us by the constitution of their crystal lattices. For example, the inclusion of *1* with acetic acid has a 2:3 host:guest ratio [79] which is unusual for this host (see Table 1, Sect. 3.2.2). The constitution of the asymmetric unit reveals that the two molecules of *1* form a H-bonded dimer with each other (Fig. 20, type I), as also observed for the 1-PrOH inclusion of *1* (cf. Fig. 19, type IIb) and those of the inclusions of *26* (see below). Such an entity binds one of the acetic acid molecules *via* mutual H-bond donor/acceptor relations forming a pseudo-dimeric arrangement. The two other independent guest molecules are arranged in tunnels in the crystal, made up by the first acetic acid complemented host matrix in the form of symmetry-center related H-bonded dimers.

A similar behavior is found in the 1:2 inclusion of *26* with formic acid [71] (Fig. 20, type II). We notice a H-bonded dimer of *26* and one of the formic acid molecules binding the host dimer in a pseudo-dimeric arrangement via the "free" —COOH groups of the host dimer. The second guest molecule is also placed into an interstitial tunnel of the dimeric host/bound-guest matrix. Here the 1:2 stoichiometry is due to the small size of the guest partner.

The structures of the acetic acid [50] and of the propionic acid [71] inclusions of *26* (Fig. 20, type III) are isomorphous to each other. The increased guest volume with respect to formic acid yields 1:1 stoichiometry, with no H-bonds between host and guest molecules in either case. The tunnel where the dimers of guests are situated (see Fig. 32a) is functionally the same as in the case of the self-dimerized pairs of the formic acid guests.

◄
different shading (hatched, dotted, and broken areas, respectively). Packing excerpt given for type III shows *26* · acetic acid (1:1). It stands for the isomorphous propionic acid clathrate als well (H-bonds as broken lines; O atoms dotted; H-atoms connected with non-heteroatoms are shown as sticks only)

All these known associates of *1* and *26* exemplify analogously structured crystals which may be regarded as either "true" clathrates (e.g. the acetic acid and propionic acid inclusions of *26*) or as "partial" coordinatoclathrates (acetic acid and formic acid inclusions of *1* and *26*, respectively). Depending on the spatial conditions and provided that the guest acids are small enough in size to fit into the intercoil clefts left free in the *1* and *26* host matrices, they intercalate and interact with the host —COOH at the same time. Obviously, acetic acid meets these criteria for *1* and formic acid for *26*. One may say that the size difference between the differing guest partners (formic acid *vs.* acetic acid) is well matched by the differences of the respective hosts *26* and *1*. The acetic acid and propionic acid guests are simply too voluminous to fit into a smaller cleft. Hence, they are placed into the more spacious tunnels formed by the *26* host matrix. This versatile manner of making inclusions of different strength has an obvious impact on the chemistry of these hosts, too.

The above class of inclusions shows unexpected features for one could well assume that guests will be coordinated strongly by the —COOH groups of host molecules (cf. alcohol inclusions). By way of contrast, in these examples all kinds of host-guest, host-host, and guest-guest H-bond contacts coexist, but exclusive host-guest contacts are never seen in these crystals. The geometry of the H-bonds may be assumed as being equally good for all types of these contacts (cf. Table 13). Hence, there is no obvious preference from these data alone as to which type of contact is the preferred one. Indications suggest that H-bonding is only a minor parameter in the mutual recognition of the carboxylic acid hosts and acidic guests. Actually the incidence of direct host-guest association is more a problem of spatial fit in the crystal than the presence of complementary functional groups, i.e. direct host-guest binding occurs

Table 13. H-bond dimensions in *26* (Fig. 15) and of its inclusion compounds with 1-BuOH (Fig. 18b) and acids (Fig. 20) as guest partners

Cmpd[a]	D—H ... A	D ... A (Å)	H ... A (Å)	D—H (Å)	D—H ... A (deg)
26	O18—H18 ... O14′	2.650(5)	1.69	0.98	167
	O15′—H15′ ... O17	2.759(5)	1.84	0.92	173
	O18′—H18′ ... O17′	2.705(5)	1.83	0.90	166
	O15—H15 ... O14	2.681(4)	[b]		
26a	O15—H15 ... O1b	2.599(5)	1.72	0.88	176
	O18—H18 ... O17	2.646(4)	1.83	0.82	173
	O1b—H1b ... O14	2.741(5)	[b]		
26b	O15—H15 ... O1f1	2.690(4)	1.62	1.07	174
	O18—H18 ... O17	2.647(3)	1.67	0.98	175
	O2f1—Hf1 ... O14	2.676(4)	1.74	0.94	172
	O2f2—Hf2 ... O1f2	2.640(5)	1.67	1.09	146
26c	O15—H15 ... O14	2.663	1.67	0.99	171
	O18—H18 ... O17	2.634	1.67	1.02	169
	O2a—H2a ... O1a	2.674	1.72	0.95	178
26d	O15—H15 ... O14	2.661(4)	1.76	0.93	163
	O18—H18 ... O17	2.651(4)	1.78	0.89	165
	O2a—H2a ... O1a	2.647(4)	1.59	1.08	167

[a] Designation: *26a* = *26* · n-BuOH (1:1), *26b* = *26* · formic acid (1:2), *26c* = 26 · acetic acid (1:1), *26d* = *26* · propionic acid (1:1). [b] These H atoms could not be located.

only when a small enough guest molecule is available and is being used for possible stabilization of the crystal. The pK-difference between acids may also add some drift towards such behavior.

4.2.2 Inclusion Compounds Involving Readily Protonizable (Proton Accepting) Guest Molecules: Salt-type Associates and Ternary Complexes of 7 and Imidazole Clathrate of 13

Salt-Type Associates and Ternary Complexes of 7

Though molecules *1* and *7* are closely connected in structure, they have totally different host properties, i.e. *1* readily forms inclusion compounds with a wide variety of guests (see Sect. 3.2.2) while *7* does not. For example, crystals of the pure host could be obtained from dimethylformamide, a solvent which is tightly held by *1*. Reasons for the different behavior of *1* and *7* have already been mentioned when the crystal structure of the free host *7* was discussed (Sect. 4.1). However, the ability of *7* to form a crystalline associate is increased, if a solvent with the property of a base is present, e.g. pyridine and substituted derivatives of pyridine (see Table 14) [80].

As confirmed by the structural studies, proton transfer from one of the carboxyl groups of *7* plays a principal role in these events (salt formation). Obviously this process is faciliated by the same factor as already seen in the structure of solvent-free *7*, namely the favorable steric placement of the two carboxyl groups. The principle of forming these salt-type inclusion aggregates is best demonstrated by the crystal structure given in Fig. 21 as type I which shows the structure of the 1:1 associate of *7* with pyridine. The same type I arrangement applies for the 1:1 associate of *7* with 2-(hydroxymethyl)pyridine. The carboxylate anion arising from the host molecule is stabilized by the internal (intramolecular) H-bond from the *syn*-positioned —OH of the neighboring carboxyl group in both cases. Besides, there is a direct contact between the charged species.

The ternary aggregate composed of *7*, pyridine, and acetic acid with 1:1:1 stoichiometry composition (Fig. 21, type II) is an even more interesting case. Apparently

Table 14. Crystal data of the inclusion associates of *7*

Compound[a]	*7a*	*7b*	*7c*	*7d*
Space group	$P2_1$	*Pbca*	$P\bar{1}$	$P1$
Unit cell				
a (Å)	8.080(1)	15.4267(4)	14.525(6)	7.569(4)
b (Å)	17.254(2)	21.1895(6)	10.481(4)	8.393(2)
c (Å)	7.715(1)	13.3727(4)	8.862(4)	8.634(1)
α (deg)	90	90	105.69(4)	93.21(2)
β (deg)	106.28(3)	90	111.10(6)	106.88(3)
γ (deg)	90	90	86.37(6)	105.17(3)
V (Å^3)	1032.4(3)	4371.3(2)	1211(1)	501.3(7)
Z	2	8	2	1
D_c (gcm^{-3})	1.356	1.372	1.321	1.359

[a] Designation: *7a* = *7* · pyridine (1:1), *7b* = *7* · 2-(hydroxymethyl)pyridine (1:1), *7c* = *7* · pyridine · acetic acid (1:1:1), *7d* = *7* · imidazole (1:1).

Type and specification	Example
I 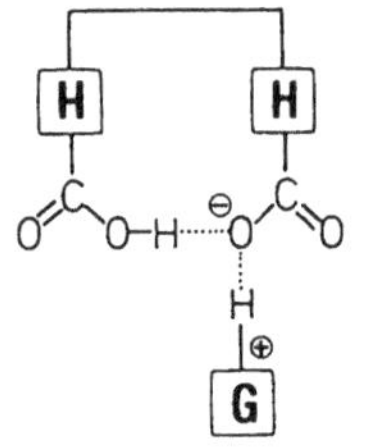	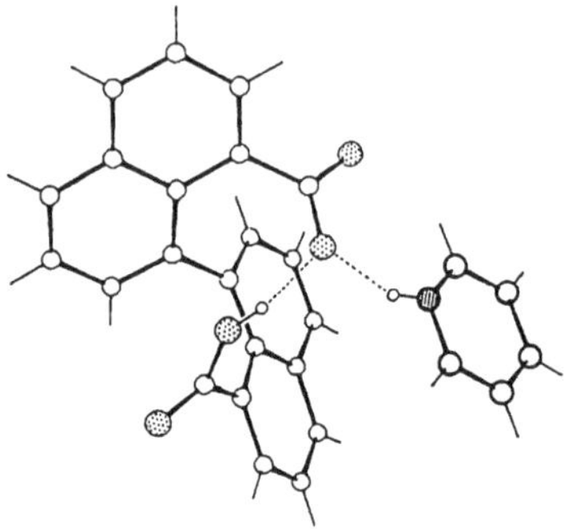 $7 \cdot$ Pyridine (1:1)
II 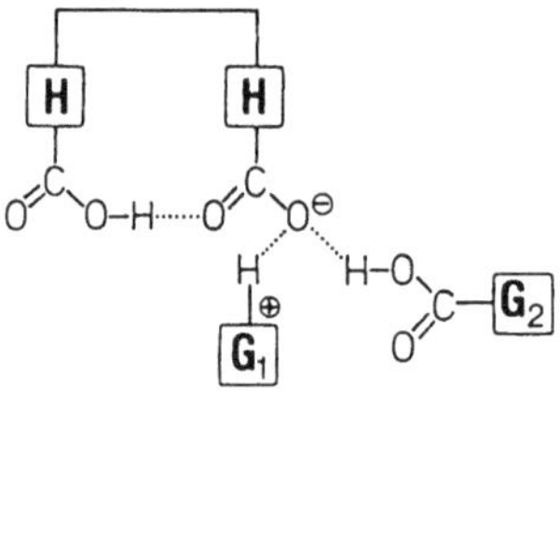	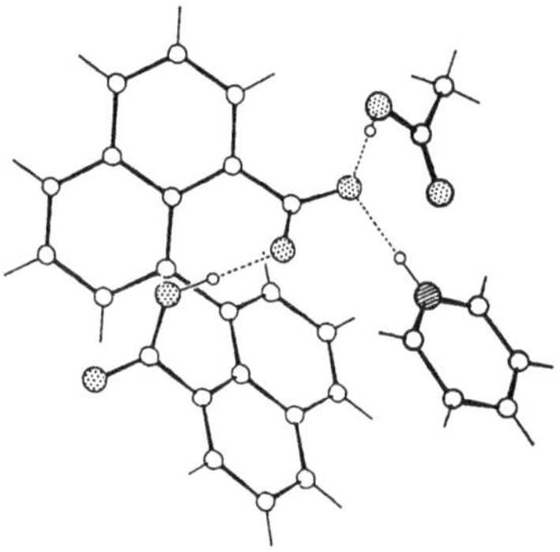

Fig. 21. Schematic and packing excerpt illustrations of the principal H-bond interactions found in the salt-type associates of 7 [80] (the bold H stands for host, G for guest; H-bonds as broken lines; O atoms dotted; N atoms hatched; H atoms connected with non-heteroatoms are shown as sticks only). The $7 \cdot$ 2-(hydroxymethyl)pyridine (1:1) associate follows nearly the same principle as shown for type I

by virtue of the basic component (pyridine), the bimolecular salt-type associate $7 \cdot$ pyridine (cf. Fig. 21, type I) became a host for a third partner (acetic acid). The intra-associate interactions may be characterized as before, involving strong H-bonds between the ionic species and internal contact of the carboxylic groups.

In summary, we may establish that the salt-type of inclusions and the ternary complex of 7 are determined by H-bonding in other respects. Here the characteristic property is the formation of strong intra-associate H-bonds (Table 15). Considering the invariable intramolecular involvement of one of the —COOH groups in the host,

Table 15. H-bond dimensions in the inclusion associates of *7* (cf. Fig. 21)

Cmpd[a]	D—H ... A	D ... A (Å)	H ... A (Å)	D—H (Å)	D—H ... A (deg)
7a	O(11)—H(11) ... O(11')	2.585(4)	0.88	1.73	163
	N(1P)—H(1N) ... O(11')	2.620(5)	0.96	1.68	166
7b	O(11)—H(11) ... O(11')	2.497(3)	0.99	1.52	172
	N(1P)—H(1N) ... O(11')	2.719(5)	0.97	1.76	171
	C(P2)—H(P2) ... O(10')[b]	3.068(5)	1.08	2.30	127
	O(P3)—H(30) ... O(10)[c]	2.739(5)	0.95	1.86	153
7c	O(11)—H(11) ... O(11')	2.576(3)	0.96	1.56	178
	N(P1)—H(N1) ... O(10')	2.692(4)	1.08	1.62	169
	O(A2)—H(A2) ... O(10')	2.623(4)	0.83	1.80	177

[a] For designation see Table 14. [b] Might be taken as a consequence of an already present H-bond. [c] Atom of a symmetry related molecule.

the major cohesion of the lattices is due to dispersion forces and van der Waals interactions. The only inter-associate H-bond exists in the structure of the *7* · 2-(hydroxymethyl)pyridine system involving —OH groups of symmetry-related solvent molecules. This associate also has a short C—H ... O contact [75b)].

Apart from the isomeric relation between *7* and *1*, the appearence of the ternary associate now showing coordinatoclathrate properties gives a reasonable motive for putting up these compounds for discussion here. The dimer formation of carboxylic acids known from the related inclusions of *1* and *26* does not occur here. Instead, one observes a well-balanced system of H-bonds between groups of different acid/base properties. It is left to future studies to find other acid/base combinations which give a comparable situation. Actually, such H-bonded systems remind one of the multiple non-bonded interactions at the active centers of enzymes.

Imidazole Clathrate of the Non-Carboxylic Host 13

The binaphthol *13* is different from *1* and *7* owing to the lower acidity of its functional groups. Therefore, crystalline complexes of *13* with amines (see Table 3) are not expected to have a salt character. The *13* · imidazole 1:2 complex (Fig. 22) [81)] was studied in the light of the general interest in this guest partner and its relation to alcohol functions in biological ensembles. The host molecule adopts ideal twofold

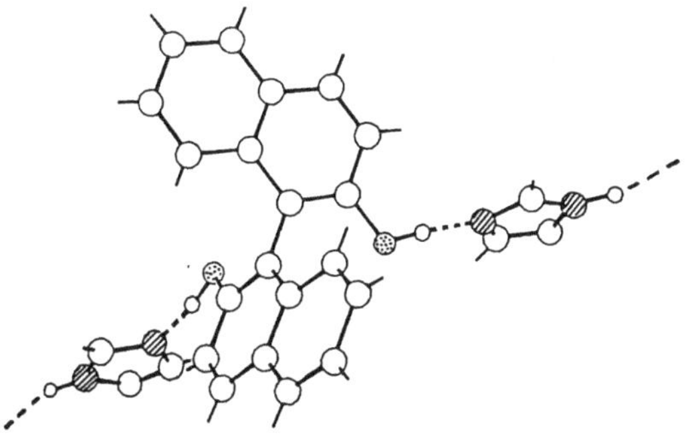

Fig. 22. H-bonding detail in the crystal structure of the *13* · imidazole (1:2) inclusion [81)] (H-bonds as broken lines; O atoms dotted; N atoms hatched; H atoms connected with non-heteroatoms are shown as sticks only)

symmetry. The imidazole molecules are linked into an infinite spiral with host molecules as H-bond donor/acceptors complementing the role of the alcohol functions (Fig. 22). The tetragonal crystal lattice consists of hosts of the same chirality (i.e. spontaneous resolution occurred). Further examples of crystal structures of inclusion compounds of *13* including enantiodifferentiation phenomena are given in Chapter 1 of this volume (see also Chapter 3 of Vol. 140 of this series).

4.2.3 Inclusion Compounds Involving Aprotic-Dipolar Guest Molecules: Dimethylformamide and Dimethyl Sulfoxide as Guest Species

The guest compounds discussed under this heading have a somewhat reduced ability to establish H-bond interactions to the host matrix. They may well act as H-bond recipients using the negatively polarized part of their dipoles, but normally their H-donor capability is low. Corresponding behavior is observed in the crystalline inclusion compounds of *1*, *20*, *22*, *25b*, *26*, *37*, and *41* with dimethylformamide and dimethyl sulfoxide. On the one hand, their crystal structures (Figs. 23 and 24) indicate the way how the hosts recognize the different guest molecules. On the other hand, an analysis of the patterns of binding between different hosts and the same guest supplies us with information on the structural conditions for recognition of a given sensor group (the —COOH group in these host compounds).

Dimethylformamide (DMF) as Guest Species

In the first instance, it is important to notice that all inclusion compounds between dimethylformamide as guest and carboxylic acid hosts so far formally studied have the same 1:1 stoichiometry (each *free* —COOH of the host binds one dimethylformamide). The schemes of the characteristic binding modes including the structures are illustrated in Fig. 23. These arrangements prove the mutual H-bond donor/acceptor ability of the sensor groups both of the host (—COOH) and of the guest. In the latter, the donor function resides in the slightly acidic C—H of the formyl group.

Closer inspection of the blocks in the crystal structure, however, reveals some interesting variation as far as the individual spatial arrangements are concerned. The most compact association is found in the 1:2 inclusion compound of the spiro host *22* with dimethylformamide [48] (Fig. 23, type Ia). This aggregate preserves a perfect twofold (C_2) molecular symmetry in the crystal lattice. The formamide moiety acts as

▶

Fig. 23. Recognition characteristics of dimethylformamide emanating from the carboxylic host molecules studied [48, 71, 82, 83]. Representative crystal structure excerpts and guest-binding schematics are shown parallel. The bold H stands for host, G for guest; A stands for acceptor, D for donor. H-bond rings coming from either host-guest or from host-host interactions (cf. type II) are indicated by different shading (dotted and hatched regions, respectively); single H-bonds are represented by a single broken line. Note that one of the DMF molecules in type Ib shows single binding only. Type Ic is characterized by an additional six-membered ring formed through an internal H-bond (O atoms dotted; N atoms hatched; H atoms connected with non-heteroatoms are shown as sticks only, except for the host molecules which are fully drawn)

Type and specification		Example
Ia	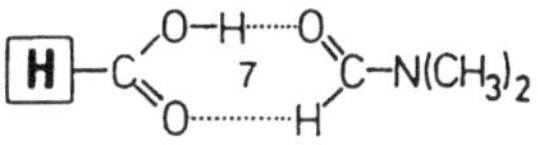 $\overbrace{D_H A_G D_G A_H}^{7}$	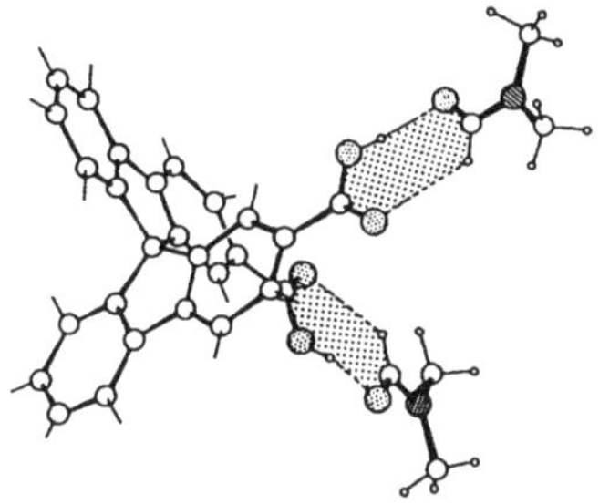
Ib	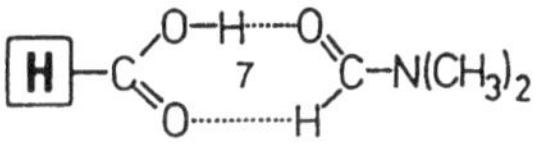	*22* · DMF (1:2)
	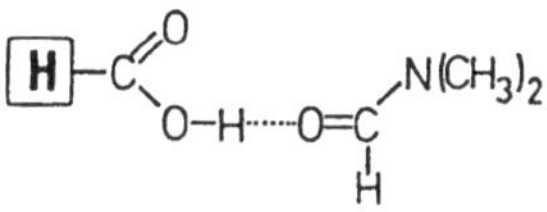 $\overbrace{D_H A_G D_G A_H}^{7} \cdot D_H A_G$	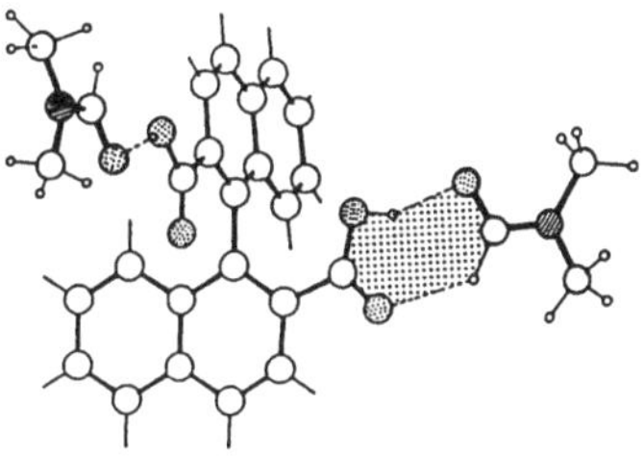 *1* · DMF (1:2)
Ic	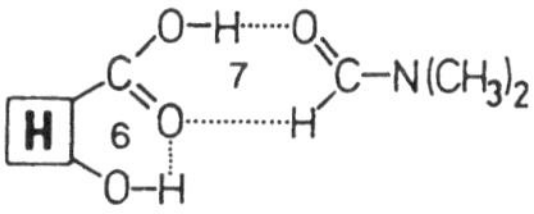 $\overbrace{D_H A_G D_G \underbrace{A_H - D_H}_{6}}^{7}$	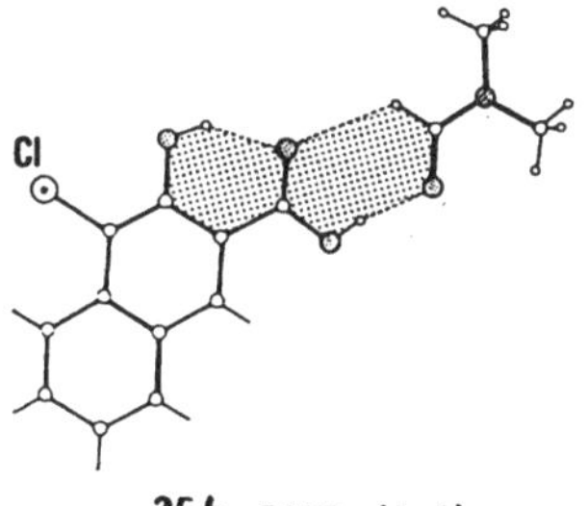 *25b* · DMF (1:1)
II	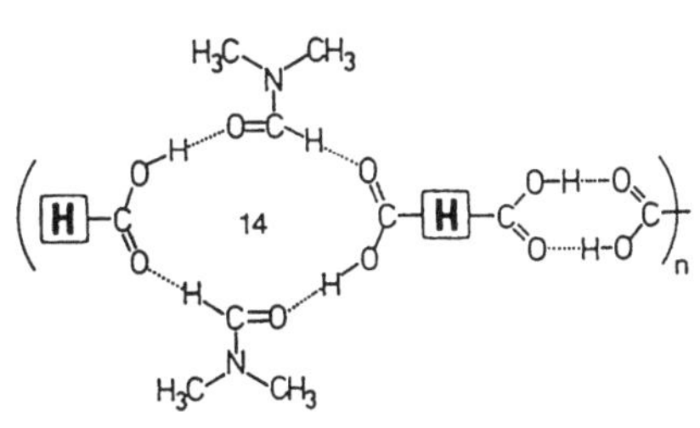 $\overbrace{(D_H A_G D_G A_H)_2}^{14} \cdot \overbrace{(D_H A_H)_2}^{8}$	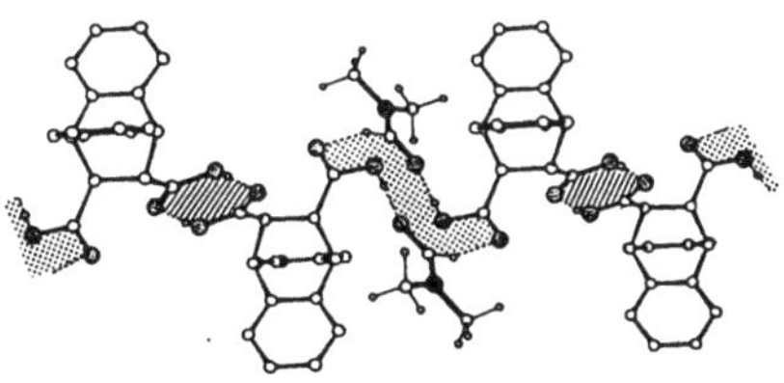 *26* · DMF (1:1)

Table 16. H-bond dimensions in the dimethylformamide inclusion of *1*, *22*, *25b*, and *26* (cf. Fig. 23)

Cmpd	D—H ... A	D ... A (Å)	H ... A (Å)	D—H (Å)	D—H ... A (deg)
1	O11—H11 ... O1d	2.692(6)	1.90	0.86	152
	O11'—H11' ... O1d'	2.613(6)	1.81	0.84	159
	C1d—H1d ... O10[a]	3.054(8)	2.26	1.08	129
22	O16—H16 ... O1d	2.593(2)	1.69	0.92	167
	C1d—H1d ... O15[a]	3.112(3)	2.42	0.99	126
25b	O11—H11 ... O1d	2.539(4)	1.74	0.82	167
	C1d—H1d ... O10[a]	3.255(3)	2.57	1.01	125
	O12—H12 ... O10	2.578(2)	1.82	0.87	144
26	O18—H18 ... O17	2.616(3)	1.76	0.87	170
	O15—H15 ... O1d	2.623(3)	1.74	0.89	179
	C2d—H2d ... O14[a]	3.240(4)	2.25	1.04	160

[a] Maintains a C—H ... O close contact.

H-bond acceptor from the carboxylic sensor of the host and also as donor making a C—H ... O type of interaction possible. In such a way, a 7-membered closed ring is formed, classified in the schematical representation as type Ia. This aggregate shows strong binding of the solvent molecule to the host as judged from the geometry parameters (Table 16).

The structure of the *1*·dymethylformamide 1:2 inclusion [82] is quite different from such a compact arrangement. One of the bound dimethylformamide molecules does not profit from keeping a cooperative C—H ... O interaction with its anchoring —COOH group here (Fig. 23, type Ib). This is due to probable packing conflicts caused by a similarly tight arrangement as in *22* (cf. Fig. 33, Sect. 4.4). One of the guest molecules is bound in a 7-membered, nearly co-planar arrangement (Tab. 16), while the other guest shows a linear binding to the anchoring carboxyl, still being co-planar with it.

Another example of dimethylformamide binding to a host is offered by the structure of the crystalline 1:1 associate with *25b* (Fig. 23, type Ic) [83]. This structure also shows the most frequent arrangement of the sensor function and the guest. The spatial correspondence between the carboxyl function and the amide moiety is also described by the dihedral angle (10.3 degrees) formed by these two moieties. The non-bound contacts (Table 16) indicate marginal (if any) interactions with the C—H moiety of the amide. This is most probably due to the presence of the —OH substituent at position 2 and the chlorine atom at position 1 in the "host" which exert substituent effects on the naphthalene moiety and give rise to an internal 6-membered H-bound ring maintained by the hydroxyl and carboxyl functions. As a result of this interaction, the carbonyl oxygen of the acid may become a weaker external acceptor.

An even more deviating pattern is found in the case of the *26* · dimethylformamide 1:1 clathrate [71] (Fig. 23, *type II*). Here, we see at first glance an arrangement that is similar to the situation found in the 1-BuOH inclusion of the same host (cf. Fig. 18b), or in the simpler alcohol inclusions of *1* (cf. Fig. 17a). A 14-membered ring is formed which formally corresponds to the sum of two 7-membered rings. The guest molecule acts as a H-bond acceptor with regard to the carboxyl group. Moreover, the formamide moiety still acts as a C—H ... O bond donor, in spite of the fact that the guest moiety is visibly no longer coplanar to either of the —COOH functions. Clearly, the system is attempting to adapt to the environment as well as possible, partly dictated by the dimerically associated hosts. The *anti* positioned —OH bond of the coordinating carboxyl group is remarkable. This bond usually adopts that position when salt formation occurs, e.g. in the salt-type associates of *7* (cf. Fig. 21). Placement of the H-atom under discussion in the common *syn* position would result in steric conflicts with the H-atom of the formamide moiety. Possibly the system tries to escape from this unfavorable situation and thus finds an energetically advantageous solution to maintain coexistence of both the C—H ... O and the O—H ... O interaction by shifting the O—H bond into the less common *anti* position. This fact also signals the importance of the C—H ... O interaction.

Summing up, recognition of dimethylformamide may be given from the information obtained in the following statements:

- the oxygen atom of dimethylformamide is always a H-bond acceptor;
- advantage is taken of electrostatically favorable C—H ... O types of contacts whenever possible;
- dimethylformamide tends to approach a coplanar conformation with the anchoring —COOH group;
- by virtue of these principles, dimethylformamide has a preference for forming a 7-membered ring partly sustained by two H-bonds of O—H ... O and C—H ... O type between the counterfacing and more or less co-planar amide and carboxyl groups.

Roughly speaking, we may say that dimethylformamide acts in a way analogously to alcohols, with the difference of having its H-bond acceptor and donor functions in sterically distant sites compared to the —OH moiety. The prevailing recognition pattern can be subject to alterations, giving an individual touch to the associate in question, nevertheless it may be traced back to its characteristic form.

Dimethyl Sulfoxide (DMSO) as Guest Species

Examples of crystalline associates where dimethyl sulfoxide is involved as one of the heteromolecular constituents are known in an appreciable number[1]. Certainly the associate between dimethyl sulfoxide and trimesic acid [84)] (cf. Chapter 5 in Vol. 140 of this series) is one of the important individual cases. Characteristic modes of association between the carboxylic hosts discussed here and dimethyl sulfoxide are illustrated in Fig. 24. Pertinent geometry data are listed in Tables 17 and 18. One may realize from Fig. 24 that the fundamental mode of association of the host acids *20*, *26*, *37*, and *41* is the formation of discrete H-bonded islands of host and (usually) one guest molecule.

The basic type I is represented by the 1:1 coordinatoclathrate of *20* · DMSO [85)] (Fig. 24). It is seen that an O—H ... O interaction occurs between host and guest. Besides, one of the methyl C-atoms is proximal to the fixing —COOH thus forming a co-planar pseudo-ring arrangement of six non-H atoms (8-membered ring including

1 Fragment search in the Cambridge Crystallographic Data Base reveals 127 hits out of 42381 entries (state May 1985) having 'dimethyl sulfoxide' as the search query and allowing for the presence of metal atoms. In absence of such elements, the number of hits is reduced to 27.

Type and specification	Example

I

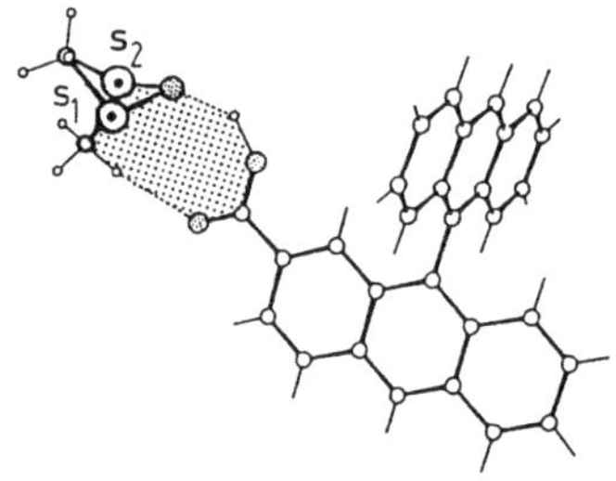

20 · DMSO (1:1)

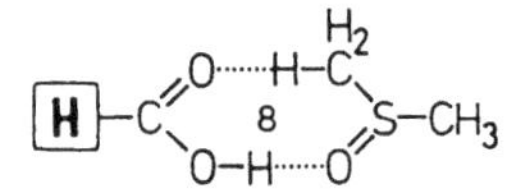

8

$D_H A_G D_G A_H$

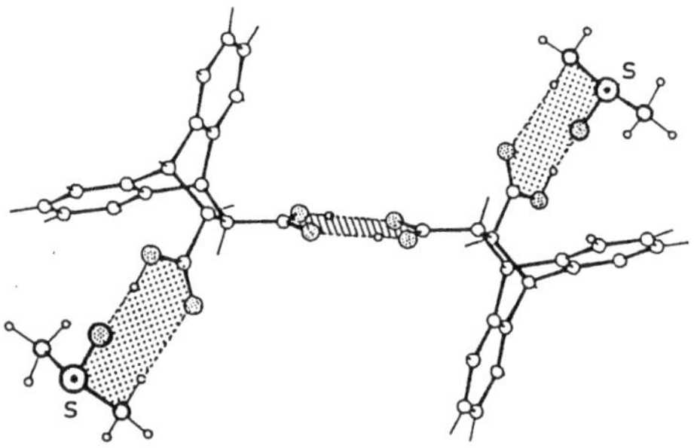

26 · DMSO (1:1)

II

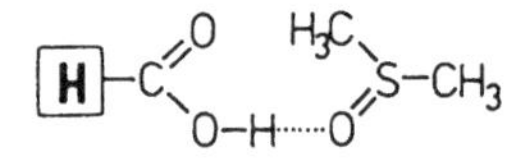

$D_H A_G$

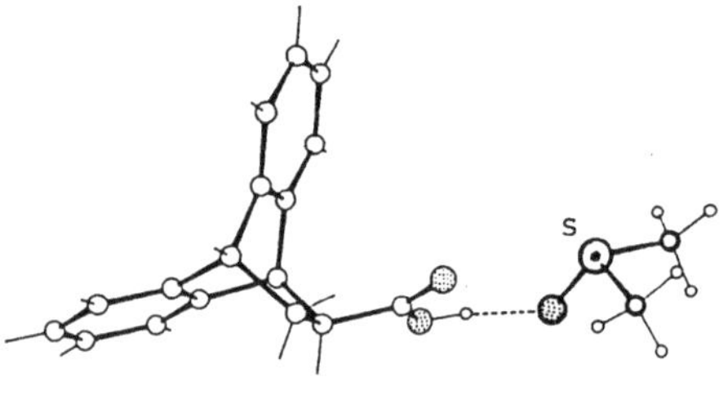

37 · DMSO (1:1)

Type and specification	Example

IIIa

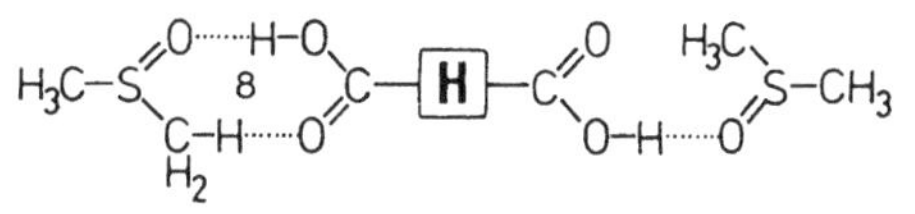

8

$[D_H A_G D_G A_H]$ $D_H A_G$

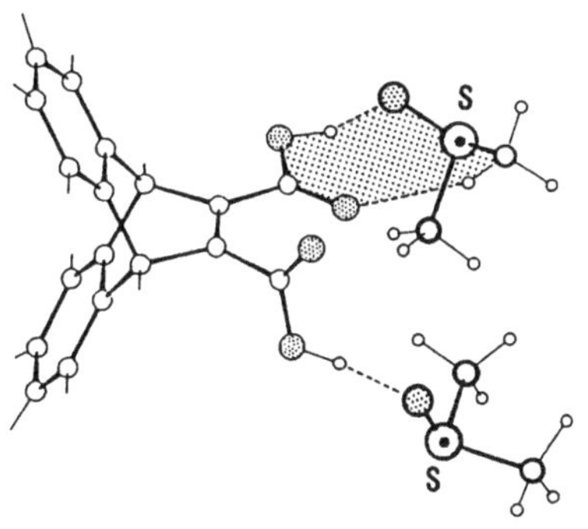

41 · DMSO (1:2)

IIIb

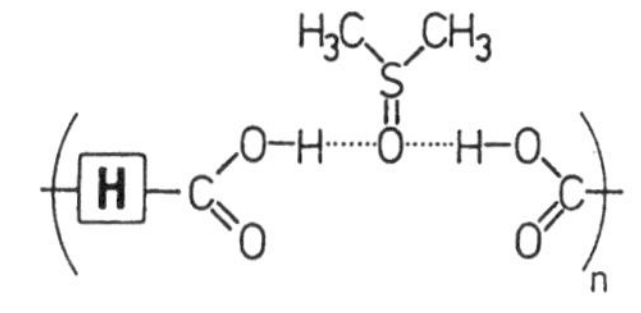

$(D_H A_G A_G D_H)_n$

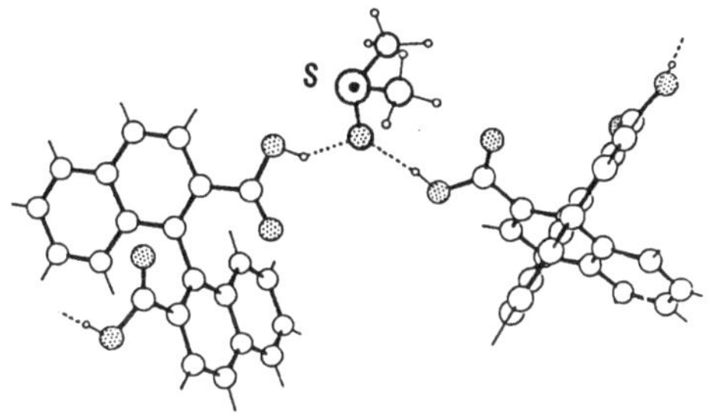

1 ·DMSO (1:1)

Fig. 24. Binding modes of dimethyl sulfoxide to roof and scissor related mono- and dicarboxylic acids [71,82,85–87]. Discrete (type I), linear (type II) and discrete/linear (or polymeric, types III) classes are distinguished. The bold H stands for host, G for guest; A stands for acceptor, D for donor. H-bond rings coming from either host-guest or from host-host interactions (cf. *26* · DMSO) are indicated by different shading (dotted and hatched regions, respectively); single H-bonds are represented by a broken line. S1 and S2 in *20* · DMF (1:1) denote two disorder sites for the sulphur atom with comparable occupancy. Detail of H-bonding in the 1 · DMSO inclusion illustrate a minor part of the chain structure and counterfacing methyl group features (O atoms dotted; S atom marked with a bold dot; H atoms connected with non-heteroatoms are shown as sticks only, except for the guest molecules which are fully drawn)

Table 17. DMSO-distances, some close contact data, and calculated densities in the DMSO-inclusions of *1*, *20*, *26*, and *41* (cf. Fig. 24)

Cmpd	S—O	S—Cl	S—C2[a]	H—O ... O=S	C=O ... C (Me)	D_c
1	1.535(5)	1.678(16)	1.770(9)*	2.652(9)	3.308(11)	1.326
				2.635(8)		
20[b]	1.430(6)	1.714(13)	1.623(13)*	2.634(7)	3.278(12)	1.312
	1.396(9)	1.720(14)	1.634(15)			
26	1.503(3)	1.772(6)	1.767(4)*	2.606(3)	3.346(5)	1.338
41	1.522(3)	1.773(4)	1.770(5)*	2.563(4)	3.288(5)	1.337
	1.504(3)	1.780(5)	1.783(5)	2.562(3)	3.383(5)	

[a] Bond marked with * maintains the C atom involved in the C=O ... C (methyl) interaction. [b] Structure has a disordered DMSO molecule in the form of mirror-imaged overlap; hence its geometry data can be considered indicative only.

H). As shown, a disordered structural model was obtained for the guest. The model comprises two mirror-related guest molecules. The oxygen atom and the proximal methyl C-atom are practically overlapping the same atomic positions in both orientations. However, the sulphur atomic positions do not average in the X-ray data and show a nearly 50/50 occupancy. As indicated by the comparison of the respective bond distances and intra-associate contact distances of the DMSO molecule (Table 17), the effect of disorder is serious (e.g. the S=O distances appear abnormally short in the *20* · DMSO instance). This precludes the possibility of assessing interaction between the O atom of the carboxyl and a methyl of dimethyl sulfoxide.

Basically the same type I binding is found in the 1:1 coordinatoclathrate of *26* · DMSO [71] (Fig. 24). That is to say, the binding of the guest is characterized by the O—H ... O hydrogen bond and by the proximity of one of the methyl groups. Besides, the dimer formation of host *26* (cf. Fig. 23, type II) appears again. In these two type I cases, the oxygen atom of dimethyl sulfoxide in the inclusion acts as a single H-bond acceptor.

Table 18. H-bond dimensions in the dimethyl sulfoxide inclusions of *1*, *20*, *26*, and *41* (cf. Fig. 24)

Cmpd	D—H ... A	D ... A (Å)	H ... A (Å)	D—H (Å)	D—H ... A (deg)
1	O11—H11 ... O1d	2.652(9)	1.78	0.98	146
	O11'—H11' ... O1d	2.635(8)	1.70	0.99	155
	C2d—H2d3 ... O10'[a]	3.308(11)	2.27	1.08	161
20	O13—H13 ... O1d	2.634(7)			
	C2d—H2d ... O12[a,b]	3.278(13)	2.47	1.11	129
26	O15—H15 ... O1d	2.606(3)	1.65	0.97	167
	O18—H18 ... O17	2.629(3)	1.80	0.83	174
	C1d—H1d ... O14[a,b]	3.346(5)	2.38	1.08	148
41	O17—H17 ... O1d	2.563(4)	1.62	0.95	172
	O15—H15 ... O2d	2.562(3)	1.52	1.07	161
	C1d2—H121 ... O18[a]	3.288(5)	2.37	1.06	144

[a] Maintains a C—H ... O close contact. [b] Guest molecules disordered over two sites; hence these data may only be considered *cum grano salis*.

Another very recent example of a somewhat deviating association pattern is provided by the 1:1 coordinatoclathrate of the monoacid *37* with dimethyl sulfoxide [86] (Fig. 24, type II). Here the guest molecule is bound to the host by the aid of a single O—H ... O interaction and both methyls are placed further off from the —COOH group.

The structure of the related 1:2 inclusion compound of *41* [87] (Fig. 24, type III a) is constructed again in a slightly different way. It may be considered as a combination of the pure types I and II in a single island of the inclusion aggregate. Both oxygen atoms act as single acceptors of H-bonds. Probably due to steric crowding, only one of the guests is allowed to adopt a "proximal methyl group" configuration to its fixing —COOH. This pattern (termed as type III a in Fig. 24) tells us that there is a degree of flexibility in the way dimethyl sulfoxide is kept by the —COOH group, depending on, e.g. local (packing) conditions.

The 1:1 inclusion compound of *1* with dimethyl sulfoxide was studied [82] in order to broaden knowledge of the binding modes of *1* to different solvents. The preferred accommodation of dimethyl sulfoxide in competition experiments with other dipolar-aprotic solvents (e.g. acetonitrile, Table 2.) suggested some kind of a strong interaction. As shown in Fig. 24 (type III b), the crystal structure of *1* · DMSO is most appropriately described as a double-acceptor guest molecule pattern. Alternating enantiomers of *1* and intercalated dimethyl sulfoxide molecules are stacked to form H-bound infite chains (see Table 18 for geometry data). These chains are repeated in the crystal by a glide plane symmetry operator in the *ac* crystallographic plane. The structure shows that both the proximal methyl group (type I) and the single-acceptor (type II) patterns coexist in this crystal. Both types I and II are combined in a way to focus these features in one *guest* molecule. This is in contrast with type III a since in the latter case combination of types I and II applies to the relations with one *host* molecule. In fact, due to the chain-like nature of the H-bonds between guest and host molecules, it is not possible to discriminate between intra- and inter-associate relations any longer. A similar chain-like involvement of dimethyl sulfoxide has recently been found for an associate composed of dimethyl sulfoxide and the adduct of hexachloroacetone with water [88].

Four out of the five associates in our studies display 1:1 host:guest stoichiometry (those of *1*, *20*, *26*, and *37*). Only the corresponding inclusion of *41* has a 1:2 stoichiometry. This difference could be linked up with the altered orientation of the vicinal carboxyl groups.

Summing up, one may conclude that formulation of the rules with reference to molecular recognition of a dimethyl sulfoxide guest by carboxyl groups is problematic. This is due to the nature of the basic attachment to the acid group: it may be seen as a flexible one-point type of binding. Concerted packing effects and such a high degree of freedom easily yield somewhat varying structural motifs.

However, there still remain some basic similarities which may be summarized as follows:

— the sulfoxide oxygen acts as a H-bond acceptor (single or double one) and these H-bonds are usually strong ones (cf. Table 18);
— dimethyl sulfoxide is inclined to give a "counter-facing methyl group", i.e. one of the methyls of dimethyl sulfoxide faces the carbonyl oxygen of the guest-binding carboxyl group.

Hence, the dimethyl sulfoxide molecule in these examples tends to reach a possibly coplanar orientation with regard to the binding carboxyl group.

A more accurate model of this orientational behavior can not be deduced from the data at our disposal due to the known problems of locating reliable H-atom positions from electron-density maps based on room-temperature X-ray data. The appearance of structural disorder, especially manifested in the *20* · dimethyl sulfoxide inclusion, requires low-temperature data. Assumptions concerning the possibility of C(methyl)-H ... O interactions might perhaps be ascertained by other H-atom position sensitive methods (e.g. neutron diffraction). As indicated by some geometry parameters (Table 17), there are data which seem to correlate with such an assumption, while others are contradicting. Calculated densities which are also listed in Table 17 for some compounds parallel the observed disorder phenomenon of the *20* · DMSO aggregate having the lowest density among all dimethyl sulfoxide inclusions.

4.3 Inclusion Compounds Without Specific Binding Contacts Between Host and Guest: Apolar Molecules as Guest Species

So far, in accord with the original idea of coordinatoclathrate formation (see Sect. 2.2), such guest molecules which have been the main topic of the discussions could more or less intimately interact with the protic sensor groups of the host matrix. Another approach where an attempt is made to get rid of such dominant interactions between host and guest will be examined briefly in the following. One could expect from Fig. 8 (Sect. 2.2) that the more geometric dependent part of host-guest interactions will now gain in importance. Actually, remembering Sect. 2.1 of this chapter, one may expect that on meeting a proper guest, a host molecule supplied with functional groups could be able to form "true" or conventional clathrates (see Fig. 6a) as well. This way of thinking also prepares the ground for a kind of conscientious engineering of crystal structures based solely on the host shape. Corresponding ideas have been incorporated into the forthcoming discussion.

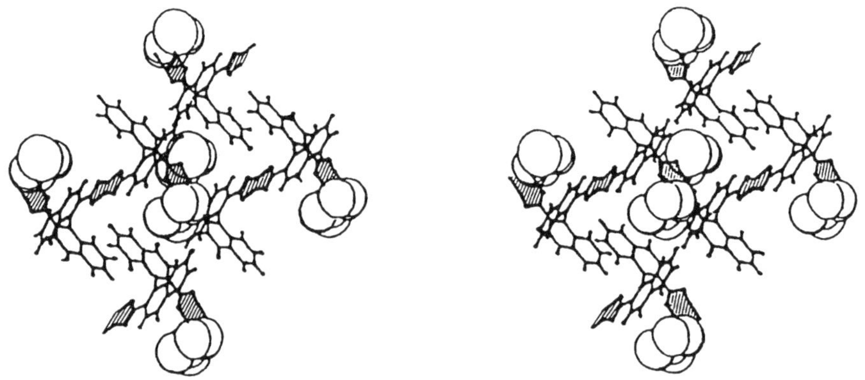

Fig. 25. Stereoscopic packing illustration of the *1* · bromobenzene (1:1) clathrate [82]. The guest molecules are shown by enlarged atomic radii (arbitrary values). H-bond rings coming from carboxylic group dimerization are indicated by hatching

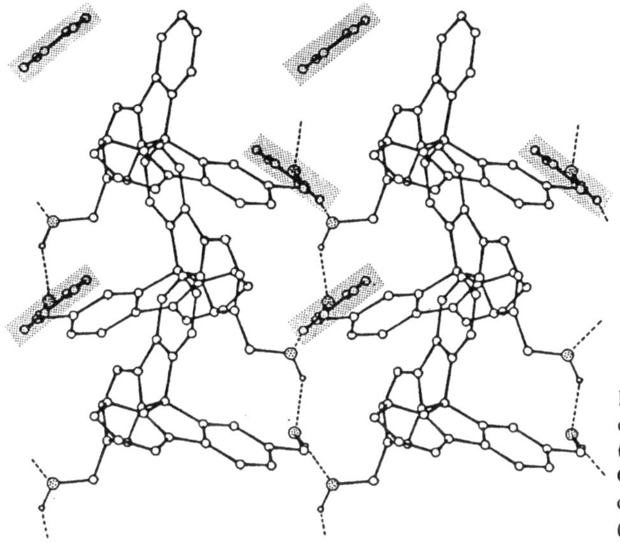

Fig. 26. Packing detail from the crystal structure of *24* · benzene (1:1) (H-bonds as broken lines; O atoms dotted; the guest molecules are marked by shading) (Adapted from Ref. 89)

We assume from Table 1 (Sect. 3.2.2) that the prototypical coordinatoclathrate host *1* forms indeed a crystalline 1:1 inclusion with bromobenzene. It is obvious, why the structure of this compound arose our curiosity. The result of the structural study [82] is shown in Fig. 25. As expected, bromobenze has no direct contacts to the encircling host matrix. It is located in channels formed between endless zigzag-like chains of host *1* with alternating chirality. The arrangement shown offers the possibility of having the guest molecule included in the apolar channels formed between such chains. Accordingly "true" clathrate formation characterized by strong host-host, but no direct host-guest contact has occurred.

Spatial accommodation of the guest evidently allows disordering of the guest molecule, another characteristic feature of "true" clathrates. This structure may serve as a general model for other possible inclusion compounds of *1* with apolar guests and also, in lack of the structure of the free host *1* (cf. Sect. 4.1), it may help to imagine a probable steric arrangement for that case.

Another example of the same building principle is found in the literature [89] for the *15*-analogous dialcoholic spiro host *24*, namely in its 1:1 inclusion compound with benzene (Fig. 26). The host molecules are bound into infinite zigzag chains by H-bonds and disordered benzene molecules appear interstitially placed between such chains.

Oddly enough, the 1:1 clathrate formed between 9,9′-bianthryl (*47*) and benzene has a very well ordered structure [64] (Fig. 27). The guest molecule has almost ideal geometry in this crystal. This is explained by the tight fit of the pairs of benzene guests into the environment maintained by the cross-shaped host (see also Sect. 4.4, Fig. 36). Unquestionably, this structure repeats basic characteristics except the functional groups of the species described before. However, unlike the aforementioned cases, both host-guest and host-host interactions are lacking here. Thus, in the strictest sense of classification (see Sect. 4 of Chapter 1 in Vol. 140), we deal with an example specified as (*a*) in Fig. 15 of Vol. 140, namely a "true" clathrate. The

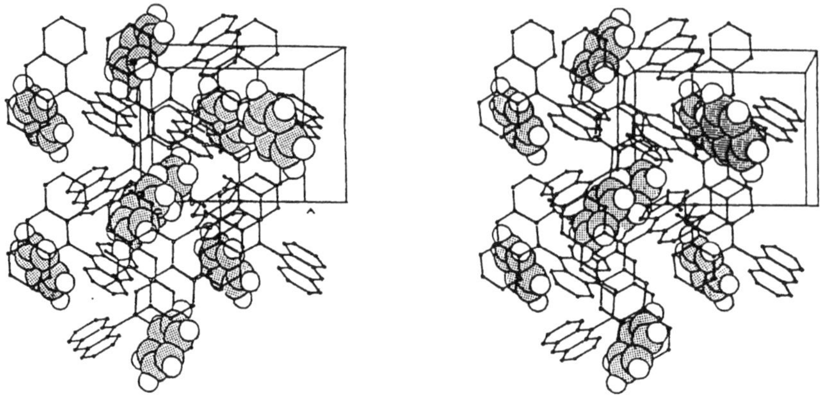

Fig. 27. Packing relations in the crystal structure of *47* · benzene (1:1) [64]. Stereo drawing of complementary stick style and space filling representations of host and guest molecules, respectively (atomic radii of the corresponding guest atoms in the space filling style are set to about half of their common van der Waals values; the H atoms of the host molecules are omitted)

former two clathrates belong to the "coordination-assisted host lattice"-type (Fig. 15b in Vol. 140), but none of all these represents a "coordinatoclathrate".

The family of "true" clathrates based on hydrocarbons only is further enriched by the inclusion compounds of *48* with benzene (1:1) and p-xylene (2:1) [90] (Table 19). Figure 28 illustrates the structure of *48* · benzene (1:1). The structure of the p-xylene clathrate shows intercalated guest molecules at centers of symmetry in the crystal lattice. Both clathrates are rather unstable at ambient temperatures and decompose easily, e.g. on exposure to X-rays (even in the presences of mother liquor). The *48* · p-xylene clathrate is unstable to such a degree that decomposition occurs at low temperature.

In accordance with this behavior, specific interactions between host and guest molecules are not indicated in the structures. The guest molecules take an inclination of approximately 45 degrees to both planes of the rectangular aromatic systems of

Table 19. Crystal data of pure hydrocarbon host-guest inclusions

Compound[a]	*47a*	*48a*	*48b*
Space group	$P2_1/n$	$P2_1/n$	$P2_1/c$
unit cell			
a (Å)	11.704	10.809	13.928
b (Å)	8.790	18.455	9.166
c (Å)	22.685	10.910	16.095
β (deg)	95.80	92.50	99.91
V (Å^3)	2321.57	2174.26	2024.09
Z	4	4	4
D_c (gcm^{-3})	1.237	1.205	1.212

[a] Designation: *47a* = *47* · benzene (1:1), *48a* = *48* · benzene (1:1), *48b* = *48* · p-xylene (2:1).

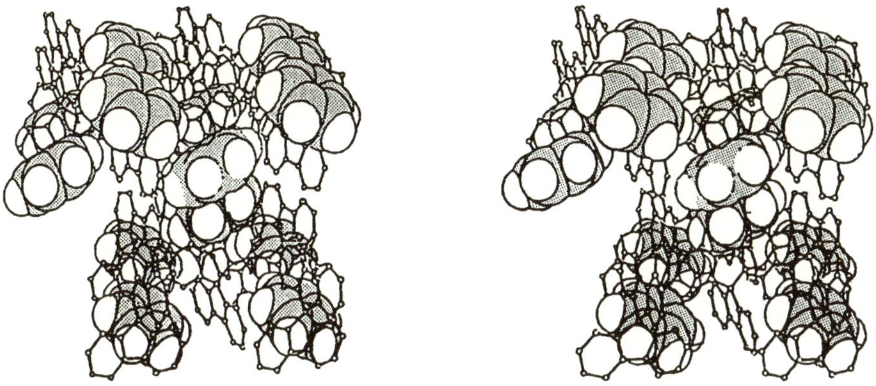

Fig. 28. Stereo view of the packing in the *48* · benzene (1 : 1) clathrate [90] (see notes of Fig. 27)

the hosts *47* and *48* in all three inclusions which excludes that π-π or similar interactions exist between the host and the guest partners.

4.4 Cavity Shapes: Steric Fit of Host-Guest Compounds

One of the aims of the crystallographic studies is to visualize the spatial conditions of non-H-bond type of interactions. Van der Waals forces (dispersion and exchange repulsion) and polarization are representatives of such interactive forces. They are governed by geometric features such as contact surfaces and volumes of the host and guest matrices.

The basic structure pattern which is noticed in the packing schemes of the crystalline inclusions between *1* and alcohols is a channel matrix formed by the host. The walls of these channels are hydrophobic in their main constitution, but they are regularly interrupted by protruding carboxyl groups where the guests bind via their polar endings (Fig. 29). Thus, depending on the lattice symmetry which is controlled by the guest shape, different arrangements of the polar/apolar segments with respect to the symmetry elements will exist. For the same reason, one may also observe that

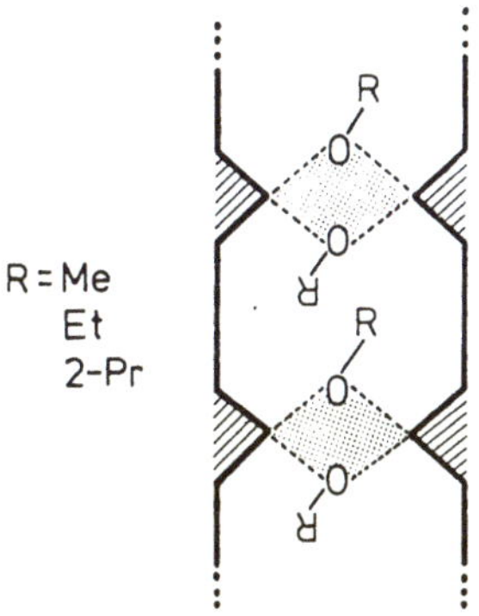

Fig. 29. Schematic representation of the longitudinal cross-section of the inclusion channel for the simple alcohol inclusions of *1* with MeOH, EtOH, and 2-PrOH [2]. Hatched triangles and dotted squares represent polar areas (cf. Fig. 19, type IIa), while the rest is of apolar property

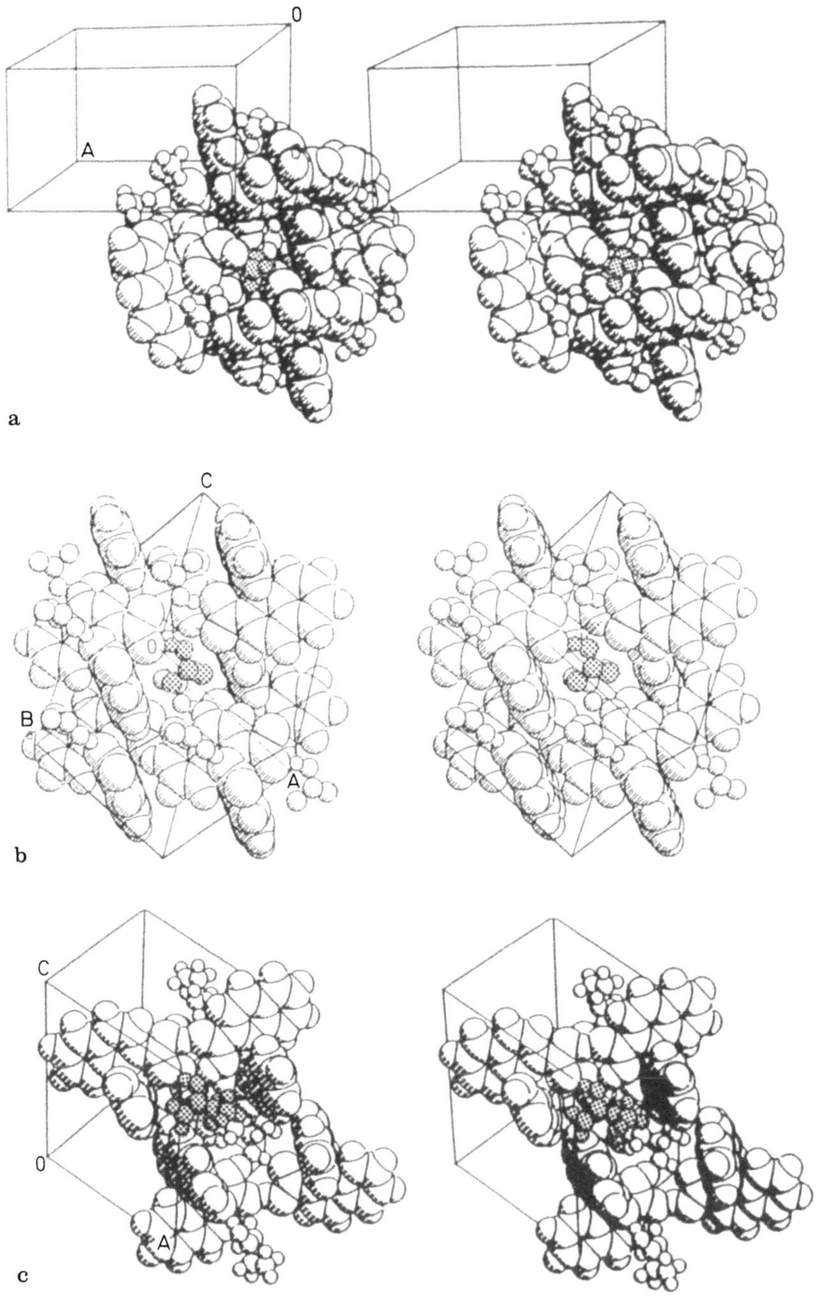

Fig. 30. Stereoscopic space filling illustrations of inclusion channels present in *1* · alcohol clathrates [2]. In each illustration, one of the guest molecules included in the channel is specified by shading (atoms of the guest molecules are shown with 20% of their van der Waals radii throughout these representations): **(a)** *1* · MeOH (1:2); **(b)** *1* · 2-PrOH (1:2) and *1* · EtOH (1:2). Due to isomorphism only the 2-PrOH structure is shown (guest H atoms are omitted for the sake of clarity); **(c)** *1* · 2-BuOH (1:1).

the size of the inclusion channel varies, although identical building blocks are used for creating H-bonds.

For instance, the arrangement of host molecules in the methanol inclusion of *1* (Fig. 30a) resembles the structure observed in the crystal of unsubstituted solvent-free 1,1′-binaphthyl *43* [72]. A difference between the basic packing rules in plain *43* and the methanol inclusion of *1* is in the following point: in case of the inclusion compound, the H-atoms at the far end of the host skeleton (with respect to the —COOH positions) approach closely the planes of neighboring aromatic rings lying nearly perpendicular to each other. In the crystal of 1,1′-binaphthyl, however, these H-atoms point to the lateral (mantle) atoms (e.g. at position 2) instead of the ring centres.

In the inclusion compounds of *1* with ethanol and isopropanol (Fig. 30b), the H-atoms under discussion are shifted even more outwards with respect to the neighboring naphthyl rings placed at right angles in the crystal space. This occurrence may also be seen as an opening of the inclusion tunnel which seems to be adapting itself to the steric requirements of the aliphatic portion of the guest molecule (cf. Figs. 30a and 30b).

In fact, the wall of the channel is lined with polar and apolar segments in a specific way. This is demonstrated by Figs. 31a and 31b showing that the polar segments are located in the corners of the channel, while the apolar ones make up the greater part of the wall. Hence the channel is tailored in a way that it corresponds to the characteristics of the substrate molecule. Hydrophobic portions (aliphatic tail of the alcohols) are placed into the hydrophobic region of the recepting channel while the hydrophilic groups (—OH functions) match the proper —COOH moieties protruding into this space (cf. Fig. 29).

A certain flexibility of this particular arrangement is obvious when considering the 2-butanol case of inclusion. As already discussed, the scheme of H-bonding changes here from a symmetric to an asymmetric ring system (cf. Fig. 19, types I vs. IIa). Nevertheless, another center of symmetry (placed between the aliphatic termini of related guest molecules) is retained, similarly to the lattices of the simpler alcohols mentioned above. The so-formed enlarged cavity around the 2-butanol guest (Fig. 30c) is still preserving the main characteristics of the former set-up. One of the polar

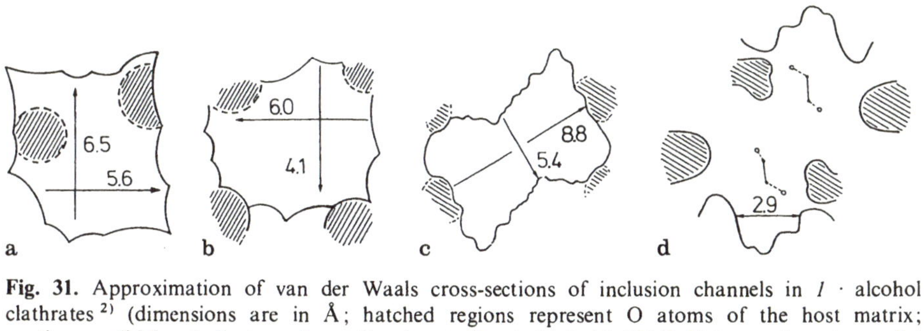

Fig. 31. Approximation of van der Waals cross-sections of inclusion channels in *1* · alcohol clathrates [2] (dimensions are in Å; hatched regions represent O atoms of the host matrix; continous solid lines indicate surfaces of apolar attribute): **(a)** *1* · MeOH (1:2) (approximately parallel to the $O_{(\alpha)}$—$C_{(\alpha)}$ vectors, cf. Fig. 17a); **(b)** *1* · 2-PrOH (1:2) (orientation as before); **(c)** *1* · 2-BuOH (1:1) (through a center of symmetry at 1,1/2,1/2, cf. Fig. 30c; non-zero electron density contours); **(d)** *1* · ethylene glycol (1:1) (in the plane of the C—C single bonds of a guest molecule, indicated by projected stick models; non-zero electron density contours)

centers (a coordinating oxygen of a carboxyl group) differs as it appears in the middle rather than in the corner of the channel (Fig. 31c). The increase of one dimension by approximately a factor of two (from 4.1 to 8.8 Å, see Fig. 31) follows the conservation of the pattern in Figs. 31a and 31b. The width of the channel in the height of the captived alcohol remains practically the same and is around 5.4–6 Å in these cases. Thus, these findings also indicate the ability of *1* to act in a systematically complementary way in order to build a stable heteromolecular crystal structure.

In the ethylene glycol inclusion of *1*, the channel structure cannot be detected any longer. An explanation is found in the fact that the gracile guest molecule,

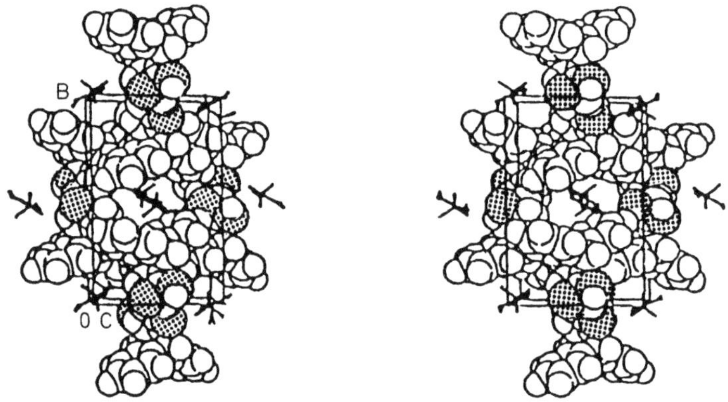

a

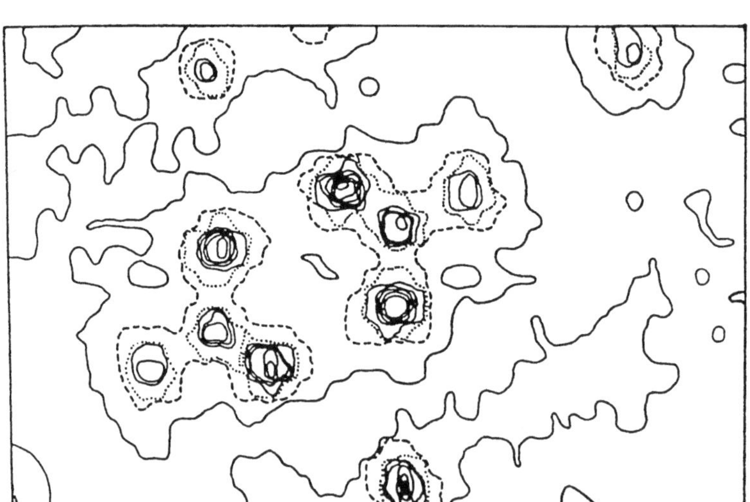

b

Fig. 32. Packing relations and steric fit of the *26* · acetic acid (1:1) clathrate (isomorphous with the corresponding propionic acid clathrate of *26*)[71]: **(a)** Stereoscopic packing illustration; acetic acid (shown in stick style) forms dimers in the tunnel running along the *c* crystal axis of the *26* host matrix (space filling representation, O atoms shaded). **(b)** Electron density contours in the plane of the acetic acid dimer[50]. First contour (solid line) is at 0.4 $eÅ^{-3}$, while subsequent ones are with arbitrary spacings of either 0.5 and 1 $eÅ^{-3}$. Density of the enclosing walls comes from C and H atoms of host molecules.

fully engaged in H-bonding, may approach the hydrophobic lap of the binaphthyl skeleton (Fig. 31 d) thus resulting in a more compact structure. This is also indicated by the calculated density and the respective packing coefficient (0.77) for this associate which is the highest among the alcohol inclusions of *1* (see Table 11).

A further example of the steric fit and thus the conditions of the second rank interactions between host and guest is illustrated by the channel structure of the acid inclusions of *26* (see inclusion compound with acetic acid, Fig. 32a). The tunnel has a mostly hydrophobic character being made up mainly from the aromatic portions of the roof-shaped host molecule. We must note that this arrangement applies possibly for the acetic acid clathrate of *1* as well.

In Fig. 32 b, showing the corresponding acetic acid clathrate of *26*, the cross-section of the electron density used to approximate von der Waals surfaces is taken in the plane of an acetic acid dimer [50]. It is revealed that by forming H-bonds only to another (symmetry related) guest acid and not to the —COOH groups of the host matrix, the roughly rectangular shaped dimer of the guest acids actually behaves as a certain hydrophobic species. This pattern is also observed in the isomorphous propionic acid clathrate of *26* and partly in the formic acid case as well. Therefore guest dimers fit the apolar channel ideally. Correspondingly, the present inclusion compounds between *26* and these acids may no longer be termed as coordinatoclathrates but "true" clathrates (cf. Sect. 2.1). Faster decomposition of the respective crystals on exposure to air is a macroscopic indication of this behavior.

As mentioned in Sect. 4.2.2, salt-type associates are the only representatives of the aggregates formed by *7*. In these crystals, the pyridinium cations appear surrounded by a rectangle-like environment maintained by *7*.

Following the order of discussion, the next type of guest molecules we examine briefly is the aprotic-dipolar class of solvents. A good example of the fit of guests and hosts is set in the packing of the *22* · DMF 1:2 inclusion [48] (Fig. 33). It shows that the orientation of the guest molecules is largely dictated by the rigid right angle shaped host framework. Such a guest orienting effect certainly contributes to the recognition pattern of the solvent joined with the effects of the sensor (—COOH) group (cf. Fig. 23, type I a).

Another example of this solvent class can be studied in the case of the *20* · DMSO

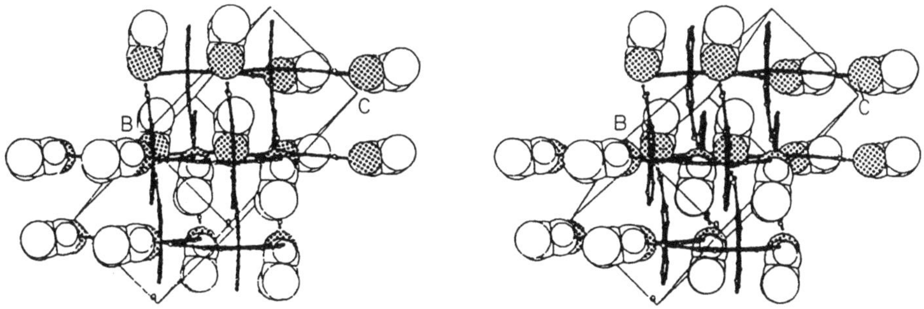

Fig. 33. Packing of the *22* · DMF clathrate [48] (stereo drawing). The joined orienting effect of the host lattice and of the sensor groups is illustrated showing the fit of the guest molecules (with 3/4 of the van der Waals radii of the composing atoms, O atoms shaded) to the host matrix (stick style)

1:1 inclusion [85] (Fig. 34). A part of a zigzag channel is seen with mostly apolar surfaces. Similarly to the channels of the alcohol inclusions of *1*, oxygen atoms of the —COOH groups appear to form polar (hydrophilic) corners. The drawing also shows ample space around the two centers of symmetry-related guest molecules in the middle pocket. This suggests a reasonable model for the disorder observed for the dimethyl sulfoxide atomic positions (cf. Fig. 24, type I, Sect. 4.2.3). Let us suppose that both DMSO-pyramids are oriented in the same way, say with their tips pointing upwards. Consequently the symmetry center which is still present in the drawing exactly in the middle of the box (at 1/2, 1/2, 1/2) would cease to exist in some of the unit cells. The empty space around the center at (1/2, 1/2, 1/2) in the crystal, enables the pyramid-shaped guest molecule to adopt both conformations with regard to the binding carboxyl group.

Another example to show that hydrophobic guest molecules favor disordering in the similarly tailored environment of the host matrix is seen in the bromobenzene inclusion of *1* [82]. The guest environment displayed in Fig. 35 by the contours of electron density both in and perpendicular to the plane of the aromatic guest molecule contains enough free space to allow for more than one orientation. In fact, it has been suggested by packing analysis [91] that there is a second orientation for the guest with ca. 1/3 population. This orientation may be derived by tilting the model from the main population by 180 degrees along an axis perpendicular to the longitudinal (the Br—C_{para}) axis of the molecule.

The cavity shapes in the case of pure hydrocarbon hosts may play an even more important role. This holds, e.g. for *47* in its inclusion compound with benzene (Fig. 36). The map of the electron density in the plane of the benzene molecule (Fig. 36a) suggests a tight envelope around the guest in the form of a hexagon. Benzene molecules are located in pairs in the crystal lattice (Figs. 36b and 36c), occupying almost completely closed cages with the edge lengths of ca. 6.5 × 7 × 12 Å.

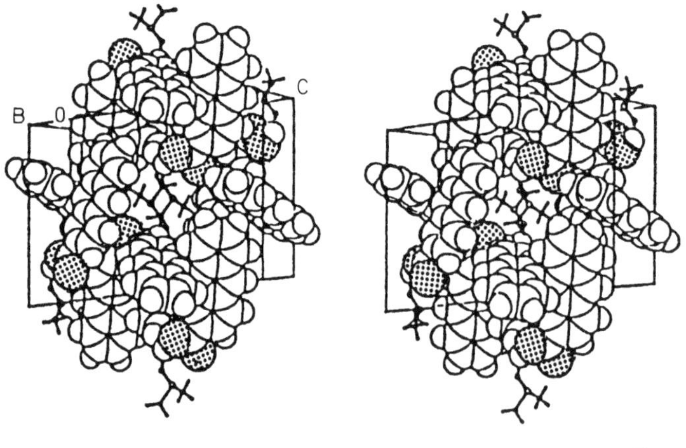

Fig. 34. Stereo drawing of the packing in the *20* · DMSO clathrate [85] (complementary space filling and stick style representations of host and guest molecules, respectively; O atoms of the host are shaded). Space around guest molecules in the center of the drawing, related by the symmetry center operator, indicates the opportunity for disorder

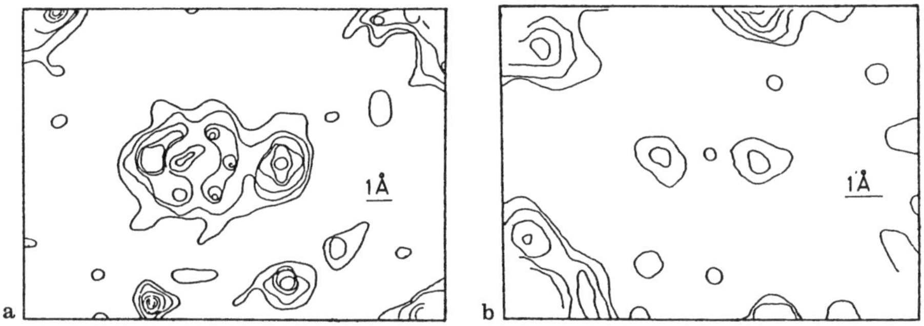

Fig. 35. Electron density distribution (arbitrary spacing) **(a)** in the plane of the guest molecule and **(b)** perpendicular to the longitudinal axis in *1* · bromobenzene (1:1)[82]. Second plane is bisecting through the middle of the Br—$C_{(\alpha)}$ bond of the main site. Plenty of empty space around the guest readily enables disordering

which are formed by the flat surfaces from eight contributing host molecules (cf. Fig. 27). The relative steric positioning of the benzene molecules arranged in pairs (Figs. 36b and 36c) suggests that base stacking between them does not exist. Figures 36b and 36c also reflect suitable conditions for the pairs of benzenes in the host cage by indicating dimensions. The cage structure of the host matrix and the extremely good spatial fit between hosts and guest are certainly responsible of the pronounced selectivity behavior of *47* and for the remarkably high thermal stability of this particular inclusion compound (see Sect. 3.5).

4.5 Interrelations Attributable to Special Host and Guest Features

With reference to hosts and a guest, molecular assemblies have to conform to certain circumstances, generally called complementary relationships. They involve both steric and electronic terms. The objects may be achieved by the use of properly chosen sensor groups and by a suitably tailored basic skeleton as exemplified by the present scissor- or roof-shaped host molecules. From the point of view of the introductory thoughts of this chapter (cf. Sect. 3.1), it is a matter of consideration to see how consistent the "scissor" or the "roof" simile is in the light of crystal structures.

4.5.1 Shape and Symmetry Considerations: Dihedral Angles of Host Compounds

The obvious thing to do is to establish a kind of link, if any exists, between the various host molecules which may seem to differ principally at first sight. Examination of hidden similarities also throws light upon possible conceptual relations and may prove useful for the future.

Shape of Host 1

As a general descriptor of the molecular shape of *1*, one may consider the dihedral angle between the chemically bound naphthalene moieties. Data of this parameter are listed in Table 20 for the studied inclusion compounds of *1* combined with a scheme that

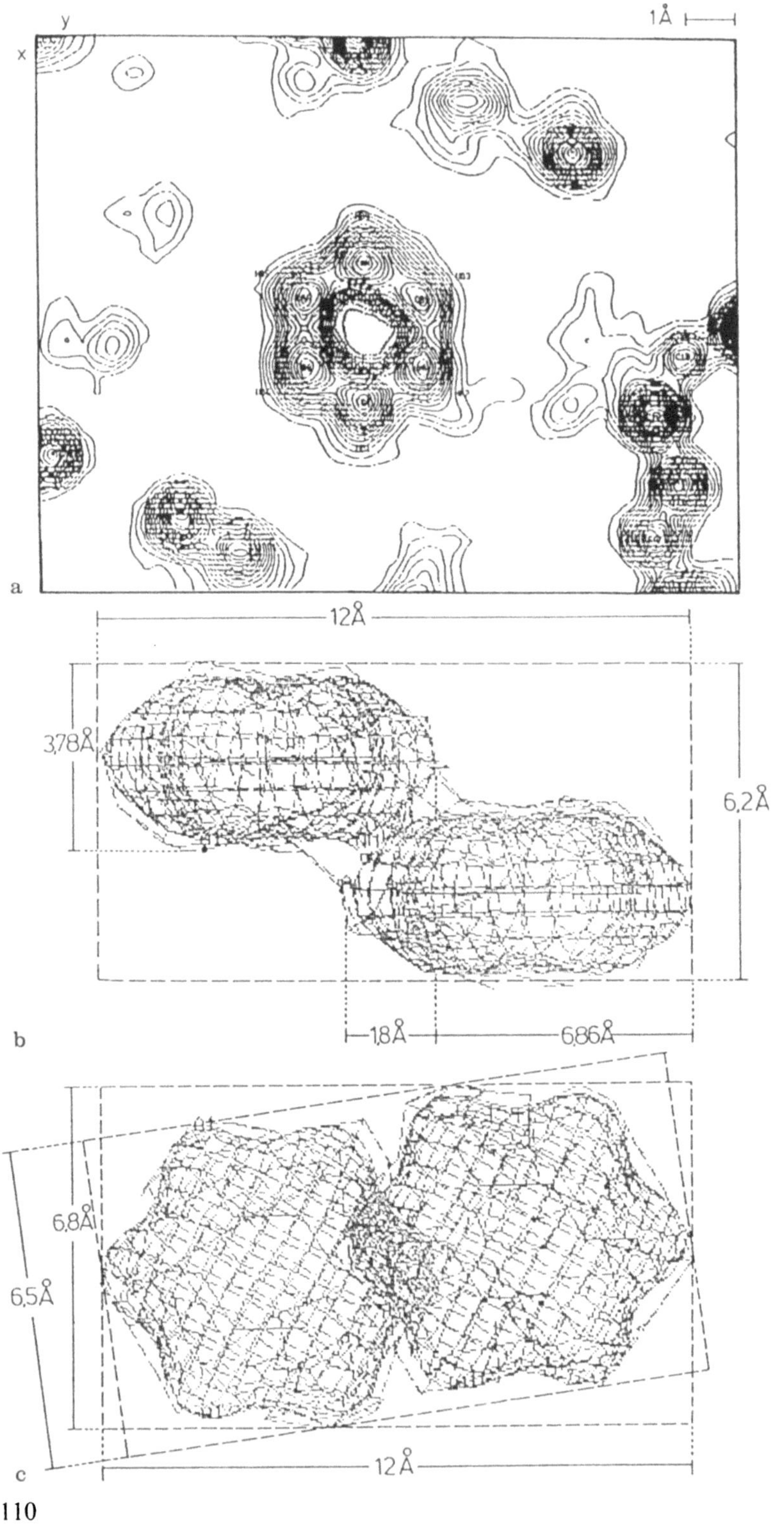

a
b
c

highlights the basic shape descriptors used in the following treatment. Characteristic shapes of the host molecule involving the aggregates of protic, aprotic-dipolar, and apolar guests are shown by selected examples for each class of compounds (Fig. 37).

Table 20. Shape of host *1* in its crystal inclusions as characterized by the virtue of dihedral angles. Attached diagram explains the angle designations[a]

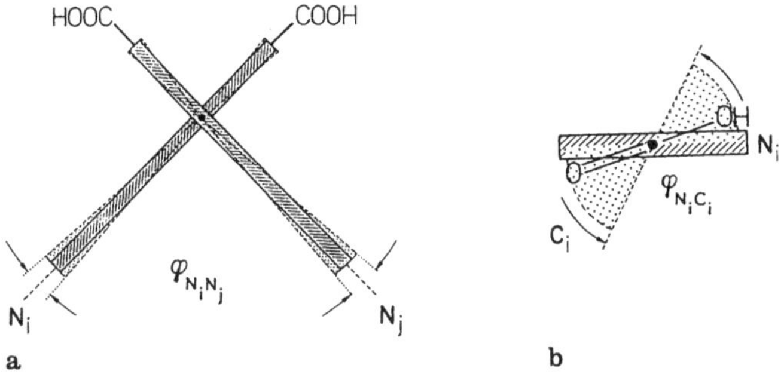

Compound[b]	N_i/N_j	N_i/C_i		C_i/C_j
1a	92.2	7.2	6.0	84.7
1b	85.3	8.5	8.5	90.0
1c	86.3	17.2	17.2	91.6
1d	89.6	7.7	2.9	85.2
	87.3	1.1	6.8	87.2
1e	89.1	25.3	14.9	82.5
1f	86.0	10.8	29.5	80.8
1e	92.0	19.8	11.5	76.1
1h	92.5	10.4	3.1	87.3
1i	98.1	9.9	33.3	74.3
1j	93.1	2.9	15.0	91.7
	87.3	25.6	0.1	73.9
1k	92.2	17.2	4.0	96.9
1l	81.4	50.8[c]	60.7[c]	71.7
1m	87.7	25.4	34.8[c]	51.9[c]
Mean[d]	89.3(4.1)	11.9(8.3)		83.9(7.7)

[a] Planes N_i, N_j are the planes of the naphthyl rings (10 atoms); planes C_i, C_j are the planes of the —COOH groups (3 atoms). (Dihedral angles calculated by Chem X, Ref. 139). Mean values include 15, 26, and 14 data, respectively. [b] Designation: *1a–1g* see table 11; *1h* = *1* · DMF (1:2), *1i* = *1* · DMSO (1:1), *1j* = *1* · acetic acid (2:3), *1k* = *1* · bromobenzene (1:1), *1l* = *1* · imidazole · H_2O (1:1:2), *1m* = *1* · imidazole (1:1). [c] Data not included in the mean value. [d] R.m.s.d. in parentheses.

◀

Fig. 36. Spatial fit between host and guest in *47* · benzene (1:1) (see Ref. 64): **(a)** Electron density in the mean plane of a benzene revealing the encasing hexagonal environment aroung the guest; and **(b)** and **(c)** van der Waals surfaces of the dimeric benzene units as seen in Fig. 27 (indicated dimensions were calculated by the aid of the CHEM X program system, see Ref. 139). The lack of extensive-enough overlapping to yield in pi-pi interactions is visible from this drawing

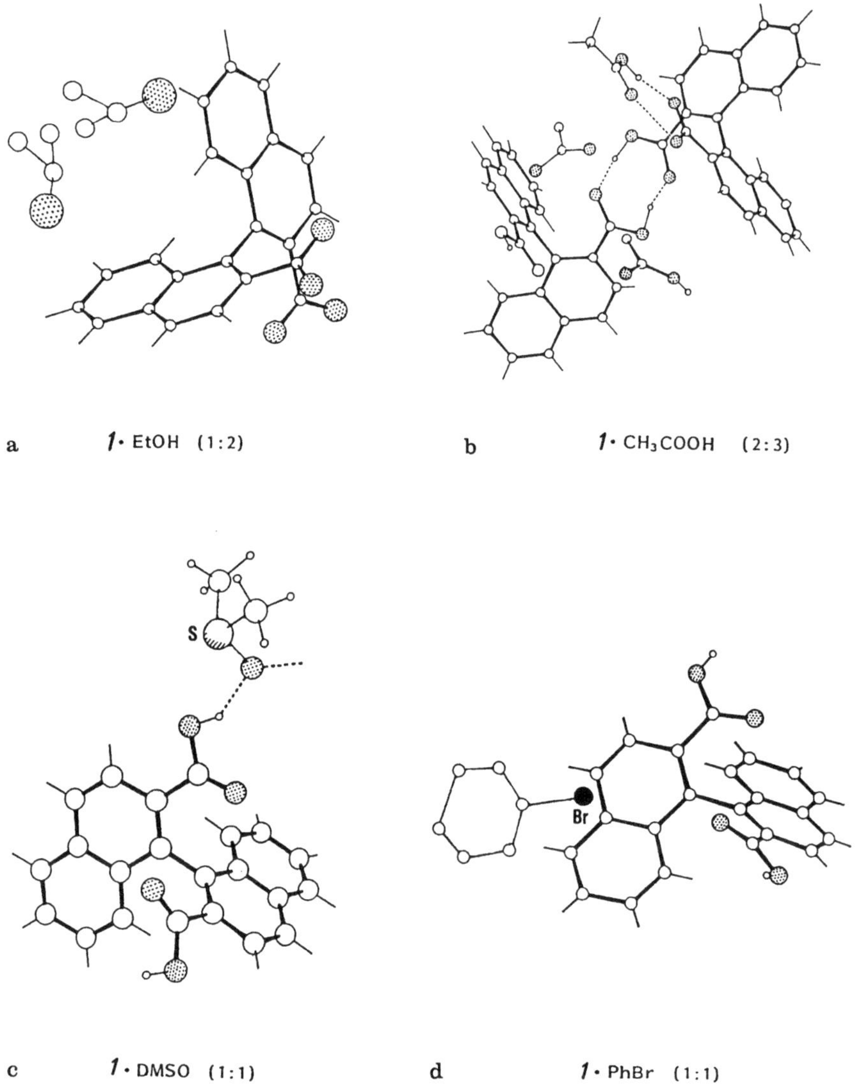

Fig. 37. Shape of host molecule *1* in some characteristic inclusion compounds [2, 79, 82] **(a–d)**, (H-bonds are indicated as broken lines; O atoms dotted; H atoms connected with non-heteroatoms of the host are shown as sticks only; disordered terminal methyl groups are found for EtOH)

We observe a rather narrow distribution in the alignment of the naphthyl planes which lie nearly at right angle to each other [mean value 89.3(4.1)°], with a slight variation only (see Table 20) showing that this molecule has an almost stable shape in the crystal structures. It is possible to draw an interesting comparison with the same parameter found for some related compounds containing the binaphthyl moiety [45, 60, 61, 72, 73, 92–99]. In these structures, the values vary between 68 and 111° indicating that this dihedral angle may be subject to alteration due to environmental effects.

Another characteristic feature associated with the structure of host molecule *1* is the inclination angle of the —COOH groups to their anchoring naphtyl moiety. The mean value of this quantity and its r.m.s.d [11.9(8.3)°] obtained from the different inclusions show a moderate inclination to the aromatic moiety and a somewhat enhanced scattering of the data (Table 20). The differing steric and electronic needs of the particular guest species seem to be reflected mainly in this parameter. As a general tendency one may deduce: the bulkier (more branched) a guest is, the larger this dihedral angle becomes (cf. Table 20.)

As a third shape descriptor of *1*, the dihedral angle between the two —COOH groups is considered. The mean value [83.9(7.7)°] (Table 20) reflects moderate deviation of these moieties from a roughly perpendicular steric positioning with respect to each other.

Shape of Host 7

Chemical facts in the case of host *7* point to rigorous steric dependence of associate formation. Though the number of the available data is rather limited, their mean values seem to be in conformity with such an assumption. The dihedral angles between the characteristic planes in these structures show that the inclination of the naphthyl planes to each other varies in a range between 58 and 68° with a mean value of 64.2(3.5)° (Table 21). The inclination angles of the —COOH groups to their respective naphthyl moiety lie also in a relatively narrow range with a mean value of 53.5(5.1)° deviating significantly from *1* (cf. Table 20) and is certainly involved in explaining the different inclusion behavior between *1* and *7*. A similar value for this angle is found, however, in the imidazole associates of *1* (see Sect. 5). The —COOH functions are at an angle of 73.8(2.9)° to each other. Two examples shown in Fig. 38 represent the particular host shape of the free [68] and of the pyridine associated form [80].

Shape of Host 26

Selected examples illustrating the shape of host *26* under various conditions are given in Figs. 39a–39c. The first parameter we choose to characterize the shape of *26* is the inclination angle of the phenyl rings readable at the gable of the roof-shaped molecule (graphical representation in Table 22). It indicates little variation [see

Table 21. Shape characteristics of host *7*

Compound[a]	N_i/N_j[b]	N_i/C_i		C_i/C_j
7	58.3	45.0	45.0	78.7
7a	64.2	53.6	56.7	73.0
7b	67.5	61.1	54.0	71.1
7c	65.1	54.5	57.0	72.5
7d	65.8	56.5	51.6	73.6
Mean[c]	64.2(3.5)	53.5(5.1)		73.8(2.9)

[a] For designation see Table 14. [b] Details in Table 20. [c] R.m.s.d. in parentheses.

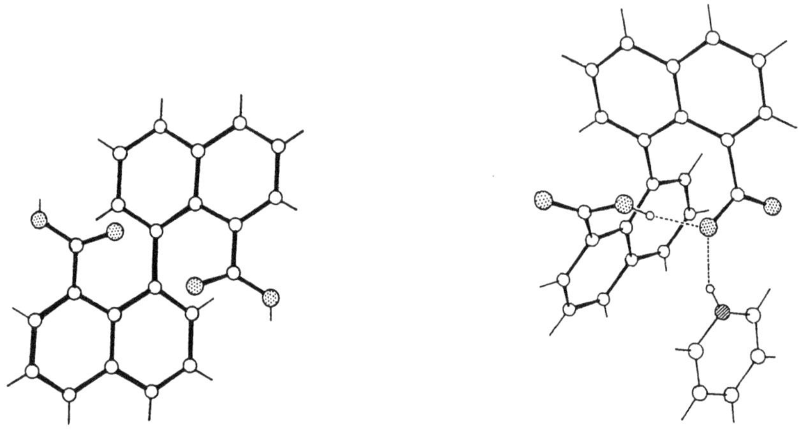

a 7 b 7 · Pyridine (1:1)

Fig. 38. Geometry of *7*: **(a)** in the free state (unsolvated crystal) and **(b)** in a typical salt-type associate (*7* · pyridine, 1:1) (H-bonds are indicated as broken lines; O atoms dotted, N atoms hatched; H atoms connected with nonheteroatoms are shown as sticks only)[80]

Table 22, mean value 123.9(2.4)°], as expected of a molecule with highly rigid constitution. Rigidity of the basic skeleton of *26* is also reflected in the symmetrical displacement of the finial (the bridging ethano-moiety) with respect to the roof-planes [mean angle 62.0(2.1)°]. The carboxyl groups show also little variation in their spatial arrangement with respect to the finial, generally they adopt a tilted arrangement with a mean angle of 50.9(6.9)°. The mutual displacement of the acid groups, with one exception, leads to a nearly perpendicular arrangement with a mean value of 86.5(8.3)° for this angle. The exceptional case is for one of the free host molecules (out of the two in the asymmetric unit, cf. Table 22) which is probably forced into this alignment on the formation of the dimer.

It is most interesting to compare these data to the respective angles obtained for *41* in its dimethyl sulfoxide inclusion compound (Table 22 and Fig. 39d). One finds a certain feature departing from those noticed above. It concerns the angles of the —COOH groups to the finial (an etheno-moiety in *41*) which now have values pointing to a coplanar/perpendicular positioning of the acidic groups with reference to the etheno-segment. However, the nearly perpendicular arrangement of the —COOH functions with respect to each other is still maintained in *41*. It seems to reflect a favored way of arranging the two carboxylic groups in space in this particular family of host compounds.

Generalization of the Shape and Symmetry of Coordinatoclathrate Hosts

The scheme of the most common hosts in this study (Fig. 40) shows that they can be considered as a homologous series of dicarboxylic acids of different chain length (1,2-, 1,4-, 1,6-, and 1,7-diacids, respectively) involving an essentially rigid central segment. A summary substantiates the similarity of the mean angles of *1*, *7*, and *26* (Table 23) and a more or less steady shape of these host molecules in the free state

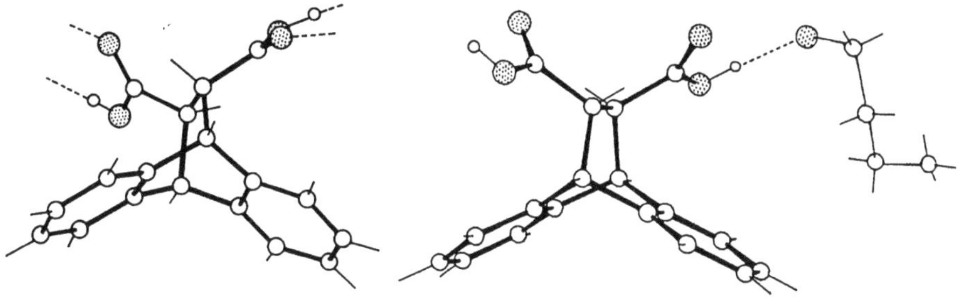

a *26* b *26* · 1-BuOH (1:1)

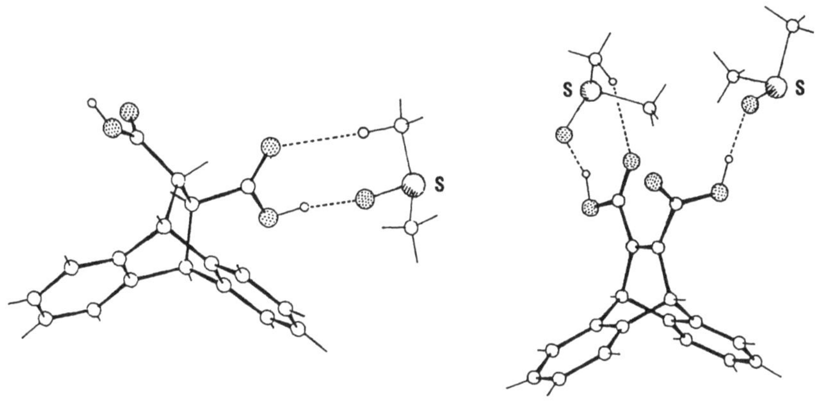

c *26* · DMSO (1:1) d *41* · DMSO (1:2)

Fig. 39. Geometry of *26* and of the related host *41* in the free state (unsolvated crystal) and in some characteristic inclusion compounds [71] **(a–d)** (H-bonds are indicated as broken lines; O atoms dotted; H atoms connected with non-heteroatoms are shown as sticks only)

and in their inclusions (see Tables 20–22) as well. Even the conformation of the mobile —COOH groups shows only a narrow range of data.

It can be concluded from the X-ray data that these hosts possess a rigid or semirigid molecular skeleton. The importance of a preformed receptor shape is well documented in the chemistry of the artificial [100, 101] and natural [102] receptors. For example, a preformed receptor was found to be helpful in reducing the activation energy for complexation that contributes to the recognition of a proper substrate [100, 101]. Such an essentially entropic effect through the preorganization of a binding site may also be effective in the inclusion formation of the scissor- and roof-shaped host molecules discussed here.

Another remarkable feature of most of the inclusion hosts of this chapter is that they possess at least approximately (i.e. noncrystallographic) twofold symmetry.

Table 22. Shape of hosts *26* and *41* in their crystal inclusions, or in the free state, as characterized by inclination angles of some planes. Attached drawing explains definition of the interplanar angles (deg)[a]

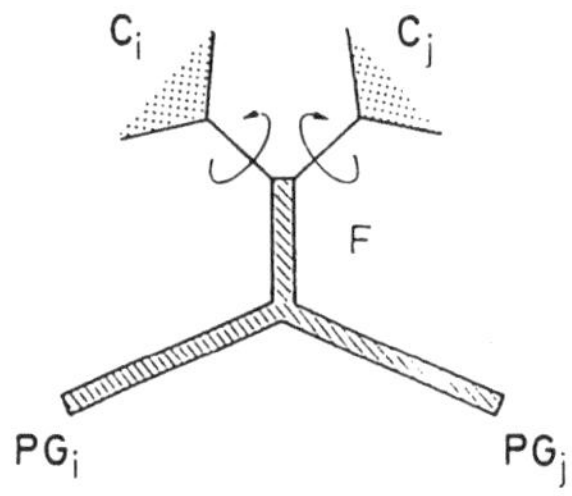

Cmpd[b]	PG_i/PG_j	PG_i/F	PG_j/F	F/C_i	F/C_j	C_i/C_j
	126.2	61.9	64.4	51.0	64.7	65.1[c]
26	123.6	62.7	60.9	51.5	49.9	83.0
26a	122.3	60.4	62.0	50.5	46.1	90.2
26b	122.0	61.9	60.1	60.2	45.5	75.7
26c	128.3	63.5	64.8	54.0	43.0	85.9
26d	121.2	57.9	63.7	43.4	55.3	84.2
26e	125.1	65.2	60.0	61.5	41.4	102.7
26f	122.5	59.5	63.1	52.0	44.8	83.8
Mean[d]	123.9(2.4)	62.0(2.1)		50.9(6.9)		86.5(8.3)
41a	117.1	60.3	56.8	7.6	85.3	95.3

[a] Planes PG_i, PG_j are the planes of phenyl rings forming the gable (6 atoms); plane F is the plane of the finial (4 atoms); planes C_i, C_j are the planes of the —COOH groups. (Dihedral angles calculated by Chem X, Ref. 139). [b] Designation: *26a–26d* see Table 13, *26e* = *26* · DMF (1:1), *26f* = *26* · DMSO (1:1), *41a* = *41* · DMSO (1:2). [c] Omitted from the mean value; if included, the mean for C_iC_j becomes 83.8(10.8)°. [d] R.m.s.d. in parentheses.

In a few instances this property becomes ideal, i.e. there is a coincidence of molecular and twofold crystallographic symmetry (see inclusion compounds *1* · EtOH, *1* · 2-PrOH, and *22* · DMF). The presence of such an internal symmetry has been recognized to bear general consequences on the constitution of crystal lattices [27,103], in particular for dicarboxylic acids [102] and hence applies also for the diacids under consideration. Actually the internal C_2 molecular symmetry axis of the host molecules is oriented perpendicular to *c* or *n* glide planes in the crystal lattices. This is the prominent feature of all the crystal structures of this study, whenever the space group permits the presence of such glide planes. The propagation of the chain build-up of H-bond fused molecules in several of the crystals behaves according to this rule and the spatial positioning of the carboxyl functions with respect to the internal symmetry element.

As shown in Fig. 40, all of the hosts have their acidic groups placed on the same side in respect of the internal C_2 symmetry counterfacing the hydrophobic region of the molecules in question. In other words, they could be described as having a "cisoid" or "Z" conformation.

It has also been noted that self-complementary objects must follow twofold sym-

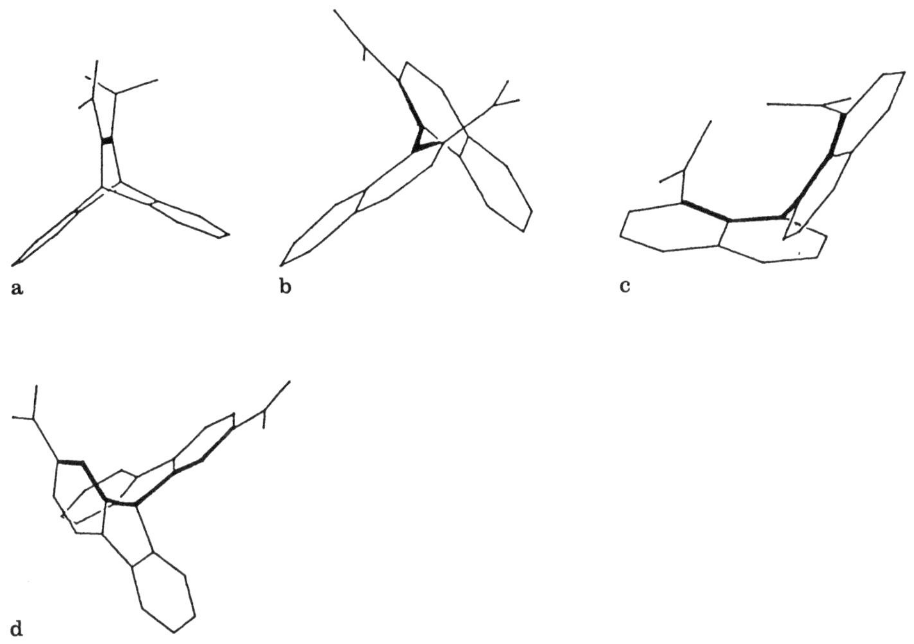

Fig. 40. Relation between host molecules of this study and dicarboxylic acids of different chain length (specified by heavy lines in the skeletal drawings): **(a)** (*41*), **(b)** (*1*), **(c)** (*7*), and **(d)** (*22*) correspond to 1,2-, 1,4-, 1,6-, and 1,7-diacids, respectively [67a)]

metry on their contact surfaces [105)]. Minor deviations from such a molecular symmetry in a crystal cause local breakdowns of the close-packing principle which remains one of the governing principles of the solid associate formation (cf. Ref. 27) and may give rise to the appearance of voids filled advantageously with guest species. Apparently gross violation of this internal molecular symmetry may reduce the ability of inclusion formation as shown for e.g. *38* and to some extent for *40*.

4.5.2 Effects Due to the Amphiphilic Nature of Host Compounds

The tendency of organic molecules to achieve the closest possible packing arrangement has been substantiated by many examples [27,103,106)]. The packing coefficients (C_k)

Table 23. Summary of shape characteristics for hosts *1*, *7*, and *26* (mean values of dihedral and inclination angles, deg, with their r.m.s.d.)[a]

Host	N_i/N_j	N_i/C_i	C_i/C_j
1	89.3(4.1)	11.9(8.3)	83.9(7.7)
7	64.2(3.5)	53.5(5.1)	73.8(2.9)
26	123.9(2.4)	50.9(6.9)	86.5(8.3)

[a] N_i, N_j are the planes of the aromatics (10 atoms for *1* and *7*, 6 atoms for *26*); C_i, C_j are the planes for the —COOH groups. (Plane angles calculated by the aid of Chem X, Ref. 139).

of the alcohol inclusion of *1* range between 0.71 and 0.77 (Table 11) thus falling close to the expected normal range [27, 103, 106]. The segregation of hydrophilic regions in the crystals, as visualized in Figs. 29–31, indicates that there are regions in the crystals linked *via* strong H-bonds (hydrophilic interactions), on the one hand, and other less interacting areas on the other hand. These latter parts of the lattices are subject to weak intermolecular forces (e.g. dispersion) and enable the guest molecules in the respective portion to be statistically distributed in space (i.e. a disordered crystal lattice results).

Another simple parameter which reflects the tightness of crystal packing (cf. Sect. 4.4) is the calculated density (D_c). The data of D_c for the alcohol inclusions of *1* (Table 11, Sect. 4.2.1) indicate a slightly different compactness of the crystal structures. The extent of the disorder observed in the aliphatic part of some guest molecules seems to be correlated with this simple quantity. For example, the somewhat lower density of *1* · EtOH with respect to the nearly identically built structure of *1* · MeOH (cf. Sect. 4.2.1) indicates more space in the crystal of the ethanol inclusion. Thus, disordering of the terminal methyl group of the ethanol guest is possible. This property is even more pronounced for *1* · 2-PrOH. The resolved disorder sites for the methyl termini in the latter two guest molecules correspond to positions rotated approximately 60 degrees apart from each other. In the case of the more extended branched alcohols (see *1* · 2-BuOH and *1* · t-BuOH), the degree of disorder is somewhat slighter and is mainly indicated by the unreasonably short terminal C—C (methyl) distances [2]. The highest density in this series of compounds found for the inclusion of *1* with ethylene glycol indicates a closely packed structure (see. Sect. 4.4).

Summing up, the density data and the slight variation of the seemingly normal packing coefficients (Table 11) may be rationalized in terms of the observed disorder pattern. It means that there are regions of different compactness in some of these crystals. This becomes visible in those instances where the repulsive forces are somewhat more pronounced due to the less cooperative aliphatic moieties in the respective guests (cf. inclusion compounds of *1* with EtOH, 2-PrOH, 2-BuOH, t-BuOH, and ethylene glycol). These structures illustrate how the ideas put forward in Sect. 3.1 are verified in the crystals. The inclusions of other hosts (*20*, *26*) also exhibit such properties: matching of regions with proper characteristics and formation of a complementary host matrix to the respective guest volume.

4.5.3 Stoichiometry and Other Thermodynamics Related Effects

Stoichiometry is just one of the consequences of the fitting requirements between a guest molecule and a host matrix. It is important to recall that the meaning of stoichiometry being used to describe the real association nature of heteromolecular aggregates (cf. Chapter 1 in Vol. 140 of this series) might be somewhat different and more extended with regard to the common usage of the term. In this sense it is applied to describe the structural conditions of aggregate formation ("building block stoichiometry"). It obviously influences kinetically controlled events. As such, it may exert a certain influence on the selectivity pattern of a given host molecule towards different solvents (cf. Table 2 in Sect. 3.2.4). Naturally, selectivity is further controlled by enthalpy (e.g. strength of H-bonding in the crystal of an aggregate) and entropy effects.

An example of the contribution of the latter, originating from the properties of such systems, is obvious from the frequent appearance of statistical disorder in the inclusion compounds (cf. Sect. 4.5.2). Let us consider a several-component system (a host matrix made up of a few molecules and possibly already bound guest molecules and two different solvent molecules). The small (0.4 kJ/mol) preference in the Gibbs free energy of *1* · EtOH (or *1* · 2-PrOH) over a single conformation in the crystal structure [2] may be a relevant factor in steering such a system towards an equilibrium of slightly lower energy, as compared to a respectively non-productive case, e.g. *1* · MeOH.

Naturally selectivity in a several-component system is primarily influenced by rather strong effects such as the presence or absence of strong H-bonding, but possibly also by much weaker interactions (e.g. of C—H ... O type). In this regard, it is interesting to note the similarity between the selectivity exerted by such simple inclusion hosts, e.g. *1*, and chiral recognition [103]. In both cases, weak interactions are of decisive importance in the final outcome of the experiments. Entropic effects have been demonstrated to play a fundamental role in enzymatic reactions [102,107]. Conceptual similarity of inclusion compounds to more complicated associates is underlined thereby.

5 Coordinatoclathrates in Active Site Modelling of Protease Enzymes: Associates of *1* with Imidazole

As we saw in the previous sections, inclusion compounds have many structural properties which relate them to other systems based on the hierarchy of non-bound interactions, like enzymes or enzyme-substrate complexes. As a matter of fact, most of the so-called "artificial enzymes" are based on well-known host molecules (e.g. β-cyclodextrin) and are designed to act partly on such bases [108,109]. Most of these models, however, take advantage of the inclusion (intra-host encapsulation) phenomena. Construction of proper covalently bound model molecules is a formidable task for the synthetic chemist [110]. Therefore, any kind of advance towards such a goal is welcomed.

Attemps at creating non-covalent models is a logical choice in this sense [111]. Many of the features of the coordinatoclathrates relate them to more complicated associates of biological molecules. As demonstrated in the preceding sections, simple hosts, like *1*, are able to maintain extended H-bound loops ("tertiary" structure formation), show in some cases nearly perfect selectivity (substrate specificity, cf. Sect. 3.2.4), and display entropic effects in their associates. Crystal growth and dissolution has also been shown to be dependent on the presence of chiroselective *inhibitors* [112]. Thus, apart from the former *static* analogies, one may find *dynamic* relationships as well. All these observations may be explained by the simple fact that the organizing forces are of the same type and approximate magnitude both for biological assemblies and for the crystals of simple organic molecules. As a consequence, the packing densities of biological assemblies (e.g. enzyme-substrate complexes) and organic crystals are also close to one another [113]. Proving by evidence, the possible similarity of simple crystal structures to the much more complicated network in protein molecules is a challenging task for the future.

5.1 Crystalline Associate of *1* with Imidazole and Water (1:1:2) [*1* · Im · 2 H_2O]

The outstanding inclusion ability and the carboxylic functions of host *1* raised the idea of co-crystallizing it with imidazole (Im) which, due to its versatile nature [114], is one of the frequently used components in enzyme active sites, generally presented by histidine. Formally, a system made of imidazole and an acid component may mimic two essential components of the so-called catalytic triad of the serine protease family of enzymes: the acid function of Asp102 and the imidazole nucleus of His57 [115] (trypsin sequence numbering). The third (albeit essential) component of the triad corresponding to the alcohol function of Ser195 was not considered in this attempt. This family of enzymes is of prime importance in metabolitic processes. By virtue of the (— + —) charge distribution of the aforementioned Asp-His-Ser triad, they are able to cleave peptide (or ester) bonds [116] (Scheme 1).

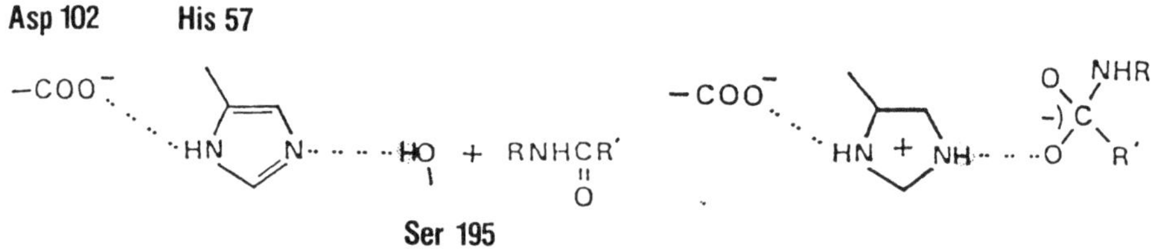

Scheme 1. Formation of the tetrahedral intermediate with the development of (— + —) charge distribution in serine proteases [111]

On intuition, a minute amount of water was added to the solvent (ethyl acetate) in the first crystallization experiment containing a molar excess of imidazole corresponding to *1*. Regularly shaped crystals were formed within one hour. Such a crystal, subjected to X-ray analysis, has the structure as shown in Fig. 41 [111]. Apart from the formation of the expected salt-type associate (carboxylate-imidazolium ion pair, cf. Sect. 4.2.2), two water molecules are present in the asymmetric unit of the crystal structure. This fact called our attention again to the family of serine protease enzymes, where water molecules are reported as being located in the close vicinity of the active sites [115–120].

5.1.1 Intra-associate Relation in *1* · Im · 2 H_2O

As already demonstrated (see Sect. 4.5.1), *1* displays a characteristic shape in its inclusions even with respect to the inclination angle of the carboxyl groups to their naphthyl planes. Coplanarity of these moieties depends partly on the bulkiness of the substrate molecule, however, the interplanar angle usually does not exceed 30° [mean value 11.9(8.3)°, Table 20].

This property is contrary to the present case, where the dihedral angle approaches a *gauche* arrangement of the naphthyl and carboxyl/carboxylate moieties with respect to each other (dihedral angles 50.8 and 60.7°, Table 20). Another remarkable conformational difference in the geometry of *1* is the presence of an *intra*molecular H-bridge between the carboxyl and carboxylate groups with the aid of a *cis* positioned O—H bond. A similar disposition of bonds has been observed in the case of the salt-type inclusion aggregates of *7* (cf. Sect. 4.2.2). It must be noted that the present arrangement is not the common mode of the steric alignment of this bond [104].

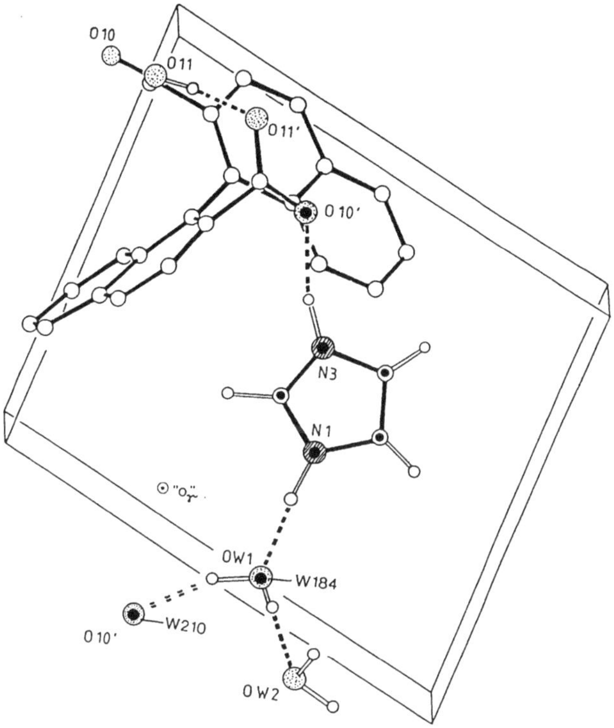

Fig. 41. Crystal structure of the *l*-imidazolium dihydrate associate [111] (O atoms dotted, N atoms hatched) showing intra-associate H-bonds (broken lines) and the resulting coinciding atomic sites from the fitting experiment with *SGPA* (bold dots). A position marked indicates the translated O10' from the anion. An expected atomic position of the O_γ atom of Ser195 (not considered as a part of the modelling experiment) is indicated merely to show the resulting would-be position executing the same transformation as for the seven fitted atoms. Only relevant H atoms are shown

Instead, the conformational characteristics are explained by the stringent requirement of the ionic interaction between the carboxylate/imidazolium ion pair coupled with the attempt of the former group to maintain as many H-bonds as possible (e.g. four H-bonds, cf. Ref. 75c). Such an attempt is obviously supported by the intramolecular H-bond in *1* · Im · 2 H_2O. The geometry of the corresponding moieties indicates the presence of strongly interacting ionic species (Fig. 42).

5.1.2 Inter-associate (Packing) Relations in *1* · Im · 2 H_2O

Packing in *1* · Im · 2 H_2O also shows some distinct features that may be related to the existence of the ionic species in the crystal. Hydrogen bonding is, of course, a primary feature (Fig. 43). An extensive network exists in this crystal which has the form of endless chains rather than that of loops usually found for the similarly double-faced (H-bond donor and acceptor) alcohols (cf. Fig. 19). As already mentioned, the carboxylate function has four connections, while its neutral —COOH neighbor maintains three H-bond contacts. The inner water molecule with respect

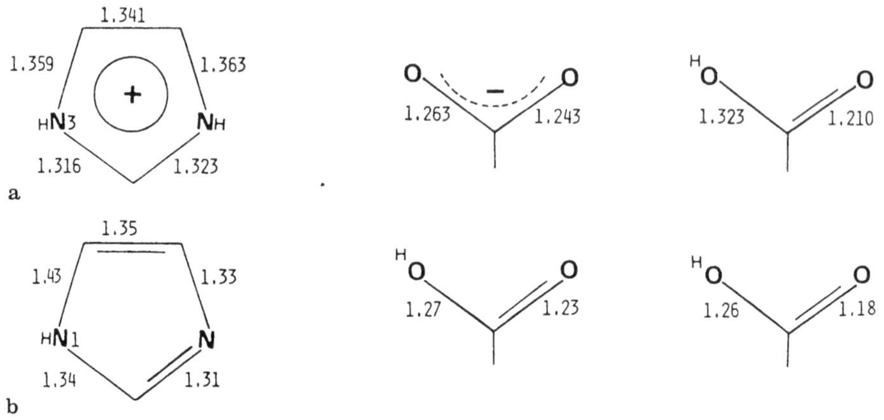

Fig. 42. Relevant dimensions **(a)** in the dihydrate and **(b)** in the anhydrous associates of *1* with imidazole [111] [e.s.d's are in the range 0.001–3 and 0.010–14 Å for **(a)** and **(b)**, respectively]

to the cation is engaged in a full binding capacity (double acceptor and donor), whereas the distant one has only three contacts (double donor, single acceptor). It is relevant to note that all but one of the H-atoms of the cationic C—H groups are involved in well-defined C—H ... O contacts as classified by Taylor and Kennard [75b] (Table 24). This possibly important feature has not been considered hitherto, e.g. in an NMR study for the especially important C(2) atom of the imidazolium moiety [121]. The H-bonds in *1* · Im · 2 H_2O have rather acceptable geometries and indicate strong interactions.

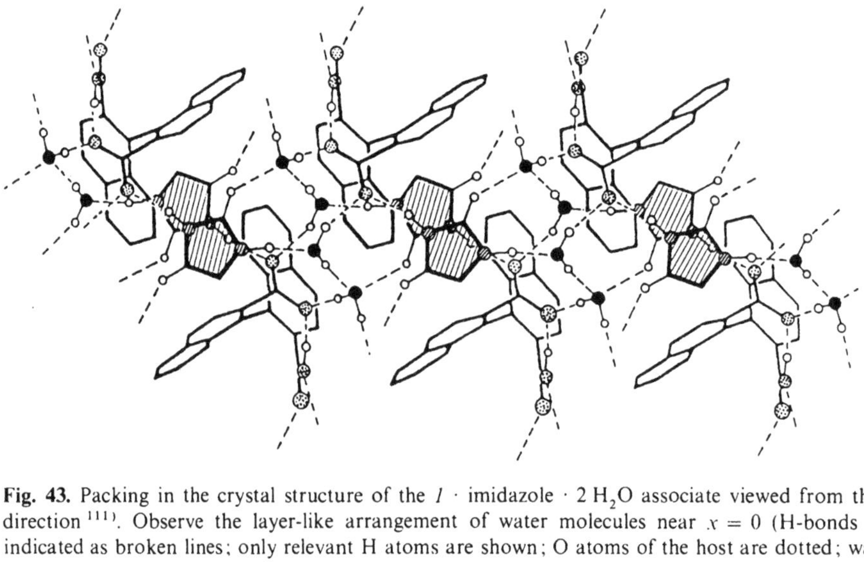

Fig. 43. Packing in the crystal structure of the *1* · imidazole · 2 H_2O associate viewed from the *c* direction [111]. Observe the layer-like arrangement of water molecules near $x = 0$ (H-bonds are indicated as broken lines; only relevant H atoms are shown; O atoms of the host are dotted; water oxygen as a bold circle; N atoms are hatched; the hatched segments signify the imidazole rings)

Table 24. H-bond dimensions in the dihydrated and unhydrated *1* · imidazole associates[a]

Donor-H	D—H (Å)	D—H ... A (deg)	A ... H (Å)	Acceptor
1 · imidazole · H_2O (1 : 1 : 2) [*1l*]				
O11—H11	1.01(3)	172(2)	1.56(3)	$O11'^{b}_{i}$
OW2—H1W2	1.01(3)	175(2)	1.81(3)	$O11'_{ii}$
OW1—H1W1	0.95(2)	157(2)	1.95(2)	$O10'_{iii}$
N3I—H3I	0.98(2)	163(2)	1.80(2)	$O10'_{i}$
OW1—H2W1	0.94(3)	170(2)	1.87(3)	$OW2_{i}$
C2I—H2I	1.05(2)	140(1)	2.35(2)	$OW2_{iv}$
OW2—H2W2	1.04(4)	171(2)	1.98(2)	$O11_{v}$
N1I—H1I	1.04(3)	175(2)	1.67(3)	$OW1_{i}$
C4I—H4I	1.00(2)	150(2)	2.28(2)	$O10_{vi}$
1 · imidazole (1 : 1) [*1m*]				
O11′—H11′	0.99	152	1.66	$O11_{i}$
O11—H11	1.12	171	1.66	$N3I_{ii}$
N1I—H1I	0.97	155	1.79	$O10_{iii}$
C2I—H2I	0.95	116	2.33	$O10'_{iv}$
C5I—H5I	0.94	164	2.35	$O11'_{iii}$

[a] E.s.d.'s are only given where appropriate. [b] Subscripts referring to symmetry operations relating H-bonded pairs of atoms are as follows: i, x, y, z; ii, $1 - x$, $1 - y$, $1 - z$; iii, $x - 1$, y, z; iv, $-x$, $1 - y$, $1 - z$; v, $x - 1$, $1 + y$, z; vi, x, $1 + y$, z for *1l*; i, $1 - x$, $-y$, $1 - z$; ii, $1 - x$, $y - 1/2$, $1/2 - z$; iii, $-x$, $-y - 1/2z - 1/2$; iv, x, $1/2 - y$, $z - 1/2$ for *1m*.

Another conspicuous packing feature is the *sheet*-like arrangement of the water molecules in the present crystal structure (Fig. 43). This may be understood as the structural manifestation of the shielding effect of the solvent molecules arranged into a layer-like pattern and effectively depolarizing the negative charges of the counter-facing carboxylate anions. Such a structural role many be invariably found in the crystal structures containing hydrated proton species [122]. Shielding of the charges arising in this crystal structure may also be affected by the counterfacing naphthyl groups protruding over the symmetry-center-related pair of imidazolium cations. The latter units are parallel and at a distance of 3.4 Å to each other. A further interesting point in the packing is the fact that the crystal structure shows the segregation of the enantiomers of *1* into homochiral strands which fit to each other through symmetry centres yielding the centrosymmetrical crystal structure.

5.1.3 Structural and Electrostatic Similarity of *1* · Im · 2 H_2O to Serine Protease Enzymes

As outlined in Sect. 5.1, the inclusion of two molecules of hydrating water enhanced the belief in the extension of the initially assumed similarity of the carboxyl-imidazole pair to the respective functions in serine proteases. The presence of two to four water molecules (or assigned as such) in this family of enzymes [115,117,118,120] and in *1* · Im · 2 H_2O parallels more recent calculations on the hydration enthalpy of bulky organic cations [123]. According to these results, the most important gain in the hydration enthalpy is encountered when the first two water molecules enter the hydration sphere of such cations. Adding more than four solvent molecules will not cause any further essential change in these systems.

Comparison between the steric arrangement in *1* · Im · 2 H_2O and an enzyme was carried out by the means of the least-squares fitting of the atoms of the imidazolium ring and of the two associated oxygen atomic sites at H-bonding distance originating from the carboxylate moiety and of W(1). For the comparison, the bacterial enzyme *Streptomyces Griseus Protease A (SGPA)* was chosen in the native form [124] since this species makes available one of the best resolved structures of the serine protease family. The comparison also relates to a peptide-aldehyde inhibited form of this enzyme [118b]. The respective atomic sites involved the imidazole ring of His57, the Oδ1 atom of Asp102, and a water molecule (W184) of the native enzyme. Results of the least-squares fitting revealed appreciable positional agreement with a mean deviation of less than 0.27 Å for the seven adjusted atomic sites (Fig. 41) [111].

Looking for further similarities, a second water site of *SGPA* (W210) was considered. This W210 is in H-bonding distance from the W184 site. Both positions are considered to mimic the carboxylate-oxygen atomic sites of a would-be transient product of the enzymic cleavage procedure. Upon transformation into the lattice of *1* · Im · 2 H_2O (Fig. 41) the W210 site falls into a place occupied by an O10′ atom within 0.4 Å. This site belongs to one of the carboxylate oxygens of *1* and is related by a unity translation along the crystallographic *a* axis to the other O10′ site which is supposed to be the Oδ1 aspartic carboxylate. Thus, in the crystal structure of *1* · Im · 2 H_2O an atom with a partial negative charge adopts the same steric position as W210 in the native *SGPA*. This fact may draw attention to the interpretation of solvent atomic sites in extremely big structures. This interpretation may be biased, similar to the results with subtilisin [115] and α-chymotrypsin [120] where the presence of even sulphate anions close to the His57 function were found in subsequent studies.

The fitting was not successful for the peptide-aldehyde inhibited form of the *SGPA* enzyme underlining the alignment difference between the active site residues His57 and Asp102 between the two forms of this enzyme (cf. Ref. 118).

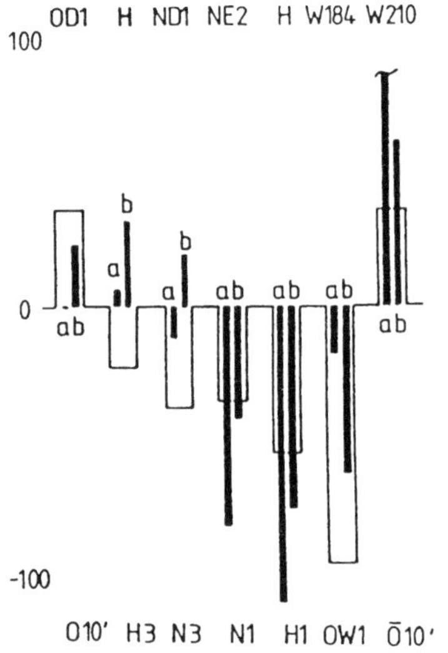

Fig. 44. Electrostatic potential (kJ/mol) pattern in *1* · imidazole · 2 H_2O (empty bars) and *SGPA* (solid lines) taken at the H-bonded atomic sites in both structures. Letters *a* and *b* for *SGPA* denote the potential values deriving from two different orientations of the H atom of the hydroxyl function of Ser195. The value of the potential in the *a* orientation is +192 kJ/mol at the W210 site (cf. Ref. 111)

The remarkable steric agreement of the respective atomic sites discussed above called for the examination of some electronic properties in the two systems being compared. Electrostatic potential patterns were chosen for their simplicity and the relative ease of accessibility [125]. The crystal structure of *1* · Im · 2 H_2O was decomposed into formic acid, water, and imidazole due to practical considerations and the resulting electrostatic potential pattern was compared to that of the native *SGPA*. Potential values at the atomic sites of the main chain of H-bonds connecting two of the catalytically important residues (Asp102 and His57) were considered. The resulting pattern reflects appreciable qualitative and quantitative similarity between the potentials (Fig. 44).

The potential pattern for the enzyme shows high sensitivity with respect to the non-trivial positions of H-atoms at the Oγ-sites of Ser195 of the catalytic machinery. The (— + —) charge distribution characteristic for the tetrahedral intermediate state is reproduced for both models. The decisive feature in this regard is the periodicity of the crystal structure in *1* · Im · 2 H_2O which results in the positive maxima due to the O10′ sites, apart from a unit cell edge. Correspondingly, it is unexpected that the value at the enzyme W210 site is even more positive (Fig. 44). A possible conclusion from the steric and electrostatic similarity is that probably a (— + —) charge distribution already exists in the native *SGPA* enzyme with the second negative charge being carried by an anion at the W210 site similar to sulphate anions in the active site of subtilisin [115].

In conformity with this reflection, trypsin [117b], elastase [126] and subtilisin [115] were found to retain sulphate anions associated with the active site over a pH-range between 5 and 7.5. Curiously, the binding of the anions is stronger on the acidic side of the pH-range. This is an especially interesting observation. It contrasts to the expectations prohibiting the presence of anions at pH values well below 7. The result may indicate that considerations based on common chemical arguments will not necessarily apply to *SGPA* [127]. One is inclined to believe that serine protease enzymes may have the ability to create a micro-environment for their active sites and bind anions more effectively than anticipated.

5.1.4 Water-Mediated Proton Transfer

The discussed analogy between the structures of the simple crystalline model *1* · Im · 2 H_2O and the active site of *SGPA* calls for an examination of the possible role and importance of the water molecules in such structures. The topology of the water molecules in the hydrated associate resembles those found for the hydrated proton structures containing H_3O^+ and $H_5O_2^+$ species [122]. Such cations probably play an essential role in the extremely fast proton transfer observed in liquid water [114,128]. Recent quantum chemical studies throw light on such events in different model systems [128c,129,130]. An H-bonded water molecule in an amidine/water system was shown to work as a mediator at a moderate energy expense [129]. The conditions compare well with the energy balance shown in Scheme 2 [111]. Another supporting observation was made from the study of a 2-pyridone-dihydrate structure [130]. In that case, the existence of a $H_5O_2^+$ ion was in fact corroborated as the proton-transferring agent, with a tempting steric similarity between the arrangement of the contributing water molecules and *1* · Im · 2 H_2O.

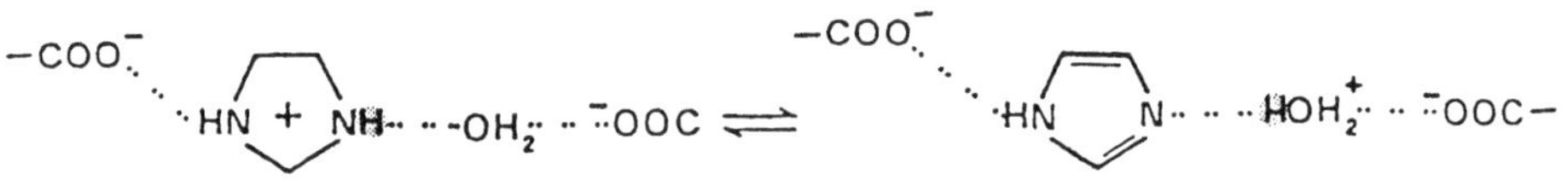

Scheme 2. Hypothetical proton removal path in the *1* · imidazole · 2 H_2O associate [111]

Water molecules or anions close to the active sites in the protease enzymes, mentioned above, may not be considered circumstantial, but may effectively contribute to the removal of the surplus proton from the imidazolium cation before the actual catalytic event. They could serve well to create the initial ion/neutral form of the Asp102-His57 couple which is important for the initial step of the catalytic process in most discussions [116,118,131]. Such a proton removal may be caused by the productive binding of a true substrate (or inhibitor) of the enzyme to the neighboring recognition clefts of the active site.

5.2 Crystal Structure of the Anhydrous 1:1 Associate of *1* with Imidazol [*1* · Im]

The importance of water in the preceding structure and theoretical considerations of its role suggested growing crystals in a water-free environment. The resulting crystals of unhydrated *1* · Im were, in general, hardly suitable for X-ray analysis. Nevertheless, out of interest, data collection from a rather small crystal was attempted. The subsequent analysis gave the structural model [111] as depicted in Fig. 45.

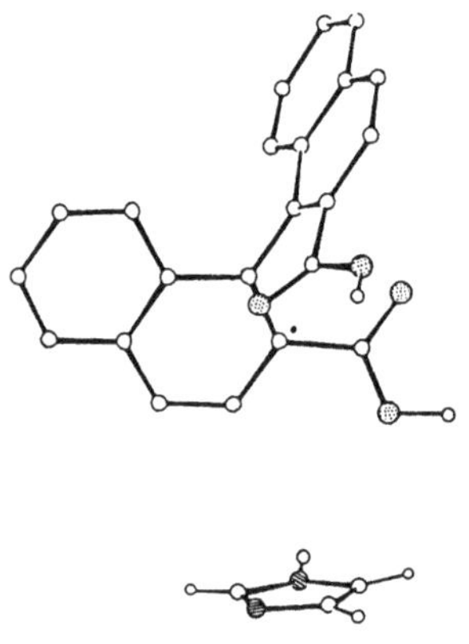

Fig. 45. Molecular structure of the anhydrous imidazole associate of *1* [111] (H atoms of the carboxyl groups indicate putative atomic positions only; O atoms dotted; N atoms hatched)

5.2.1 Intra-associate Relation in *1* · Im (1:1)

It is of importance to discuss briefly the present *1* · Im structure in comparison with the preceding hydrated analogue and with other inclusion ensembles of *1*. Some differences between corresponding dimensions of *1* and imidazole in *1* · Im · 2 H_2O

are shown in Fig. 42. The data given for *1* · Im supports the presence of a neutral rather than a salt-like associate. This important difference can also be noted in the alignment of the —COOH groups with respect to their naphthyl moieties and to each other (see Table 20 in Sect. 4.5.1). The values of the former parameter clearly relate this anhydrous imidazole associate to the common neutral host-guest ensemble of *1*. Dihedral angles in the given range fall under the inclusions of voluminous guest molecules (cf. *1* · 2-BuOH, *1* · t-BuOH, or *1* · DMSO). The mutual orientation of the —COOH groups are also different between the hydrated and the unhydrated forms. The internal H-bridge between the carboxylate and the neutral acid group in *1* · Im · 2 H_2O (cf. Fig. 41) cannot form in the unhydrated *1* · Im due to the unfavorable —COOH alignment (Fig. 45).

5.2.2 Packing and Energy Relations in *1* · Im (1:1): Structural Model for Logic Circuits at the Molecular Level

The H-bonding in the anhydrous *1* · Im (Table 24) has topologic properties (Fig. 46) similar to those in the alcohol coordinatoclathrates of *1* with 1:2 host:guest stoichiometry (cf. Fig. 17a). Assuming a perfectly ordered crystal lattice, the resulting central loop of H-bonds should appear to have homodromic directionality with the donor/acceptor functions separated in space. This contrasts to the behavior in the dihydrated *1* · Im where no such characteristic loops are formed. Involvement of the C—H hydrogen atoms of the imidazole molecule, however, is similar in both cases.

An interesting aspect of the present arrangement arises in connection with the poor quality of the data set and at the same time the reliability of H-atom positions. These are included in the scattering model with a fair amount of ambiguity in their positions, more than usual in X-ray experiments. Certain abnormalities in the geometry of the carboxyl groups may be understood as a result of conformational

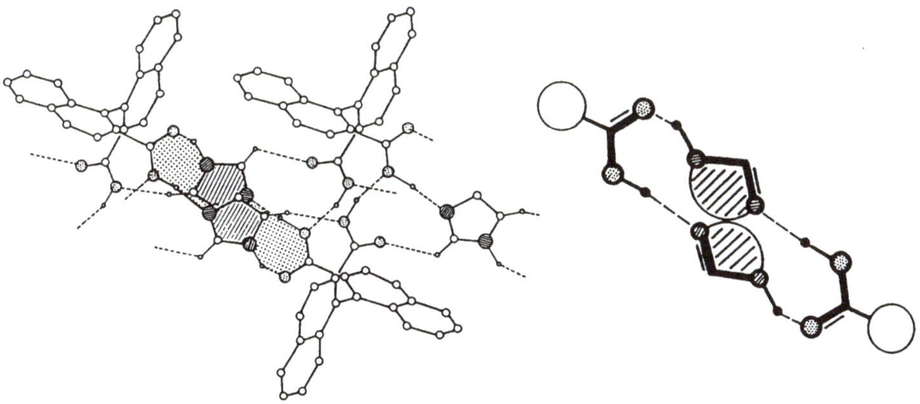

Fig. 46. Characteristics of the packing arrangement in unhydrated *1* · imidazole with a separate schematics emphasizing the central loop topology [111] (H-bonds are indicated as broken lines; backbone H atoms are omitted; O atoms dotted; N atoms hatched; the hatched segments in the schematic drawing signify the imidazole rings)

disorder resulting in the irregular interchange of C—O and C=O moieties at symmetry equivalent positions of the lattice. Averaging of C—O and C=O bonds in the crystal of unhydrated *1* · Im may lead to bond lengths as in Fig. 42. Such dimensions may, of course, arise when at least some of the molecules are in the ionized (salt-type) state. It is virtually impossible to distinguish these effects from each other from X-ray data unless crystals of high quality are provided.

Another peculiarity in the arrangement of the central loop (Fig. 46) is the close resemblance to the arrangement proposed for a theoretically constructed molecule to be used as a molecular flip-flop gate [132]. Such an imaginary molecule would consist of two hemiquinone moieties bound covalently to a central bisimidazolyl nucleus. The central H-bond loop satisfies most of the formal steric and electronic requirements for the proposed proton switching machinery. Indeed, theoretical calculations for this model indicate relatively little energy difference between the ionized and the neutral forms, which may be overcome by the energy of thermal motion [111]. Thus, the apparently small energy differences may also contribute to the assumed coexistence of salt-type and neutral forms which account for the low order in the structure of anhydrous *1* · Im.

6 Conclusions and Prospects

In short, the principle of coordinative assistance in clathrate formation whose possible advantages compared with the classical clathrate type are unfolded in Sect. 2.2 stood the test. It was shown that molecules related to the geometry of scissors and roofs, suitably equipped with functional groups, are nearly perfect host systems providing many clathrates of high stability and with designed selectivity. Hosts derived from other geometric figures and using other functional groups, still connected with the discussed principle, are imaginable in almost any number. In this respect, the coordinatoclathrates described here have added a new dimension to the inclusion chemistry with crystals [7], namely the relationship of polar site (functional group) complementarity between host and guest molecules (cf. Refs. 100 and 101).

On considering future prospects, it is advisable to visualize main aspects to be learned from this study in a more detailed way. The first conclusion is that closely homologous host compounds do form solid heteromolecular associates which, as documented by their crystal structures, are built in a systematic and nearly analogous manner. In other words, this means that in much the same way chemical reactions involving analogously functionalized compounds occur in an analogous (and predictable) manner (homology principle), the crystal chemistry of heteromolecular associates also possesses this feature, extending beyond the boundaries of individual molecules. Moreover, it is seen that there is a definable recognition in heteromolecular assemblies [67a], corresponding to the main structural motifs that one observes. In accord with the term *supramolecular chemistry* coined by Lehn for mostly *ionic* types of host/guest assemblies (cryptates, coronates, speleates, etc.) [100], we are now fairly confident in predicting a further expansion of the boundaries of this new discipline into the more puzzling and promising region of the associates of *neutral* molecules as well [133]. The odds are in our favor that it will soon become possible to predict the outcome of co-crystallization experiments [112].

Another promising point deduced from this study is that structures having a "Gestalt" like those illustrated in this chapter may also be helpful in the design of intramolecularly encapsulating hosts. The simple idea behind this is to translate a fixed multimolecular section of an inclusion lattice into a covalently linked arrangement of building blocks yielding the host. A recent example from another laboratory [134] illustrates the applicability of the crystal model as a useful tool to aid host engineering. The interesting synthetic work performed led to a cross-linked dimer of *26* having virtually the same hydrophobic tunnel geometry as seen in Fig. 32. Many other clathrate structures of this study wait for exploitation in an analogous sense (see also Chapter 4 of this book).

A further subject not fully exploited is the use of crystalline associate formation in artificial enzyme modelling [135]. Certainly, the disclosed structural and electronic relations between the hydrated *1* · imidazole co-crystal and a serine protease enzyme are only the beginning of an imaginable research field. Another proof of this kind has recently been found in the structure of the imidazole associate of *7* [136]. In agreement with the markedly different inclusion behavior of this binaphthyl derivative (see Sect. 4.2.2), *7* will not include water like *1* under the same circumstances. The so formed associate *7* · imidazole (Fig. 47) also deviates in structure and in packing from that of the imidazole associate of *1*. In spite of the lack of water, *7* · imidazole displays a salt-type structure. The resulting crystals are enantiomorphous (space group *P*1) indicating spontaneous resolution on crystallization which is a further difference to *1*. Moreover, the so-formed aggregate will not match when attempts are made to fit it to the respective atomic sites of *SGPA*. It fits, however, with a mean deviation of 0.6–0.7 Å, to some atoms of the subtilisin active site [136]. It was noticed that the similar active sites of the serine proteases are by no means identical [115]. Hence,

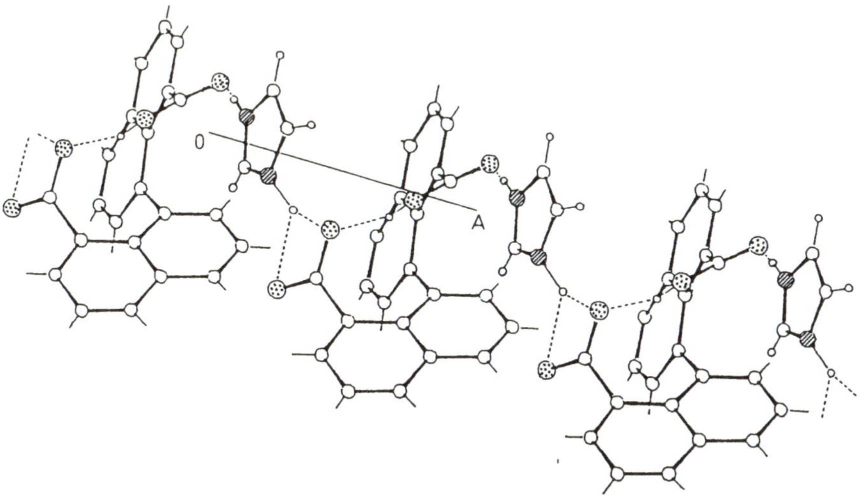

Fig. 47. Molecular structure and packing in the crystalline *7* · imidazole (1:1) associate with an indication of the H-bonding network [136] (H-bonds as broken lines; backbone H atoms of the host are shown as sticks only; O atoms dotted; N atoms hatched)

another challenge arises, namely, to attempt to understand minor alterations occuring in such related systems.

Conglomerate crystallization in the above case indicates that the inclusion approach may be further extended into the realm of the salt-type associates. Such an attempt is especially interesting due to the obvious role in enantiomer separation which relies heavily on the solubility difference of the enantiomeric salts under certain circumstances[137].

It goes without saying, that all the aspects pointed out give rise to a great number of applications in industrial and academic sectors[138]. In future, the use of specific computer programs[139] will help to reach some of the goals more easily.

7 Acknowledgements

The authors are very much indebted to Dr. I. Csöregh (University of Stockholm, Arrhenius Laboratory) for intensive and fruitful cooperation over many years and to Prof. P. Kierkegaard (Stockholm), and Prof. A. Kálmán (Budapest) for their continous interest in and support of this research. The authors would also like to thank Dipl. Chem. M. Hecker and Dipl. Chem. W. Seichter for drawing the figures, and Mrs. M. Weber for the trouble with the typing of the manuscript. The work at Bonn was financially supported by the Deutsche Forschungsgemeinschaft and the Fonds der Chemischen Industrie.

8 References

1. Hall, D. M., Turner, E. E.: J. Chem. Soc. *1955*, 1242
2. Weber, E., Csöregh, I., Stensland, B., Czugler, M.: J. Am. Chem. Soc. *106*, 3297 (1984)
3. An experimental note of Ref. 1 only vaguely suggests possible "solvate formation"
4. Encyclopaedia Britannica USA, Instant Service Report R-1519, Britannica Centre, 310 South Michigan Ave, Chicago, IL 60604
5. Brown Jr., J. F.: Sci. Am. *207*, 82 (1962)
6. Saenger, W.: Umschau *74*, 635 (1974)
7. Atwood, J. L., Davies, J. E. D., MacNicol, D. D. (eds.): Inclusion Compounds, Vols. 1–3, London, Academic Press 1984[8]
8. Recent and most comprehensive representation of this topic. A complete reference list of monographies and reviews on clathrate compounds is found in Chapter 1 of Vol. 140 of this series (Molecular Inclusion and Molecular Recognition — Clathrates I)
9. Schlenk Jr., W.: Fortschr. Chem. Forsch. *2*, 92 (1951)
10. Cramer, F.: Angew. Chem. *64*, 437 (1952)
11. Takemoto, K., Sonoda, N. in: Vol. 2 of Ref. 7, p. 47
12. Smith, A. E.: Acta Crystallogr. *5*, 224 (1952)
13. Giglio, E. in: Vol. 2 of Ref. 7, p. 207
14. Popovitz-Biro, R., Chang, H. C., Tang C. P., Shochet, N. R., Lahav, M., Leiserowitz, L.: Pure Appl. Chem. *52*, 2693 (1980)
15. Powell, H. M. in: Non-stoichiometric Compounds, Mandelcorn, L. (ed.), New York—London, Academic Press 1964, p. 438
16. MacNicol, D. D. in: Vol. 2 of Ref. 7, p. 1
17. Bhatnagar, V. M.: Clathrate Compounds, New Delhi, S. Chand 1968, p. 16
18. Jeffrey, G. A. in: Vol. 1 of Ref. 7, p. 135
19. Weber, E., Josel, H.-P.: J. Incl. Phenom. *1*, 79 (1983)

20. Hyatt, J. A., Duesler, E. N., Curtin, D. Y., Paul, I. C.: J. Org. Chem. *45*, 5074 (1980); Rahman, A., van der Helm, D.: Cryst. Struct. Commun. *10*, 731 (1981); Thierbach, D., Huber, F.: Z. Anorg. Allg. Chem. *477*, 101 (1981); Mac, T. C. W.: J. Chem. Soc., Perkin Trans. 2, *1982*, 1435; Mentzafos, D., Terzis, A., Filippakis, S. E.: Cryst. Struct. Commun. *11*, 71 (1982); Pickering, M., Small, R. W. H.: Acta Crystallogr. B *38*, 3161 (1982); Shiel, H. S., Hoard, L. G., Nordmann, C. E.: Acta Crystallogr. B *38*, 2411 (1982)
21. "Fourth International Symposium on Inclusion Phenomena and Third International Symposium on Cyclodextrins", July 20–25, 1986, Lancaster (cf. Coll. Abstr.)
22. Davies, J. E. D., Kemula, W., Powell, H. M., Smith, N. O.: J. Incl. Phenom. *1*, 3 (1983)
23. Fischer, E.: Ber. Dtsch. Chem. Ges. *27*, 2985 (1894)
24. Cram, D. J. in: Applications of Biochemical Systems in Organic Chemistry, Part II, Techniques of Chemistry, Vol. X, Jones, J. B., Sih, C. J., Perlman, D. (eds.), New York, Wiley Interscience 1976, p. 815
25. E.g. Allcock, H. R., Allen, R. W., Bissell, E. C., Smeltz, L. A., Teeter, M.: J. Am. Chem. Soc. *98*, 5120 (1976); Löhr, H.-G., Vögtle, F., Schuh, W., Puff, H.: J. Chem. Soc., Chem. Commun. *1983*, 924; Herbstein, F. H., Mak, T. C. W., Reisner, G. M., Wong, H. N. C.: J. Incl. Phenom. *1*, 301 (1984)
26. Kitaigorodsky, A. I.: Order and Disorder in the World of Atoms, The Heidelberg Science Library, Vol. 3, New York, Springer Verlag 1967
27. Kitaigorodsky, A. I.: Molecular Crystals and Molecules, New York—London, Academic Press 1973
28. Kitaigorodsky, A. I.: Mixed Crystals, Berlin—Heidelberg—New York—Tokyo, Springer Verlag 1984
29. With reference to graphic arts, see: Die Welten des M. C. Escher, Hersching, Manfred Pawlak Verlagsgesellschaft 1971, 3rd ed.
30. Recent examples of host molecules with C_2-symmetry from other laboratories: Chan, T. L., Mak, T. C. W., Trotter, J.: J. Chem. Soc., Perkin Trans. 2, *1980*, 672; Mann, B. J., Paul, I. C., Curtin, D. Y.: J. Chem. Soc., Perkin Trans. 2, *1981*, 1583; Zheng Huang, N., Mak, T. C. W.: J. Chem. Soc., Chem. Commun. *1982*, 543; Bishop, R., Dance, I. G., Hawkins, S. C.: J. Chem. Soc., Chem. Commun. *1983*, 889; Radcliffe, M. D., Gutiérrez, A., Blount, J. F., Mislow, K.: J. Am. Chem. Soc. *106*, 682 (1984)
31. Baker, W., Gilbert, B., Ollis, W. D.: J. Chem. Soc. *1952*, 1443
32. Grasselli, J. G. (ed.): Atlas of Spectral Data and Physical Constants for Organic Compounds, Cleveland, Ohio, The Chemical Rubber Co. 1973; see also Davies, J. E. D. in: Vol. 3 of Ref. 7, p. 37
33. Guest selectivity properties to this extent are out of the ordinary, cf. Ref. 7
34. Reichardt, C.: Solvent Effects in Organic Chemistry, Weinheim, Verlag Chemie 1979
35. Hall, D. M., Ridgewell, S., Turner, E. E.: J. Chem. Soc. *1954*, 2498
36. Optically resolved compound, see: Mislow, K., Glass, M. A. W., O'Brien, R. E., Rutkin, P., Steinberg, D. H., Weiss, J., Djerassi, C.: J. Am. Chem. Soc. *84*, 1455 (1962)
37. Weber, E., Ahrendt, J., Finge, S.: unpublished result (1986)
38. Pummerer, R., Prell, E., Riede, A.: Ber. Dtsch. Chem. Ges. *59*, 2159 (1926)
39. Weil, K., Kuhn, W.: Helv. Chim. Acta *27*, 1648 (1944); Barber, H. J., Gaimster, K.: J. Appl. Chem. *2*, 565 (1952)
40. Cf. Akimoto, H., Yamada, S.: Tetrahedron *27*, 5999 (1971)
41. Brass, K., Sommer, P.: Ber. Dtsch. Chem. Ges. *61*, 997 (1928)
42. Koukotas, C., Schwartz, L. H.: J. Chem. Soc., Chem. Commun. *1969*, 1400; see also Bell, F., Waring, D. H.: J. Chem. Soc. *1949*, 2689
43. Haas, G., Prelog, V.: Helv. Chim. Acta *52*, 1202 (1969)
44. Toda, F., Tanaka, K., Nagamatsu, S.: Tetrahedron Lett. *25*, 4929 (1984); Toda, F., Tanaka, K., Mak, C. W.: Chem. Lett. *1984*, 2085
45. Bromobenzene solvate of optically resolved dimethylester of *16*, see: Akimoto, H., Iitaka, Y.: Acta Crystallogr. B *25*, 1491 (1969)
46. Schwenk, A.: Chem.-Ztg. *53*, 335 (1929)
47. With current knowledge, possible clathrate formation of *22* with ethanol may be anticipated from the original synthetic description in Ref. 43
48. Czugler, M., Stezowski, J. J., Weber, E.: J. Chem. Soc., Chem. Commun. *1983*, 154

49. Neupert-Laves, K., Dobler, M.: Helv. Chim. Acta *64*, 1653 (1981)
50. Czugler, M., Weber, E., Ahrendt, J.: J. Chem. Soc., Chem. Commun. *1984*, 1632
51. Mowry, D. T.: J. Am. Chem. Soc. *69*, 573 (1947); Hurd, D. C., Tockman, A.: J. Am. Chem. Soc. *81*, 116 (1959); Baumgartner, P., Hugel, G.: Bull. Soc. Chim. Fr. *1954*, 1005; Walborskiy, H. M.: Helv. Chim. Acta *36*, 1251 (1953)
52. Barnett, E. D. B., Goodway, N. F., Lawrence, C. A.: J. Chem. Soc. *1935*, 1102; Scheibler, H., Scheibler, U.: Chem. Ber. *87*, 379 (1954); see also Ref. 51
53. Bachmann, W. E., Scott, L. B.: J. Am. Chem. Soc. *70*, 1458 (1948)
54. Bachmann, W. E., Cole, W.: J. Org. Chem. *4*, 60 (1939); Johnson, W. K., Patton, T. L.: U.S. 2,938,049, May 24, 1960 [Chem. Abstr. *54*, 19628i (1960)]
55. Figeys, H. P., Dralants, A.: Tetrahedron Lett. *28*, 3031 (1972); Huebner, C. F.: Ger. Off. 1,914,998, Oct. 30, 1969 [Chem. Abstr. *72*, 78769p (1970)]
56. Sakellarios, E., Kyrimis, T.: Ber. Dtsch. Chem. Ges. *57*, 322 (1924); Wilson, K. R., Pincock, R. E.: J. Am. Chem. Soc. *97*, 1474 (1975)
57. Maigrot, N., Mazaleyrat, J. P.: Synthesis *1985*, 317; Tamao, K., Minato, A., Miyake, N., Matsuda, T., Kiso, Y., Kumada, M.: Chem. Lett. *1975*, 133
58. Bergmann, F., Eschinazi, H. E., Neeman, M.: J. Org. Chem. *8*, 179 (1943)
59. Ueji, S., Nakatsu, K., Yoshioka, H., Kinoshita, K.: Tetrahedron Lett. *23*, 1173 (1982)
60. Badar, Y., Cheung King Ling, C., Cooke, A. S., Harris, M. M.: J. Chem. Soc. *1965*, 1543
61. Kress, R. B., Duesler, E. N., Etter, M. C., Paul, I. C., Curtin, D. Y.: J. Am. Chem. Soc. *102*, 7709 (1980)
62. Newman, M. S.: J. Am. Chem. Soc. *62*, 1683 (1940)
63. Bell, F., Waring, D. H.: J. Chem. Soc. *1949*, 267, 1579
64. Weber, E., Ahrendt, J., Czugler, M., Csöregh, I.: Angew. Chem. *98*, 719 (1986); Angew. Chem., Int. Ed. Engl. *25*, 746 (1986)
65. Dale, J.: Stereochemie und Konformationsanalyse, Weinheim, Verlag Chemie 1978
66. Clarkson, R. G., Gomberg, M.: J. Am. Chem. Soc. *52*, 2881 (1930)
67. (a) Czugler, M.: Transactions of the "Symposium on Molecular Structure: Chemical Reactivity and Biological Activity", Beijing, China 1986, Oxford, University Press, to be published (1988)
(b) For example, a ternary aggregate was reported for a *32*-related host compound [code ANTTCN in the Cambridge Crystallographic Database: 9,10-dihydro-9,10-ethanoanthracene-11,11,12,12-tetracarbonitrile-tetracyanoethylene complex methylene chloride solvate. $8(C_{20}H_{10}N_4):(C_6N_4):2(CH_2Cl_2)$]. See Karle, I. L., Fratini, A. V.: Acta Crystallogr. B *26*, 596 (1970)
68. Crystal structure of 1,1'-binaphthyl-8,8'-dicarboxylic acid: Czugler, M., Weber, E.: Unpublished result (1985)
69. Crystal structure of bis-β-naphthol: Gridunova, G. V., Furmanova, N. G., Shklover, V. E., Struchkov, Yu. T., Ezhkov, Z. I., Chayanov, B. A.: Kristallografiya *27*, 477 (1982)
70. Crystal structure of 2,2'-bis(hydroxymethyl)-1,1'-binaphthyl: Czugler, M., Csöregh, I., Weber, E.: Unpublished result (1984)
71. Csöregh, I., Czugler, M., Weber, E.: Transactions of the "Symposium on Molecular Structure: Chemical Reactivity and Biological Activity", Beijing, China 1986, Oxford, University Press, to be published (1988)
72. Racemic 1,1'-binaphthyl: Kerr, K. A., Robertson, J. M.: J. Chem. Soc. B *1969*, 1146
73. Optically active 1,1'-binaphthyl: Kuroda, R., Mason, S. F.: J. Chem. Soc., Perkin Trans. 2, *1981*, 167. See also Ref. 61
74. Crystal structure of 9,9'-spirobifluorene: Csöregh, I., Czugler, M., Weber, E.: Unpublished result (1985)
75. Recent literature on H-bonds:
(a) Taylor, R., Kennard, O.: Acc. Chem. Res. *17*, 320 (1984)
(b) Taylor, R., Kennard, O.: J. Am. Chem. Soc. *104*, 5063 (1982) (C—H ... X type H-bonds)
(c) Taylor, R., Kennard, O., Versichel, W.: ibid. *105*, 5761 (1983)
(d) Taylor, R., Kennard, O., Versichel, W.: ibid. *106*, 244 (1984)
(e) Taylor, R., Kennard, O., Versichel, W.: Acta Crystallogr. B *40*, 280 (1984)

(f) Murray-Rust, P., Glusker, J. P.: J. Am. Chem. Soc. *106*, 1018 (1984)
(g) Berkovitch-Yellin, Z., Ariel, S., Leiserowitz, L.: J. Am. Chem. Soc. *105*, 765 (1983)
76. Saenger, W.: Nature (London) *279*, 343 (1979); Lindner, K., Saenger, W.: Acta Crystallogr. B *38*, 203 (1982) and references therein
77. Crystal structure of *1* with 1-propanol (2:1): Czugler, M., Weber, E.: Unpublished result (1983)
78. Clementi, E.: "Structure of water and counterions for nucleic acids in solution", in: Structure and Dynamics: Nucleic Acids and Proteins, Clementi, E., Sarma, R. H. (eds.), New York, Adeline Press 1983
79. Crystal structure of the acetic acid inclusion of *1*: Csöregh, I., Weber, E.: Unpublished result (1985)
80. Crystal structures of the *7* · pyridine (1:1), *7* · 2-(hydroxymethyl)-pyridine (1:1), and *7* · pyridine · acetic acid (1:1:1) aggregates: Csöregh, I., Czugler, M., Weber, E.: Unpublished results (1984)
81. Crystal structure of *13* · imidazole: Czugler, M., Weber, E.: Unpublished result (1985)
82. Csöregh, I., Sjögren, A., Czugler, M., Cserzö, M., Weber, E.: J. Chem. Soc., Perkin Trans. 2, *1986* 507
83. Crystal structure of *25b* · DMF (1:1): Czugler, M., Weber, E.: Unpublished result (1983)
84. (a) Bernstein, F. H., Marsh, R. E.: Acta Crystallogr. B *33*, 2358 (1977)
(b) Bernstein, F. H., Kapon, M.: ibid. *34*, 1608 (1978)
(c) Bernstein, F. H., Kapon, M., Wasserman, S. ibid. *34*, 1613 (1978)
(d) Bernstein, F. H., Kapon, M.: ibid. *35*, 1614 (1979)
85. Crystal structure of *20* · DMSO: Csöregh, I., Czugler, M., Weber, E.: Unpublished result (1985)
86. Csöregh, I.: Personal communication (1987)
87. Crystal structure of *41* · DMSO (1:2): Czugler, M., Csöregh, I., Weber, E., Ahrendt, J.: Unpublished result (1986)
88. Gold, V., Stahl, R., Wassef, W. N., Kuroda, R.: J. Chem. Soc., Perkin Trans. 2, *1986*, 477
89. Neupert-Laves, K., Dobler, M.: Helv. Chim. Acta *64*, 1653 (1981)
90. Crystal structures of *48* with benzene (1:1) and p-xylene (2:1): Czugler, M., Weber, E., Csöregh, I.: Unpublished results (1986)
91. Gavezzotti, A.: Personal communication (1985). See also Gavezzotti, A., Simonetta, M. in: Organic Solid State Chemistry (Studies in Organic Chemistry, Vol. 32), Desiraju, G. R. (ed.), Amsterdam—New York, Elsevier 1987, p. 391
92. Gridunova, G. V., Furmanova, N. G., Shklover, V. E., Struchkov, Yu. T., Ezhkova, B. A., Chayanov, B. A.: Kristallografiya, *27*, 477 (1982)
93. Harata, K., Tanaka, J.: Bull. Chem. Soc. Jpn. *46*, 2747 (1973)
94. Pauptit, R. A., Trotter, J.: Can. J. Chem. *61*, 69 (1983)
95. Gridunova, G. V., Shklover, V. E., Struchkov, Yu. T., Chayanov, B. A.: Kristallografiya *28*, 87 (1983)
96. Wells, J. L., Trus, B. L., Johnston, R. M., Marsh, R. E., Fritchie, C. J., Jr.: Acta Crystallogr. B *30*, 1127 (1974)
97. Pauptit, R. A., Trotter, J.: Can. J. Chem. *59*, 1149 (1981)
98. Pauptit, R. A., Trotter, J.: ibid. *59*, 1149 (1981)
99. Fink, R., van der Helm, D.: Cryst. Struct. Commun. *9*, 97 (1980)
100. Lehn, J.-M.: Science, *227*, 849 (1985); Lehn, J.-M.: Angew. Chem. *100*, 91 (1988); Angew. Chem., Int. Ed. Engl. *27*, 89 (1988)
101. Cram, D. J.: Angew. Chem. *98*, 1041 (1986); Angew. Chem., Int. Ed. Engl. *25*, 1039 (1986)
102. Lipscomb, W. N.: Acc. Chem. Res. *15*, 232 (1982)
103. See Chapters 2 and 3 in Vol. 140 of this series (Molecular Inclusion and Molecular Recognition — Clathrates I)
104. For a comprehensive review of the packing modes of aliphatic carboxylic acids, see: Leiserowitz, L.: Acta Crystallogr. B *32*, 775 (1976)
105. On the symmetry of self-complementary surfaces, see e.g.: Morgan, R. S., Miller, S. L., McAdon, J. M.: J. Mol. Biol. *127*, 31 (1979)
106. Gavezzotti, A.: J. Am. Chem. Soc. *105*, 5220 (1983)
107. Sturtevant, J. M.: Proc. Natl. Acad. Sci. U.S.A. *74*, 2236 (1977)
108. Recent examples of artificial enzyme models based on the β-cyclodextrin skeleton:

(a) Breslow, R., Trainor, G., Ueno, A.: J. Am. Chem. Soc. *105*, 2739 (1983)
(b) LeNoble, W. J., Srivastava, S., Breslow, R., Trainor, G.: ibid. *105*, 2748 (1983)
(c) D'Souza, V. T., Hanabusa, K., O'Leary, T., Gadwood, R. C., Bender, M. L.: Biochem. Biophys. Res. Commun. *129*, 727 (1985)
(d) Tabushi, I.: Acc. Chem. Res. *15*, 66 (1982)

109. Cram, D. J., in: Chemistry for the Future, Grünewald, H. (ed.), Oxford—New York, Pergamon Press 1984
110. Cram, D. J., Katz, H. E.: J. Am. Chem. Soc. *105*, 135 (1983)
111. Czugler, M., Ángyán, J. G., Náray-Szabó, G., Weber, E.: ibid. *108*, 1275 (1986)
112. Addadi, L., Berkovitch-Yellin, Z., Weissbuch, I., van Mil, J., Shimon, L. J., Lahav, M., Leiserowitz, L.: Angew. Chem. *97*, 476 (1985); Angew. Chem., Int. Ed. Engl. *24*, 466 (1985)
113. Cf. the crystal structure of 15,15-bis(dodecyloxymethyl) [16] crown-5 · NaSCN. This complex has a density almost identical with that of water (D_c = 1.02 g cm^{-3}): Czugler, M., Weber, E., Kálmán, A., Stensland, B., Párkányi, L.: Angew. Chem. *94*, 641 (1982); Angew. Chem., Int. Ed. Engl. *21*, 627 (1982)
114. Eigen, M.: Angew. Chem. *75*, 489 (1963)
115. Matthews, D. A., Alden, R. A., Birktoft, J. J., Freer, S. T., Kraut, J.: J. Biol. Chem. *252*, 8875 (1977)
116. (a) Kraut, J.: Annu. Rev. Biochem. *46*, 331 (1977)
(b) Polgár, L., Halász, P.: Biochem. J. *207*, 1 (1982), and references therein
117. Some reports referring to the occurence of two to four water molecules in the active site of such enzymes are:
(a) Birktoft, J. J., Blow, D. M.: J. Mol. Biol. *68*, 187 (1972)
(b) Bode, W., Schwager, P.: ibid. *98*, 693 (1975)
118. (a) Sielecki, A. R., Hendrickson, W. A., Broughton, C. C., Delbaere, L. T. J., Brayer, G. D., James, M. N. G.: ibid. *134*, 184 (1979)
(b) James, M. N. G., Sielecki, A. R., Brayer, G. D., Delbaere, L. T. J.: ibid. *144*, 43 (1980)
119. Trypsinogen, the inactive proenzyme form of trypsin has no water molecules in its unordered "active site"; cf. Fehlhammer, H., Bode, W., Huber, R.: ibid. *111*, 415 (1977)
120. Blevins, R. A., Tulinsky, A. J.: J. Biol. Chem. *260*, 8865 (1985)
121. Hunkapiller, M. W., Forgac, M. D., Whitaker, D. R., Richards, J. H.: Biochemistry *12*, 4732 (1973)
122. Cf. layering of water molecules in higher hydrates of HCl: Taesler, I.: Acta Univ. Upps. *S91* (1981); Taesler, I., Lundgren, J.-O.: Acta Crystallogr. B *34*, 2424 (1978)
123. Meot-Ner (Mautner) M.: J. Am. Chem. Soc. *106*, 1257 (1984); Meot-Ner (Mautner) M.: Acc. Chem. Res. *17*, 186 (1984)
124. Coordinates of the *SGPA* active site residue and of the two water molecules W184 and W210 are from a further refinement of the published (1.8 Å resolution) structure (Ref. 118b) with a resolution extending beyond 1.7 Å for the native enzyme: James, M. N. G., Sielecki, A. R.: Private communications (1983, 1984)
125. Ángyán, J., Náray-Szabó, G.: J. Theor. Biol. *103*, 349 (1983)
126. Shotton, D. M., White, N. J., Watson, H. C.: Cold Spring Harbor Symp. Quant. Biol. *36*, 91 (1971)
127. The native *SGPA* crystals have been prepared from a phosphate buffered solution at a pH of 4.3 (cf. Ref. 118)
128. (a) Lengyel, S., Conway, B. E. in: Comprehensive Treatise of Electrochemistry, Vol. 5, Conway, B. E., Bockhus, J. O. M., Yeager, E., (eds.), New York, Plenum Press 1983
(b) Wang, J. H.: Proc. Natl. Acad. Sci. U.S.A. *66*, 874 (1970)
(c) Ángyán, J., Allavena, M., Picard, M., Potier, A., Tapia, O.: J. Chem. Phys. *77*, 4723 (1982)
129. Zielinski, T. J., Poirier, R. A., Peterson, M. R., Csizmadia, I. G.: J. Comput. Chem. *4*, 419 (1983)
130. Field, M. J., Hillier, I. H., Guest, M. F.: J. Chem. Soc., Chem. Commun. *1984*, 1310
131. (a) Kossiakoff, A. A., Spencer, S. A.: Biochemistry *20*, 6462 (1981)
(b) Stein, R. L.: J. Am. Chem. Soc., *105*, 5111 (1983)
132. Aviram, A., Seiden, P., Ratner, M. A., in: Molecular Electronic Devices, Carter, F. L. (ed.), New York—Basel, Marcel Dekker 1983

133. Cf. Weber, E. in: Synthesis of Macrocycles — The Design of Selective Complexing Agents (Progress in Macrocyclic Chemistry, Vol. 3), Izatt, R. M., Christensen, J. J. (eds.), New York, Wiley 1987, p. 337
134. Sheppod, T. J., Petti, M. A., Dougherty, D. A.: J. Am. Chem. Soc. *108*, 6085 (1986). See also Wilcox, C. S., Greer, L. M., Lynch, V.: ibid. *109*, 1865 (1987)
135. Dugas, H., Penney, C.: Bioorganic Chemistry, New York, Springer-Verlag 1981; Green, B. S., Ashani, Y., Chipman, D. (eds.): Chemical Approaches to Understanding Enzyme Catalysis, Amsterdam—New York, Elsevier 1982
136. Czugler, M.: Unpublished results (1986)
137. Jacques, J., Collet, A., Wilen, S. A.: Enantiomers, Racemates, and Resolutions, New York, Wiley-Interscience 1981
138. J. Scheffer (ed.): Organic Chemistry in Anisotropic Media (Tetrahedron Symposia-in-Print, Number 29), Tetrahedron *43*, 1197 (1987); Desiraju, G. R. (ed.): Organic Solid State Chemistry (Studies in Organic Chemistry, Vol. 32), Amsterdam—New York, Elsevier 1987
139. Davies, I. K.: CHEM X, a program system for manipulating and displaying molecules, Chemical Design Ltd., UK., Oct. 1986

New Types of Helical Canal Inclusion Networks

Roger Bishop and Ian G. Dance

School of Chemistry, University of New South Wales, Kensington, NSW 2033 Australia

Table of Contents

Over recent years a variety of new helical canal inclusion systems have been discovered and characterised, and an increasing awareness has developed of the roles played by canal topology and helicity in the function of certain biological systems.

In this review the definition, symmetry and structural requirements for formation of helical canal inclusion networks are introduced from a crystallographic viewpoint and then individual chemical systems are examined in detail.

Multimolecular helical inclusion networks formed by rigid alicyclic diols, urea, deoxycholic acid, and tri-o-thymotide are described and contrasted, followed by discussion of DNA intercalates, amylose compounds, and other inclusion systems formed by helical polymers.

The article concludes by examining recent research involving transmembrane ion channels particularly those molecular aggregates and macromolecular systems where helicity is involved in construction of the channel.

1 Introduction

Among the secondary literature on inclusion phenomena there is no review devoted to molecular systems where the host structure has helical topology [1]. In fact, for a relatively long period there was known only one chemical system where helical canal inclusion compounds could be generally prepared and utilised, namely the urea host system. However, new helical canal inclusion networks have been discovered and characterised in recent years, and it is becoming more apparent that some biological aggregates of macromolecules involve canal topology and incorporate helicity. Therefore the purpose of this review is to assemble and assess information about the involvement of helical canals in inclusion phenomena. We have taken a broad view, covering both chemical and biochemical systems. In some cases the experimental data provides unambiguous definition of helical canals, while for others the available information and its interpretation are less clearcut. We also consider systems where helical molecular arrangements aggregate to produce centrally located canals.

We concentrate here on the structural aspects of helical canal inclusion compounds, primarily because this field of chemical inclusion is still at the relatively juvenile stage of establishing geometry and geometrical variables. Comments on structure-property relationships for the chemical systems and on structure-function relationships for the biochemical systems are made wherever possible.

Comment is required on the proliferative terminology for the geometry. The terms "canal", "channel", "tube" and "tunnel" have been used by various authors to describe host cavities extended in one dimension without restriction on cross-sectional shape. We prefer the terms proposed by Weber and Josel [2], namely "helical tubuland" for the host structure and "helical tubulate" for the host-guest complex, but since the words "canal" and "channel" predominate in the literature we have arbitrarily adopted "canal" throughout this article, except for discussion of membrane transport where the the term channel is widely used. We will distinguish unimolecular inclusion compounds where the host is a single molecule, and multimolecular inclusion compounds which require many host molecules for construction of the canal.

In the next section we consider some general questions about the occurrence, properties and experimental investigation of helical canal inclusion compounds, in relation to inclusion compounds with different topologies. In subsequent sections we describe the properties of specific systems.

2 General Analyses

2.1 General Requirements for Helical Canal Inclusion Networks

The formation of host-guest inclusion compounds in crystals (or in highly ordered membranes or other biological structures) is a phenomenon more organised than simple co-crystallisation of a pair of substances. The special characteristic of inclusion crystallisation is a constant identity of one member in a series of pairs, this member thereby being identified as host, and a constancy of structure type (but not necessarily dimensions) for the host. The higher molecular organisation involved in inclusion compounds can allow in some cases the existence of the host lattice with the same structure even in the absence of guest molecules. The occurrence of a *helical canal* topology for the host system is manifestation of an even higher degree of organisation, the sources of which should be examined. An alternative expression of the question here is "what intermolecular interaction mechanisms and energy balances allow some host compounds to form series of helical canal inclusion compounds, and what factors cause disruption of this structure type?"

In the helical canal topology there is some analogy between unimolecular and multimolecular host systems. In the former long host molecules wind around the canal. Generally this host molecule will be a polymer, and the helical canal conformation would be expected to be the result of specific attractive interactions, along the pitch of the helix, between non-sequential monomer residues. A multimolecular helical canal contains corresponding series of distinct host molecules maintained in similar topology by inter-host attractions around and along the helix.

It would be expected that the more energetic and directional of intermolecular forces, namely hydrogen bonds, would be involved as the host-host interactions in helical canals, and this is observed. No helical canal structures are known with only London dispersion forces between host molecules. It can be reasonably postulated that substantial host-host hydrogen bonding is necessary for the formation of a guest-free helical canal lattice. However it is not a sufficient condition, as some very well established non-canal host structures also depend on host-host hydrogen bonding networks. Host-guest hydrogen bonding is relatively rare in helical canal inclusion compounds.

This requirement for strong host-host hydrogen bonding in the maintenance of a helical canal lattice need not be restrictive on the shape or dimensions of the canal. Hydrogen bonding dimensions, particularly angles, are variable, and host-host hydrogen bonding networks can possess a substantial flexibility. The design of helical canal host networks therefore requires understanding and control of inter-host hydrogen bonding.

2.2 Definition and Symmetry of Helical Canals

A helical canal is one in which parts of the wall have the characteristics of a helix, that is a uni-directional coupling of rotation about, and translation along, the canal axis. These helix operations are usually symmetry operations between equivalent objects. However a helix can also arise as a collection of vectors derived from molecular topology. These two types of helicity are represented by the screw and the pitched turbine, respectively. Helical canals commonly occur along crystallographic screw axes, which are symbolised n_m and rotate in increments $2\pi/n$ in conjunction with translational increments m/n of the crystallographic repeat length. The rotational and translational increments are coupled in the right-handed sense. The crystallographic screw axes are 2_1, 3_1, 3_2, 4_1, 4_2, 4_3, 6_1, 6_2, 6_3, 6_4, 6_5. Where $m \neq n/2$ the helices occur as right- and left-handed enantiomeric pairs: n_{n-m} is the enantiomer of n_m because the fractional translation $(n - m)/n$ is equivalent to the supplementary translation $-m/n$ in the opposite direction. The screw axes 2_1, 4_2, 6_3 can be regarded as creating helices which are however not chiral, because the two senses of coupling of the rotation and translation are equivalent.

The turbine type of helicity in canals is a toroidal turbine, and can occur with proper rotation axes and with 2_1, 4_2, 6_3 screw axes, and depends on the object rotated. It is necessary that a distinctive principal plane or axis of the host molecule or molecular fragment be canted in the cylindrical surface of the canal so that it is neither parallel to or perpendicular to the axis of the canal. No unequivocal instance of this type of helical canal has been reported. The cyclodextrin unimolecular hosts [3,4] might be

Table 1. Some crystallographic space groups supporting helical canals [a]

Space group	Descriptive comments	Symmetrical sites on the canal axis
A. Screw axes without intersecting axes.		
$P2_1$		none
$P3_1$ ($P3_2$)[b]	hexagonal array of parallel 3_1 (3_2)[b] screw axes only	none
$P4_1$ ($P4_3$)	4_1 (4_3) axes surrounded only by parallel 2_1 axes	none
$P4_2$	4_2 axes surrounded by parallel 2 axes	none
$P6_1$ ($P6_5$)	6_1 (6_5) axes surrounded by parallel 3_1 (3_2) and 2_1 axes	none
$P6_2$ ($P6_4$)	6_2 (6_4) axes surrounded by parallel 3_2 (3_1)[c] and 2 axes	none
$P6_3$	6_3 axes surrounded by parallel 3 and 2_1 axes	none
$I4_1$	4_1 axes surrounded by parallel 2 and 4_3 axes[d]	none
$R3$	parallel 3_1, 3_2 and 3 axes[d]	none
B. Screw axes with intersecting twofold axes.		
$P3_121$ ($P3_221$)	hexagonal array of parallel 3_1 (3_2) axes in two sets, one pierced by lateral 2 axes at intervals of $c/6$, the other without intersecting axes	none
$P3_112$ ($P3_212$)	as in the previous case, but differing in direction of the 2 axes	none
$R32$	as for $R3$, with screw axes intersected by 2 axes at intervals of $c/6$	32 sites separated by $c/2$
$P4_122$ ($P4_322$)	array of parallel 4_1 (4_3) axes with perpendicular 2 axes at intervals of $c/8$	none
$P4_222$	as for $P4_2$, with 2 axes intersecting 4_2 axes at intervals of $c/4$	222 positions at intervals of $c/4$
$I4_122$	as for $I4_1$, with 2 axes intersecting 4_1, 4_3 axes at intervals of $c/4$	none
$P6_122$ ($P6_322$)	as for $P6_1$ ($P6_5$), with 2 axes intersecting the 6_1 (6_5) axes at intervals of $c/12$, and 2 axes intersecting 3_1 (3_2) axes at intervals of $c/6$	none
$P6_222$ ($P6_422$)	as for $P6_2$ ($P6_4$), with 2 axes intersecting the 6_2 (6_4) and 3_2 (3_1) axes at intervals of $c/6$	222 sites at intervals of $c/3$ along 6_2 (6_4)
C. Screw axes with intersecting screw axes		
$P4_12_12$ ($P4_32_12$)	as in $P4_1$ ($P4_3$), with lateral 2_1 axes at intervals of $c/4$	none

a) These descriptions are not complete because they are intended to portray only aspects relevant to helical canals; the symbols 2 and 3 represent proper twofold and threefold axes;
b) the enantiomorphic space groups and their characteristics are listed in parentheses;
c) note the reversed sense of these screw axes;
d) crystals in this space group can be regarded as a racemic mixture of helical canals

considered in this category, particularly as they are derived from the extended starch screw helix, by excision of one turn, followed by cyclisation, but there is no distinctive section of the glucose unit canted in the cylindrical surface. This example illustrates the point that assignment of turbine helicity is a disputable matter of interpretation of molecular topology, and is not necessarily related to chirality of a toroidal molecule. Cyclodextrins can function as enantiomerically selective hosts, but not because they are helical.

Crystallographic space groups in which screw axes could describe helical canals are outlined in Table 1, according to the occurrence of surrounding symmetry axes and symmetry axes which intersect the screw axes.

Structures which are unambiguously helical in topology can also occur in non-helical space groups. Examples are found in many of the macromolecular helical canals described in Section 7.2 and 7.3 where the space group is $P2_12_12_1$ with non-intersecting 2_1 axes parallel to the orthogonal lattice axes.

2.3 The Locations of Guest Molecules, and Diffraction Measurements

The characteristic of inclusion compounds is the containment of guest molecules in empty spaces in a host lattice with weak host-guest interaction forces and stronger host-host interaction forces. One consequence of this is that the host molecules constitute a regular crystallographic lattice while the array of guest molecules can be, and frequently is, disordered in some fashion. Crystals of inclusion compounds are usually considered and treated as interpenetrating arrays of the two components, host and guest, and then two questions arise: the degree of disorder in the guest lattice; and the degree of registry between host and guest lattices.

When the topology of the inclusion compound involves closed cavities, the guest disorder (type I) can arise by variation of: I(a) guest occupancy of the cavity; I(b) position of the guest in the cavity; I(c) orientation of the guest in the cavity; I(d) vibrational motion of the guest in the cavity. However, where the cavity is extended, as in canal inclusion compounds (and lamellar compounds which undergo intercalation) there are additional degrees of freedom and additional types of disorder. These canal disorders (type II) can arise due to variation of: II(a) the position of guest molecule along the canal, referenced to the host lattice; II(b) separation of guest molecules along the canal; II(c) registry of guest molecule positions in adjacent canals; II(d) translational motion of guest molecules along canals. Indeed, with relatively small and inert guest molecules which are not anchored sterically or electronically to the host canal, canal disorders of types II(a) and II(c) at least are to be expected.

Another factor is pertinent for crystals of helical canal inclusion compounds. The host molecules which comprise the canal helix are usually quite small. Consequently the helix is of relatively high order n (usually 6) and the asymmetric unit of the host canal is then a $360°/n$ angular segment of length $1/n$ times the canal repeat length. For example in the alicyclic diol hosts (Section 3) the asymmetric unit is a 60° sector of length only 1.2 Å; in urea the length of the 60° sector is 2 Å. Consequently, congruence of lattice symmetry between host and guest requires either a guest molecule that is sufficiently small to fit into the asymmetric unit of the helical canal, or a guest molecule that can adopt the symmetry *and* dimensions of the canal helix. Either requirement is likely to be satisfied in exceptional cases only. In general in helical canal inclusion compounds it will be impossible for the guest lattice to be congruent with the host lattice.

These characteristics of crystalline canal inclusion compounds have implications for structure determination by diffraction analysis. The measured diffraction is the sum of the scattering by the host and guest arrays. Irregularities in the guest array lead to diffuseness of their scattering, with concomitant loss in measurable intensity: when the guest contains only light atoms the scattering due to the guest can be very difficult to detect. In canal compounds where there is a guest lattice and it is incommensurate with the host lattice, the guest scattering will not overlap the host diffraction pattern: in closed cavity complexes where the guest lattice corresponds to or is a superlattice of the host lattice there will be some overlapping of the guest and host diffraction patterns. All of this means that the routine use of automatic diffractometers to measure the sharp diffraction patterns of crystals, and computation with automatic structure solution software, are very likely to ignore diffraction information about guest molecules in canals.

The X-ray diffraction observed for crystalline canal inclusion compounds is usually either (i) a sharp pattern due to the host alone [the guest molecules being so disordered (types I and II) and their scattering so diffuse as to be undetectable], or (ii) a sharp host pattern with diffuse streaks due to the guests. The type (i) observation is more prevalent with small guest molecules, while type (ii) usually occurs when the guest molecules are long. The streaks are diffuse along their length and occur parallel to the layer lines on photographs where the crystal is rotated about the canal axis. Streaks of this type indicate that the guest molecules form fairly regular one-dimensional arrays along the canals, but that there is no registry of these arrays between different canals. That is, the guest molecules constitute a set of equivalent one-dimensional sequences which are totally uncoordinated in directions normal to the canals.

Measurements of the streak separations yield the repeat distances of the guest molecules along the canal. These measurements have been made [5], and the results generally correspond with the known length of the guest molecule in extended conformation. See also Section 6.

3 Helical Tubulands formed by Rigid Alicyclic Diols

3.1 Description of the Structures

In 1979 the bicyclic diol *exo*-2,*exo*-6-dihydroxy-2,6-dimethylbicyclo[3.3.1]nonane (*1*) was prepared and observed to co-crystallise with various solvents, including ethyl acetate, chloroform, toluene, dioxane, and acetone. A crystal structure determination of the ethyl acetate compound revealed the occurrence of a helical canal host structure, containing ethyl acetate as guest (with 3:1 diol:ethyl acetate stoichiometry), and that spontaneous resolution had occurred on crystallisation of the multimolecular inclusion compound [6].

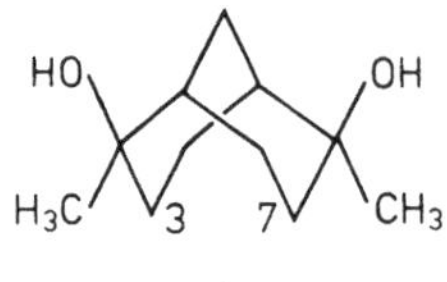

1

2

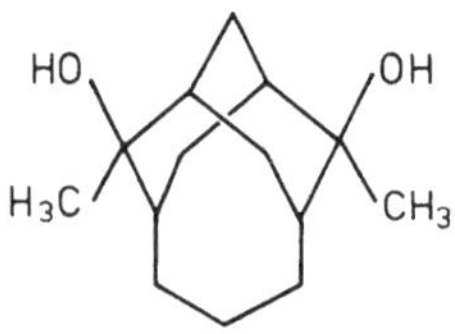

3

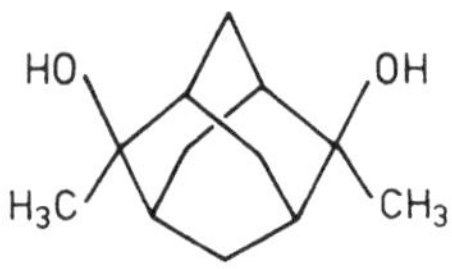

4

5

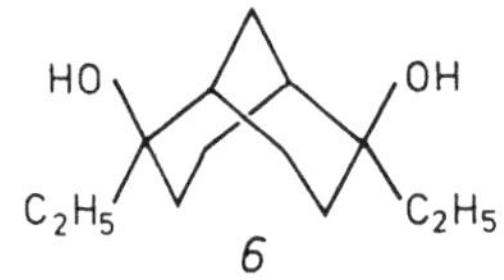

6

7

8

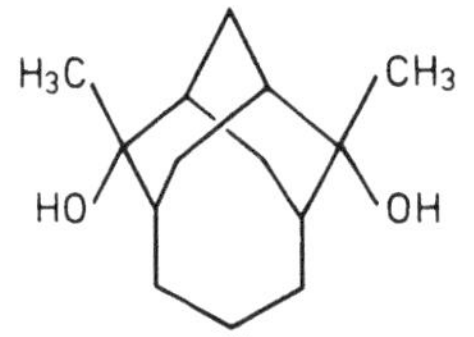

9

10

The space-group of this trigonal lattice ($a \approx 12.2$ Å, $c \approx 7.0$ Å) is $P3_121$ (or its enantiomorph $P3_221$) in which there are two types of threefold screw axes parallel to c. One type of threefold screw axis (at x, y = 0, 0) is intersected by lateral twofold axes, at intervals of $c/6$, specifically two parallel to a at z = 1/3, 5/6 two parallel to b at z = 1/6, 2/3, and two parallel to the ab diagonal at z = 0, 1/2. The canals are coaxial with these 3_1 axes at x, y = 0, 0. The diol molecules which surround the canals straddle the twofold axes at z = 1/6, 1/2, 5/6, and are connected by hydrogen bonds, which occur in infinite helical sequences ... OH ... OH ... OH ... (see Scheme 1) around the other type of 3_1 axes at x, y = 1/3, 2/3 and 2/3, 1/3. The essential elements of the host crystal structure are these helical spines of hydrogen bonds and the diol molecules which radiate from and interconnect the spines: there are six spines around each canal and three canals around each spine. Fig. 1 is a projection of the structure along the c axis, with the spines and canals marked as circles and triangles, while Fig. 2 diagrams the molecular components of one canal in relation to the symmetry elements.

```
   C      C      C      C      C
   |      |      |      |      |
...O–H...O–H...O–H...O–H...O–H...
```

Scheme 1. The infinite sequence of hydrogen bonds connecting diol molecules

```
The sequence of diol molecules around a canal is ...HOC–––COH...OC–––CO
                                                                |     |
                                                                H     H
...HOC–––COH...OC–––CO...HOC–––COH...OC–––CO...(where C–––C
               |    |                |    |
               H    H                H    H
```

represents the diol core and the dots are hydrogen bonds) such that in the sequence

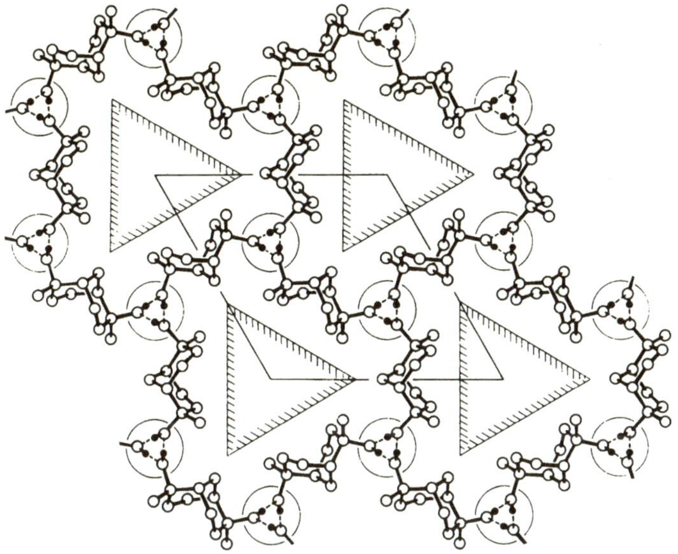

Fig. 1. Projection view, parallel to the three-fold screw axes, of the diol host network. The filled circles and dotted lines represent OH hydrogen atoms and hydrogen bonds, respectively; other hydrogen atoms are omitted for clarity. The hydrogen-bonded spines are circled, and the canals are outlined as triangles [11]

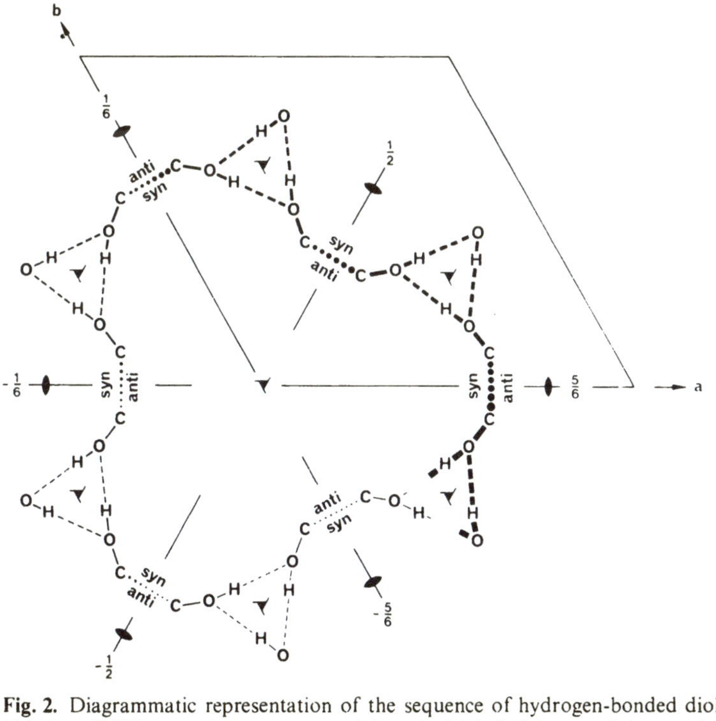

Fig. 2. Diagrammatic representation of the sequence of hydrogen-bonded diol molecules (denoted HOC . . . COH) comprising one turn of the spiral (pitch $2c$) around the canal. Lateral twofold axes at the z-coordinates marked bisect the diol molecules [8]

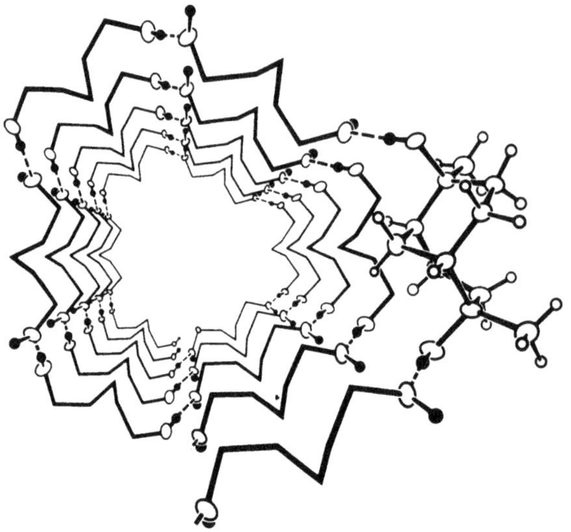

Fig. 3. Exaggerated perspective view of the helical sequence of hydrogen-bonded diol molecules *1* in one canal: all except one diol molecule are represented diagrammatically as the bridge linkage of the two OH groups [6]

each diol functions as a double hydrogen bond donor or a double hydrogen bond acceptor. A complete turn of each spiral chain contains six diol molecules. The pitch of each chain is $2c$, and, as a consequence of the c repetition of diol molecules, the host helix is comprised of *two* separate helical chains of the above type, without *direct* hydrogen bond linkage between the helices. The double helix which envelops a tube is shown diagrammatically in exaggerated perspective in Fig. 3. There is translation of $c/3$ at each hydrogen bond, and the hydrogen bonded spines are single helices of pitch c. The crystallographic asymmetric unit of the host structure in $P3_121$ is half of a diol molecule, with one acceptor and one donor hydrogen bond, while the asymmetric volume of a canal is a 60° sector between twofold axes, with height c.

In this diffraction analysis, electron density due to the ethyl acetate guest was detected, but was not resolvable into individual atoms. This is not surprising, because the $P3_121$ diffraction symmetry observed for the crystal requires that the one guest molecule in the cell be at least sixfold disordered in the canal. It is impossible for even the smallest asymmetric guests to conform with the $P3_121$ symmetry of this host lattice, because the twofold axes which pierce the canal are separated by merely $c/6 \approx 1.2$ Å. Nevertheless, the unobstructed cross-sectional area available to the (disordered) guest species in this inclusion structure is relatively large, being approximately triangular with edge 6.3 Å. Comparative analysis of the sizes and shapes of the guest canals in a series of these host lattices is made below.

This host network, termed the *helical tubuland structure type* [7,8], is unique. The walls of the canals are lined only by aliphatic hydrocarbon, and the hydrogen bonded spines are insulated from the guest canals. Powder diffraction and IR measurements indicate that when *1* is crystallised from acetonitrile the same host crystal structure occurs, but devoid of guest. This has the unusually low calculated density, 1.02 g cm^{-3}.

The melting point of the empty host crystals is 189–191 °C. Therefore it was apparent that this is an independently stable host structure containing large yet chemically inert guest canals, in which guest species cannot interfere with the strong hydrogen bonds which maintain the host network.

A further attribute of this host system is that each crystal contains only one enantiomer of the diol, and is therefore an example of the relatively uncommon conglomerate class of crystals [9].

This discovery was explored further, with emphasis on investigation of the extent to which the host molecule can be modified or substituted in order to modify the shape, size and chemical functionalities of the canal, without destabilising the helical tubuland host structure. It was noted that the C_3–C_7 separation across the open face of *1* is close to that of an ethano bridge. Therefore the homoadamantane derivative *anti*-2, *anti*-7-dihydroxy-2,7-dimethyltricyclo[4.3.1.1^{3,8}]undecane (*2*) [10] was prepared, and found to have the same helical tubuland crystal structure although with appreciably different lattice dimensions (*a* increased by 1 Å to 13.2 Å), and with substantial encroachment of the host ethano bridge into the canal (see Section 3.2 below for details). There is infrared evidence for tight inclusion of small amounts of ethyl acetate in the canals of *2*, but no significant electron density was detected in the diffraction analysis. The canals are now constricted helices, with the capacity for immobile inclusion of small molecules [7].

The homologous compound with a propano bridge, *anti*-2,*anti*-8-dihydroxy-2,8-dimethyltricyclo[5.3.1.1^{3,9}]dodecane (*3*), also adopts the same structure. The crystallographic twofold axis through the molecule is satisfied by disordering of the propano bridge equally between two conformations.

If the C_3–C_7 separation of *1* is reduced by introduction of a smaller CH_2 bridge, that is the dimethyl adamantanediol, 2,6-dihydroxy-2,6-dimethyltricyclo[3.3.1.1^{3,7}]-decane (*4*), the host crystallises from ethyl acetate with a different, non-including, layer structure. The layers are comprised of orthogonal kinked strands as shown diagrammatically in Fig. 4(a), with the molecules connected along the strands and between the strands by cycles of four hydrogen bonds, as in Scheme 2.

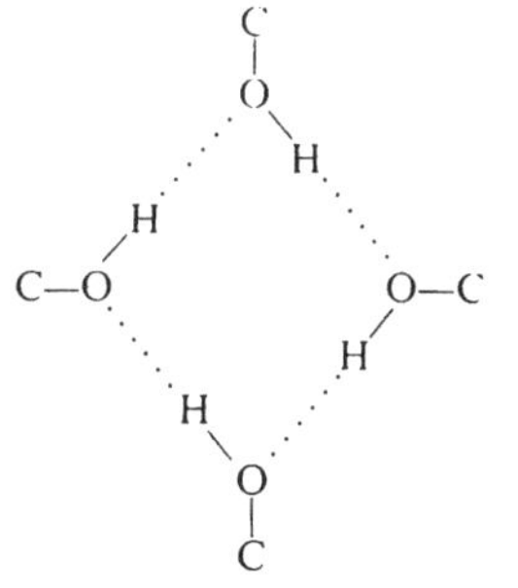

Scheme 2. The cyclic sequence of hydrogen bonds in diol structures

The role of the methyl substituents in the diol host molecule was assessed by preparation of the homologues of *1* with hydrogen [*exo*-2,*exo*-6-dihydroxybicyclo-[3.3.1]nonane, (*5*)] and ethyl [*exo*-2,*exo*-6-dihydroxy-2,6-diethylbicyclo[3.3.1]nonane, (*6*)] substituents [11]. Neither compound showed evidence of inclusion, or forma-

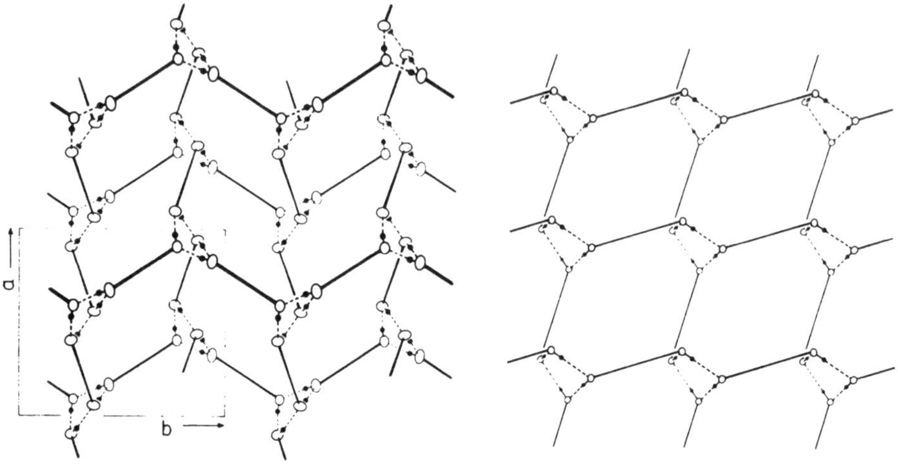

a b

Fig. 4. Diagrammatic representation of layered crystal structures of rigid diols, with closed hydrogen bonding cycles. Open circles are oxygen atoms, filled circles hydrogen atoms, and the solid lines represent the connecting diol. Hydrogen bonds are shown as broken lines. **(a)** The structure of 2,6-dihydroxy-2,6-dimethyltricyclo[3.3.1.1^{3,7}]decane (*4*). **(b)** The structure of *endo*-2,*endo*-6-dihydroxy-2,6-dimethylbicyclo[3.3.1]nonane, (*7*)

tion of low density crystals. The crystal structure of *5* (space group $P2_1/c$) contains Scheme 1 type hydrogen bonding, which can be regarded as a spiral, but there is no significant similarity with the helical tubuland structure. Crystalline *6* is close packed in space group *Fdd*2, with Scheme 2 type hydrogen bonds. It is not altogether clear why *5* or *6* does not crystallise with the helical tubuland structure, or why *1* does not crystallise in the lattices of *5* or *6*. The variation of alkyl substituent does not affect the geometry of the host molecule, and the substituent does not appear to be influencing the directions of the hydrogen bonds relative to the host molecular framework.

A significant variable of the host diol molecule is the configuration of the tertiary alcohol functions, which can be changed at one or both alcohols, with loss or retention of the twofold molecular symmetry, respectively. The double epimers of *1*, *2* and *3* are *7*, *8*, and *9*, respectively. The rationale for expecting that these double epimers with twofold molecular symmetry could crystallise with the helical tubuland structure derives from the role of the diol host as a connector between hydrogen bonded spines, that role being to provide two C—OH functions in rigid mutual orientation. Relative to the diol molecule the essential geometrical characteristics of the pair of C—O bonds in the helical tubuland structures already described are shown diagrammatically in Fig. 5. The C—O bonds lie in opposing planes parallel to the twofold axis of the host, and are directed at approximately 50° to a plane perpendicular to the twofold axis. A similar condition obtains in *7*, *8* and *9*, with the simple difference that the methano bridge is on the opposite side of the molecule. It is necessary to distinguish and define the two faces of the host molecule, relative to the *lattice-determining* directions of the C—O bonds, as the *syn* and *anti* faces as shown in Fig. 5. Thus the effect of double

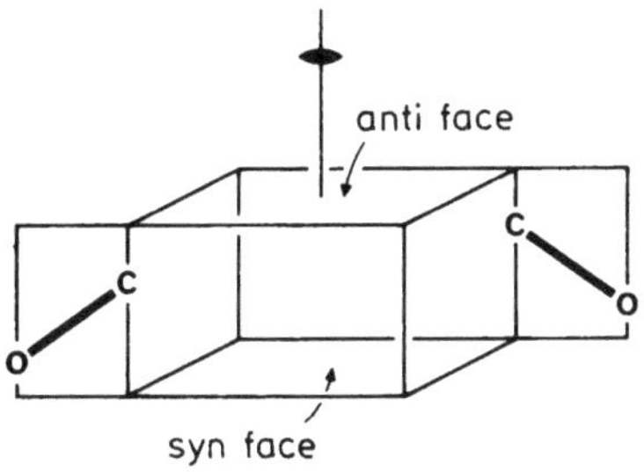

Fig. 5. Diagrammatic representation of key characteristics of the diol host molecules, with two-fold symmetry, C—O bonds in parallel planes, and faces *syn* and *anti* to the pair of C—O bonds [8]

epimerisation of a host molecule is simply interchange of the bridges on the *syn* and *anti* faces. These *syn* and *anti* faces of the host molecules, and their alternation around the host helix, are marked on Fig. 2.

Compound *7*, *endo*-2, *endo*-6-dihydroxy-2,6-dimethylbicyclo[3.3.1]nonane [12], crystallises (from tetrahydrofuran) not in the helical tubuland lattice type, but with a layer structure in which strands of hydrogen-bonded diol molecules are woven as shown diagrammatically in Fig. 4(b). Again cycles of four hydrogen bonds as in Scheme 2 provide the connections within and between the strands, with considerable overall similarity to the layer structure of *4* [Fig. 4(a)]. However, compound *8*, *syn*-2, *syn*-7-dihydroxy-2,7-dimethyltricyclo[4.3.1.1^{3,8}]undecane, when crystallised from ethyl acetate, does adopt the helical tubuland structure [7, 13]. Detailed comparison of the structures of *2* and *8* presented in Fig. 6 confirms the constancy of the orientations of the C—O bonds in the lattice structure, and the consequent inversion of the host molecule faces. The unobstructed cross-sectional area of the canals is large and trilobed (Fig. 6), and again there is no diffraction evidence for (disordered) atom positions for the ethyl acetate which is at least partly included. Powder diffraction data show that inclusion complexes of *8* with carbon tetrachloride, diethyl ether, and o- and p-xylene also possess the trigonal helical tubuland structure. Compound *9*, *syn*-2, *syn*-8-dihydroxy-2,8-dimethyltricyclo[5.3.1.1^{3,9}]dodecane, also adopts the helical tubuland structure, in which the host molecules are inverted relative to the structure of 3. The canal cross-sectional area is now even larger, and hexalobed (Fig. 6).

Therefore both *symmetrical* configurations of the hydroxy functions in the host molecule allow the helical tubuland structure. The *unsymmetrical* epimer *anti*-2, *syn*-7-dihydroxy-2,7-dimethyltricyclo[4.3.1.1^{3,8}]undecane (*10*), the hybrid of *2* and *8*, does not possess a molecular twofold axis (or pseudo twofold axis) or the conformation of C—O bonds of Fig. 5, and would not be expected to fit into the helical tubuland structure framework. Its crystal structure is indeed different, with infinite zig-zag sequences of hydrogen bonds constructing a non-including lattice [14].

Compound *8*, when crystallised instead from benzene, adopts a different lattice structure (space group $I4_1/acd$) which is an achiral multimolecular inclusion compound with benzene guests [13]. This is not a helical tubuland structure maintained by helical spines of host-host hydrogen bonds. The hydrogen bonding connections are cyclic (Scheme 2) such that the host molecules surround hydrocarbon-lined ellipsoidal cavities separated by $c/2$ (9.4 Å) along crystallographic twofold axes parallel to c. The snug fit of the benzene molecules into these cavities, which are connected and rotated along constricted canals, is shown in Fig. 7. Benzene translation along the

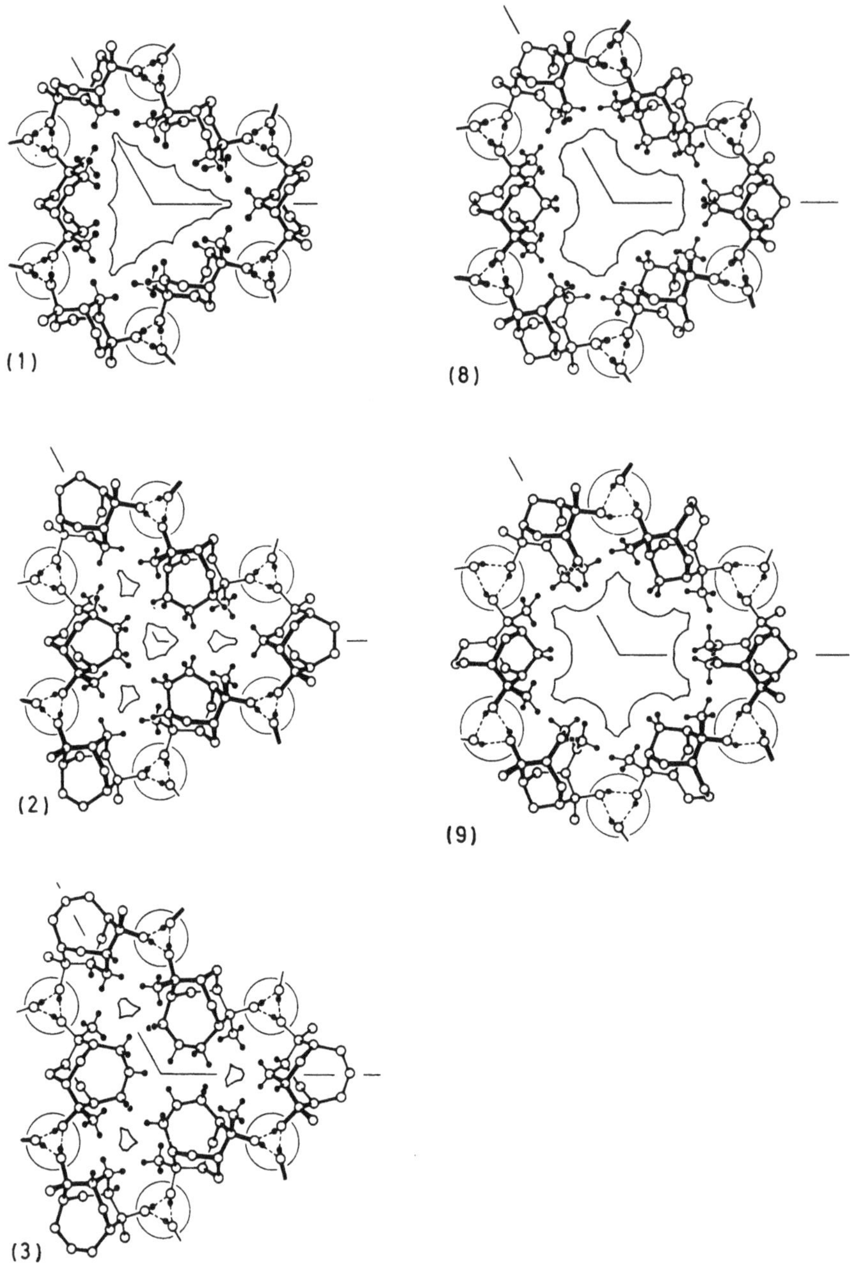

Fig. 6. Comparative projections along the *c* axis of the diol molecules and the canals they enclose in *1*, *2*, *3*, *8* and *9*. The bond thickening signifies depth in individual molecules only, because the helical characteristic is absent from these projections of the lattice. The canal boundaries are marked as the intersecting projected van der Waals spheres of the hydrogen atoms which line the canals. All five diagrams are presented on the same scale. Significant hydrogen atoms are marked as filled circles, and the spines are circled

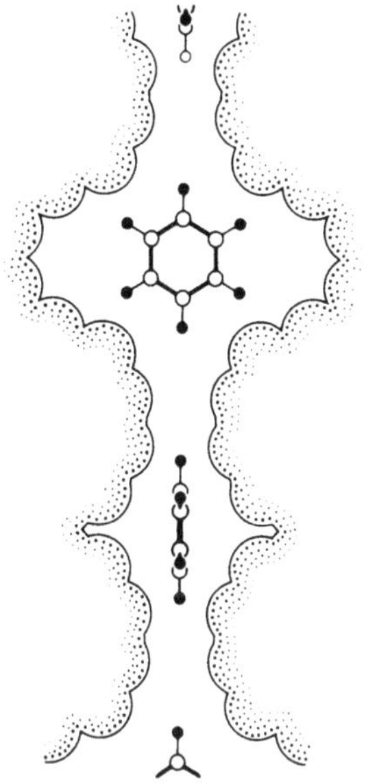

Fig. 7. Cross-sectional representation of the cavities in *8* when crystallised with benzene. The cavities, which are linked along two-fold axes parallel to *c*, are shown as the van der Waals surfaces due to the hydrogen atoms of the host molecules. Benzene guest molecules are shown in their major orientation [13]

canals is not possible. When crystallised from wet acetonitrile, compound *8* crystallises with a layer structure.

The diethyl substituted version of compound *8* similarly does not crystallise with the helical tubuland structure.

3.2 Variables of the Helical Tubuland Structure

In the following analysis of the helical tubuland structure type we distinguish (i) the host variables, and (ii) the lattice variables, and examine the correlations between them [8]. For these purposes we represent the essence of the host molecule as in Scheme 3.

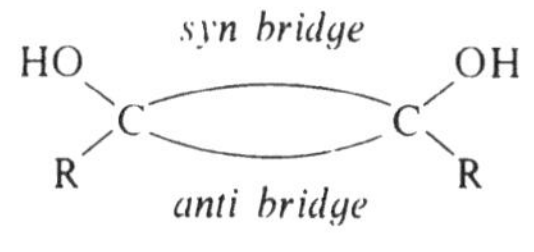

Scheme 3.

The necessary structural feature of the host molecule is clearly the rigid pair of C—OH functions, in *syn* orientation. In all instances of the hosts which form the helical tubuland structure the separation of the two hydroxy functions is almost invariant, ranging only 5.54 to 5.68 Å. The mutual angular orientations of the C—O bonds are assessed by the C′ ... C—O angle, which ranges only 122 to 126°, and by the O—C to C′—O′ torsional angle which is smaller than 90° (73.5°, 79.0°, 71.3°) in *1*, *2*, *3*, respectively, and larger than 90° (94.3°, 97.2°) in *8* and *9*.

Although the substituent R is a variable, it appears so far that R = Me is a necessary requirement. The more important variable features of the host are the *syn* and *anti* bridges. These bridges affect the sizes and shapes of the canals in two ways: one is directly by the extent of protrusion into the canal space; the other is indirectly via intermolecular inter-host interactions in the lattice. These latter interactions between host molecules affect the lattice variables. In order to assess these *lattice* variables we examine the degrees of freedom in the space group. There are two degrees of freedom (in space group $P3_121$) for the host diol molecules on the lateral twofold axes, namely twist about and displacement along these axes. Both changes affect the dimensions of the hydrogen bonds and the dimensions of the canals, and serve to illustrate an observed dimensional variability of the complete host lattice structure, under invariant symmetry.

The imagery of the wire coil mattress has been invoked in description of the essential components of the lattice structure. The coiled springs, representing the hydrogen bonded spines, are interconnected in hexagonal array by stiff linkages (the diol molecules) with some flexibility at the linkage-coil connections (the C—OH ... O hydrogen bonds). This analogy illustrates one form of variability of the lattice structure, namely compression or elongation of the spines concomitant with twisting of the diol molecules about their twofold axes. It has been estimated [8] that the stacking thickness of the diols can differ by not more than ±1 Å from 7 Å without disruption of the hydrogen bonding spines.

In fact there is substantial structural variability among the five instances of the helical tubuland structure described above, in the form of variable placements of the diol hosts along the twofold axes, and variable *a* dimensions of the lattice, both of which have marked influence on the size and shape of the canals. Full analysis of these effects is provided elsewhere [8] and the major results only are reported here. As a consequence of steric repulsions between the bridge on the *syn* face and methyl substituents R on adjacent molecules presenting an *anti* face to the canal, as shown in Fig. 8, the molecules presenting the *anti* face are moved along the twofold axes, further

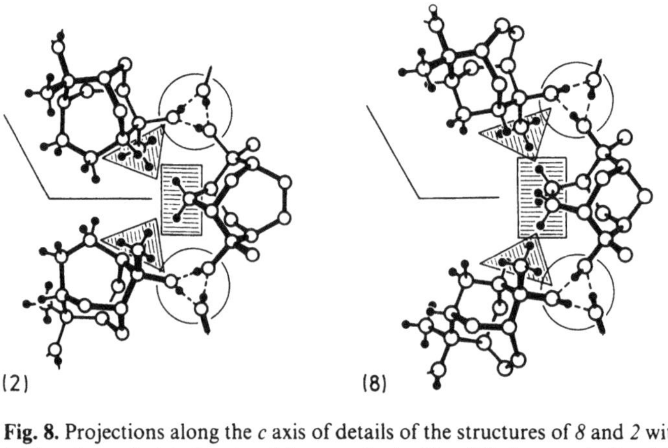
(2) (8)

Fig. 8. Projections along the *c* axis of details of the structures of *8* and *2* with ethano and methano *syn* bridges (cross-hatched rectangle), respectively. The differing proximity of the *syn* bridge to the methyl substituents (cross-hatched triangles) on the flanking *anti* walls of the canal are apparent [8] from the canal centre. These repulsions and shifts are more pronounced in *8* and *9*, causing elongation of the *a* axes, and expansion of the hexagonal area between the spines (and hence expansion of the canal area): the relative areas (a^2) are 1.00, 0.96, 1.03, 1.18, 1.28, for *1*, *2*, *3*, *8*, *9*. In addition the relative positions of the diol molecules along the twofold axes change, as can be seen on the projection diagrams of Fig. 6, with pronounced changes in the shapes of the cross sections of the canals.

Superimposed on these indirect influences on the shapes of the canals are the direct differences due to bridges on the *syn* and *anti* faces. The *anti* faces are generally closer to the axis of the canal and have greater influence on shape. It is significant that all of these effects can be recognised, and probably controlled, without recourse to considerations of the shape of the guest molecule.

3.3 Canal Capacities

Fig. 6 provides projection views of the lattice structure of each of the crystals *1*, *2*, *3*, *8*, *9*, and shows the differences of unobstructed cross sectional area A^{un} available to guests. In *1* $A^{un} = 22.4$ Å^2 and is close to triangular; in *2* there are four thin unobstructed lobes with a total cross-section of only 4.7 Å^2; in *3* even less, 2.7 Å^2; in *8* $A^{un} = 30.2$ Å^2 and is trilobed; in *9* $A^{un} = 34.7$ Å^2, and is hexa-lobed. However it is clear from examination [8] of stereo space-filling diagrams that the walls of the canals in these structures contain substantial indentations generally able to accommodate at least CH_3 or CH_2 groups of guest species, and the projection diagrams do not reveal the full capacities of the canals, and exaggerate the differences between them.

In summary, these rigid diol host molecules generate a unique helical canal inclusion system, maintained by a network of strong inter-host hydrogen bonds, presenting a hydrocarbon-lined canal surface to guest species, and with variable and controllable size and shape of the canal. The direct and indirect effects of the stereochemistry and the bridge identities in the host diol molecules has been evaluated. The significance of the hydroxyl-to-hydroxyl length of the diol, and the full details of the including capabilities of these molecules are under investigation.

4 Urea, Thiourea and Selenourea Inclusion Compounds

The fortuitous discovery that urea forms inclusion compounds with many unbranched organic molecules was first reported by Bengen in 1940 [15]. Structural studies by Schlenk [16] and by Smith [17, 18] determined the now familiar canal structure of these materials. The chemistry of these compounds has been reviewed on a number of occasions [19–21], as have specific applications in chromatography [22], their use in inclusion polymerisation [23, 24], and studies of their thermodynamic [25] and spectroscopic [26] properties.

The crystal structure of urea itself has been studied on a number of occasions using X-ray [27–30] or neutron diffraction methods [31–33] and has the tetragonal space group $P\bar{4}2_1m$. In contrast crystals of the urea inclusion compounds [16–18] are usually in the hexagonal space group $P6_122$.

Uncomplexed thiourea crystallises in the space group *Pnma* [34]. Uncomplexed selenourea crystallises in the trigonal space group $P3_2$ with nine crystallographically distinct selenourea molecules per unit cell [35,36], although it appears that the crystal approaches the higher symmetry of space group $P3_221$. The crystal structure comprises tight threefold spiral chains parallel to *c*, held together by N–H---Se hydrogen bonds.

Thiourea and selenourea both form canal inclusion compounds. The selenourea compounds, reported only briefly [37], appear to be isostructural with the rhombohedral thiourea compounds, although with larger dimensions which are more susceptible to the sizes of guest molecules. For urea, thiourea and selenourea the crystal densities of the compounds are significantly less than those of the hosts: values (g cm^{-3}) for the pairs host/complex are 1.30/1.20; 1.40/1.10; 2.08/1.60–1.65 for urea, thiourea, selenourea, respectively [37].

4.1 Description of the Canal Structures of Urea and Thiourea Inclusion Compounds

In the normal inclusion compounds of both urea and thiourea there are canals with hexagonal cross section. The urea/thiourea molecules are almost coplanar with the walls of the hexagonal prism, and the prisms are packed into a distinctive honeycomb lattice. The lattice is maintained by the maximum possible number of hydrogen bonds: two donor by each NH_2, and four acceptor by each O or S atom. All hydrogen bonds also lie virtually within the walls of the hexagonal prisms. Therefore both host structures are remarkable and unique in that all host molecules and inter-host hydrogen bonds project onto an hexagonal host array with minimal cross sectional area, and with an unusually large proportion of the projection area available to guest species.

Despite these similarities, and similarities in local hydrogen bonding connections and geometry, there are significant differences in crystallographic symmetry between the urea and thiourea host lattices. The urea host structure [18] has the hexagonal space group $P6_122$ (labelled $C6_1$ in Smith's paper [18]) in which a 6_1 screw axis passes along the axis of the canal. The urea molecules occur in columns, and radiate from the 3_1 screw axes, with the O atoms close to the axis. Additional symmetry elements are 2_1 screw axes parallel to and midway between adjacent 3_1 axes, and lateral twofold axes pierce the canals and the 6_1 axis at intervals of $c/12$.

Figure 9(a) shows the arrangement of urea molecules and hydrogen bonds. The sequence of urea molecules around the canal axis constitutes a triple helix.

In contrast, the thiourea host crystal structure is centrosymmetric in the trigonal space group $R\bar{3}c$. The thiourea molecules radiate from threefold screw axes, which alternate left- and right-handed around the canal (this is the main difference from the urea complex structure, where the threefold screw axes are all of the same hand). The axis of the thiourea canal is a threefold (instead of 6_1) axis, with sites of $\bar{3}$ (S_3) and 32 (D_3) point symmetry alternating along the canal separated by $c/4$. Figure 9(b), analogous to Figure 9(a) for the urea structure, shows the arrangement of thiourea molecules and the $\bar{3}$ and 32 sites along the canal. The important result is that the thiourea canal is not helical, and not chiral.

Canals in the urea inclusion compounds have a diameter of approximately 5.25 Å.

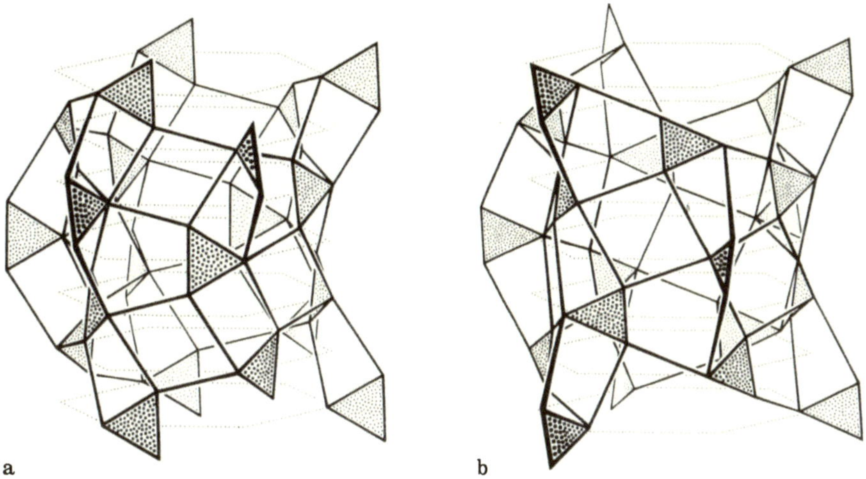

a b

Fig. 9. Schematic representations of the canal structures of **(a)** urea and **(b)** thiourea, drawn to emphasise the similarities and differences between them. Each triangular host molecule is denoted by a stippled triangle, with the NH_2 functions at the ends of the vertical edge. Oxygen or sulfur atoms lie in layers, outlined by the dotted hexagons. The thick inter-host lines are hydrogen bonds: each O or S atom is involved in four hydrogen bonds, and each NH_2 function is involved in two

Linear guest molecules are included along these canals in an extended planar zig-zag conformation. Branched molecules are generally excluded unless branching occurs near the end of a long linear chain, but aromatic derivatives can be included if they have a long alkyl chain [38]. The review article by Takemoto and Sonoda [21] contains an excellent survey of the types of molecules known to form urea inclusion compounds and of the means used to study their detailed conformations and thermal motion.

4.2 Thermal Changes and Structural Modifications of Urea Inclusion Compounds

Formation of the urea inclusion compounds is exothermic to the extent of about 6.7 kJ mol^{-1} per carbon atom of the guest molecule [39]. On heating, the hexagonal inclusion compounds decompose below the melting point of urea, which then regains its tetragonal structure [40]. Investigating the C_8–C_{12} range of paraffins, Dyadin [41] has determined values of the energy of activation for decomposition of the C-odd hydrocarbon complexes at 155–159 kJ mol^{-1} and of the C-even ones at 134 to 163 kJ mol^{-1}.

Chatani [42,43] has shown that if the hexagonal inclusion compounds are cooled many of them undergo a reversible transformation into a low-temperature form with the orthorhombic space group $P2_12_12_1$. These phase transitions take place over a narrow temperature range and are the consequence of deformation of the original hexagonal canals caused by the guest molecules adopting definite orientations in the canals. The reduction in symmetry occurs only in the plane normal to the canal axis, and the host-host hydrogen bonds are not disrupted. Similar results have been reported with the urea-polyethylene (av. MW 59,000) inclusion compounds [44,45]. The urea inclusion compound with 1,3-butadiene at low temperature also has the space group $P2_12_12_1$ but after γ-irradiation the resulting polymer inclusion compound is hexagonal at room temperature [46].

Lenne and Schlenk [38,47,48] have shown that urea can also give inclusion compounds in the rhombohedral space group $R\bar{3}c$ with α,ω-disubstituted alkanes 15 to 39 Å in length as guests. These results are particularly interesting since this is the crystal arrangement generally favoured by inclusion compounds of thiourea. The urea-vinyl chloride complex, studied at reduced temperature, also had a canal structure of a rhombohedral nature. Inclusion polymerisation of the monomer caused disintegration of the crystal lattice with formation of polyvinyl chloride and tetragonal polycrystalline urea [49]. The urea inclusion compound with trioxane is also rhombohedral at room temperature [48], and distorts to lower symmetry in the plane normal to the canal axis at low temperatures [50].

Yet another type of canal structure has been reported for the urea inclusion compound of 1,4-dichlorobutane [51]. Even though the canals are pseudo-hexagonal in dimension, there is a significant difference in their symmetry. The host lattice is orthorhombic, space group *Pbcn*. The difference lies in the directions of the six pseudo-3_1 helices of host molecules around the walls of each canal: the sequence is *cyclo*-RRRLLL, as opposed to *cyclo*-RRRRRR in the hexagonal inclusion compounds and *cyclo*-RLRLRL in the rhombohedral. This orthorhombic host structure probably occurs also in the urea inclusion compounds with 1,5-dichloropentane and 1,6-dibromohexane [51].

A considerable amount of work has been directed towards the study of detailed molecular orientations and motions of guest molecules in urea canal inclusion compounds and structural changes such as those described above. Methods used include infrared [52] and Raman spectroscopy [52,53], esr [54-56], nqr [50,57], and nmr (^{1}H, ^{2}D, ^{13}C) [57-59].

4.3 Exploitation of Urea Inclusion Compounds

Well-developed applications [19-21] of urea inclusion compounds have been in the areas of detergents, dewaxing and petroleum chemistry. Some recent novel advances include the use of the peroxydodecanoic acid-urea inclusion compound in a laundry bleach product [60] and in the purification of insect pheromones [61,62].

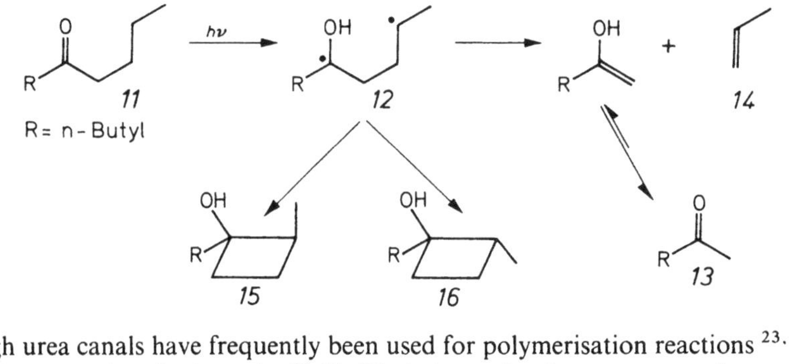

Although urea canals have frequently been used for polymerisation reactions [23,24] and investigation of radical species [26,54-56] their applications as an environment for other types of reaction has hardly been exploited. Recently de Mayo and Scaiano [63] have studied the photolysis of saturated ketones such as 5-nonanone (*11*). Normally, in solution, the principal reaction is Norrish Type II photodecomposition yielding 2-hexanone *13*, propene *14* and the two isomeric cyclobutanols *15* and *16*. This reaction also took place on irradiation of the urea inclusion compound, but along with *13* and *14* essentially only *one* cyclobutanol isomer (*15*) was formed.

Observation of the Norrish Type II reaction presents some difficulty in that generation of the biradical intermediate *12* requires a six-membered transition state and this is in conflict with the linear guest arrangement normally expected in the channel. However, as noted earlier, accommodation of planar six-membered rings in urea inclusion complexes has been observed [38]. It appears that in this case the necessary six-membered transition state can be produced in the channel without destruction of the crystal structure.

4.4 Structural Details of Thiourea Inclusion Compounds

Thiourea canal inclusion compounds [19–26] have a wider diameter than those formed by urea, such that n-alkanes are not included but that molecules of cross-section approximately 5.8–6.8 Å are trapped [64]. Thus many inclusion compounds have been reported between thiourea and branched alkanes or cyclic molecules. Of special interest are the inclusion compounds with cyclohexane derivatives and the recent studies carried out on the preferred conformation(s) of the ring in the restricted environment of the thiourea canal.

The chloro- and bromo-cyclohexane inclusion compounds have been extensively examined by infrared (4000–30 cm^{-1}) spectroscopy [65–68] and by Raman ($< 1000\ cm^{-1}$) spectroscopy [69]. In the canals both guests are found to exist exclusively in the chair conformation with an axial halogen substituent, while iodocyclohexane [65,68,69] adopts both axial and equatorial conformations in the canal. These results should be contrasted with the familiar situation in the liquid phase where the equatorial arrangement is the lowest energy conformer and is present to the extent of about 65–70% at room temperature.

It is believed that equatorial substituents such as chlorine or bromine would increase the guest diameter beyond the allowed values (assuming that the guest molecules stack roughly parallel to the canal [68]). Support for this comes from the study of fluorocyclohexane where the population of the axial conformer is not enhanced to any major extent [70]. Nitro- [71] and cyano-cyclohexane, *trans*-1,2-dichloro-, *trans*-1,2-dibromo-, *trans*-1,4-dichloro-, *trans*-1,4-dibromo-, and *trans*-1-bromo-4-chloro-cyclohexane all pack most efficiently in the thiourea canals as the axial or diaxial conformer [68,72]. *Trans*-2,3-dichloro-1,4-dioxane behaves similarly [73]. In contrast isocyanato-, *trans*-1,4-diiodo-, *trans*-1-bromo-4-iodo-, and *trans*-1-chloro-4-iodo-cyclohexane are present as mixtures of the axial/equatorial or diaxial/diequatorial conformations as appropriate [68,72]. The reason for this anomalous behaviour of the iodosubstituted cyclohexanes is not clear.

It should be noted that equatorial or diequatorial conformers of substituted cyclohexanes could be accommodated in the thiourea canals by the stacking of guests at an oblique angle or parallel to the canal axis. X-ray structural data is so far unavailable on these systems. In the light of the recent results of Gerdil and Frew [74] for the cage complex of tri-o-thymotide with chlorocyclohexane such studies would be well worthwhile provided that the guest selected is sufficiently ordered in the thiourea canals.

The thiourea-cyclohexane inclusion compound was originally investigated by Lenne using X-ray diffraction methods [75]. Although this allowed the host structure to be determined in detail, the guest molecules were disordered. The guest orientations and behaviour have recently been investigated by Clement et al. [76] and by Meirovitch et al. using ^2H nmr methods [77], over a range of temperatures. Above 270 K rapid ring inversion occurs.

Metallocenes (but only with unsubstituted cyclopentadienyl rings) also form thiourea inclusion compounds. X-ray diffraction and Mössbauer data on the ferrocene inclusion compound show that at 295 K the guest molecules are orientationally disordered at the sites of 32 symmetry. A phase change at 162 K generates a more ordered structure [78].

Esr [79–81] and ^{13}C MAS nmr methods [82] have also been used in the study of the thiourea inclusion compounds in addition to the techniques already described in detail.

4.5 Exploitation of Thiourea Inclusion Compounds

Size and conformational effects such as those described above have been exploited in a wide range of separation procedures. In a patented method for the separation of *cis-* and *trans-*1-methyl-4-*iso*hexylcyclohexane carboxylic acids only the *trans* isomer forms a thiourea inclusion compound [83]. In solution the *cis* isomer always has an axial, equatorial arrangement in the chair conformation, while the *trans* isomer can exist as the diequatorial or diaxial chair conformers. It is not reported which of these is favoured or how the guest packs in the canals.

Thiourea inclusion compounds have also found recent applications as diverse as the separation of liquid crystal isomers [84], isolation of petroleum constituents [85,86], and the recovery of squalene during olive oil refining [87], to cite just a few examples.

The use of thiourea canals to conduct polymerisation reactions has been the subject of several recent reviews [21,23,24] and consequently only new directions of investigation will be discussed here. One of the monomers frequently studied has been 2,3-dimethylbuta-1,3-diene. Recently the crystal structure [88] of the thiourea canal inclusion compound of this monomer has been determined at -130 °C and found not to have the usual rhombohedral structure but instead to have the monoclinic space group $P2_1/a$. The host molecules enclose hexagonal shaped canals where the diene molecules are situated in a longitudinal fashion. Since the cross-section of the guest across the canal is roughly 4×7 Å, the resulting canal is markedly distorted from a regular hexagon.

Ichikawa and Nakao [89] have investigated the additive effect of 2,3-dimethylbutane on the polymerisation of 2,3-dimethylbuta-1,3-diene in thiourea canals. Although yields were reduced sharply, polymerisation was found still to occur in the presence of surprisingly high percentages of alkane. Molecular weights were reduced in this procedure; for example the maximum yields of dimers were obtained with an 80% alkane content.

Recently Sergeev et al. [90,91] have developed a low temperature condensation method for the formation of inclusion compounds of thiourea with reactive and volatile guests, avoiding the use of solvents. The two guests in the joint inclusion compound of thiourea with 1,3-cyclopentadiene and maleic anhydride underwent Diels-Alder addition at 170 K. These two substances do not react at this low temperature unless they are present in the thiourea complex: the usual *endo* isomer of the product is formed. Apart from copolymerisation reactions this appears to be the first use of the thiourea canal to study reactions between different materials.

5 Deoxycholic Acid and Derivatives

Deoxycholic acid (DCA) (*17*) and apocholic acid (ACA) (*18*) are typical examples of the bile acid family of materials, but with the unique property of forming inclusion compounds with a wide variety of guest molecules [92]. Partly due to the *cis* ring junction between rings A and B, and partly due to the conformation of the steroidal side chain these compounds present a convex hydrophobic β-face and a concave hydrophilic α-face, as shown for DCA (*19*), a classical aid to the formation of inclusion compounds [93].

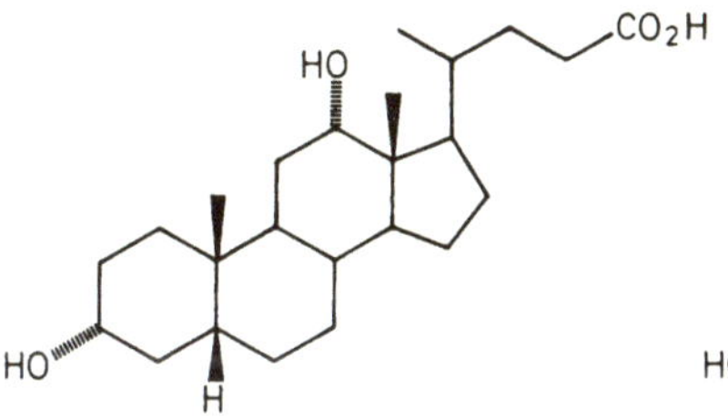

17: Deoxycholic acid (DCA)

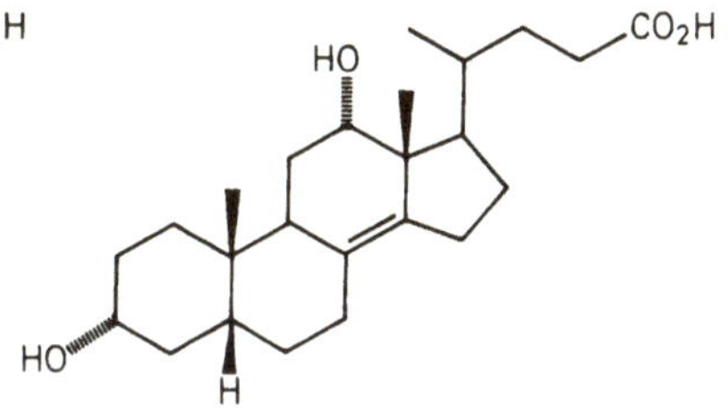

18 : Apocholic acid (ACA)

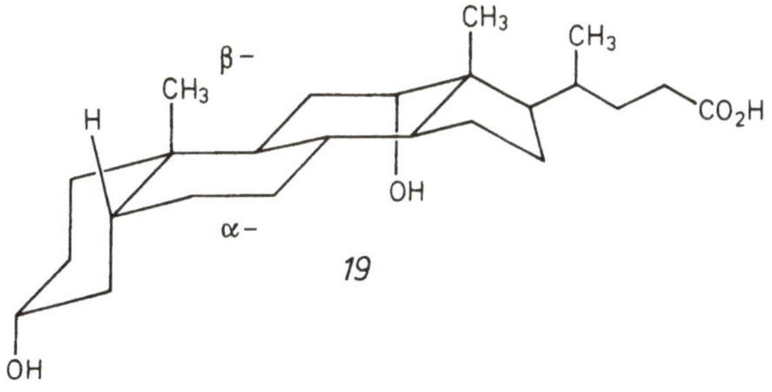

19

DCA forms canal inclusion compounds, known as choleic acids, which most frequently have the orthorhombic space group $P2_12_12_1$, or less frequently $P2_12_12$. In such crystals the DCA molecules hydrogen bond to each other to form an extended bilayer structure, thereby creating a hydrophobic canal between adjacent bilayers. The guest molecules present in these canals therefore tend to be non-polar or moderately polar molecules such as aromatic compounds, alkenes, ketones and certain carboxylic acids [92]. Since the bilayers are held together only by van der Waals forces the canals are able to adopt different dimensions to accommodate the variety of guest molecules. However such canals are not helical and the crystal space group of such compounds is achiral. Two recent crystal structure determinations which are typical of this class of inclusion compounds are those of DCA with quadricyclane [94] and with pinacolone [95].

With smaller, more polar molecules tetragonal ($P4_12_12$) and hexagonal ($P6_5$) crystal systems may be produced. In the former systems the guest molecules such as water and ethanol are present in narrow canals running inside the bilayers and hydrogen bonding with their hydrophilic interior. It is especially interesting that DCA can include a mixture of ethanol/water either as the tetragonal system (DCA:ethanol:water = 2:1:1) or as the hexagonal system (3:2:1). The latter structure is favoured at higher temperature [92].

When crystallised from hydrogen bonding liquids, DCA can form a helical tubuland structure with a helical canal of roughly circular cross section (diameter 4 Å). The space group is $P6_5$ (or $P6_1$) with a sixfold helical sequence of DCA molecules along an axial repeat of 18.7 Å, and a sixfold helical sequence of guests inside the host helix. There is host-host and host-guest hydrogen bonding in this structure. Two isomorphous cases only have been reported: one in which the guest molecules are ethanol and water in ratio 4:2 [96] and a second with dimethyl sulfoxide (DMSO) and water 3:3 [97]. The structure is shown diagrammatically in Fig. 10, emphasising the hydrogen bonding sequences. The outer sequences link three different host molecules, while the inner helix of hydrogen bonds zig-zag between host and guest oxygen atoms. In

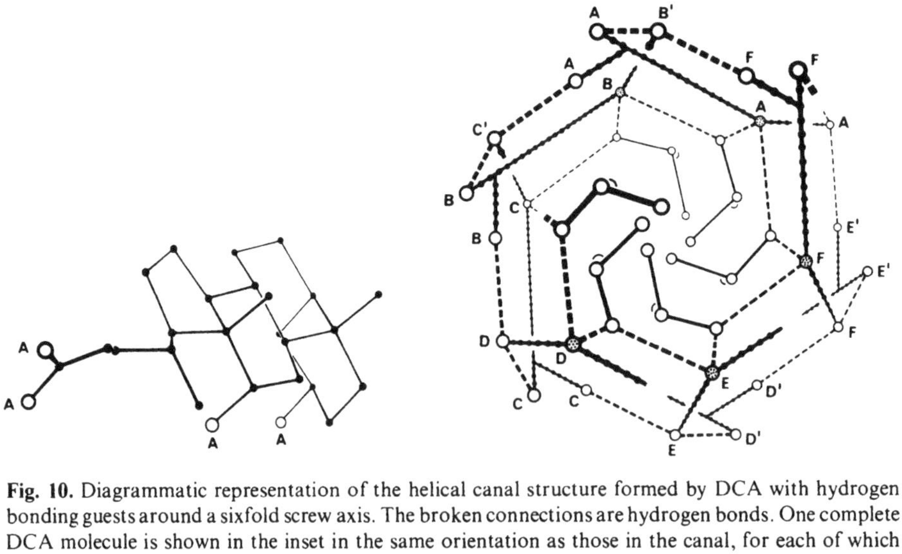

Fig. 10. Diagrammatic representation of the helical canal structure formed by DCA with hydrogen bonding guests around a sixfold screw axis. The broken connections are hydrogen bonds. One complete DCA molecule is shown in the inset in the same orientation as those in the canal, for each of which only the four oxygen atoms of the two hydroxy groups and the carboxylic acid function are drawn. The letters A to F identify the groupings of oxygen atoms for each of the six DCA molecules, with primed letters signifying unit cell translation along the canal. It can be seen that each DCA molecule extends around a ca. 100° are of the canal, with the consequence that the sets of hydrogen bonds around the periphery connect three different DCA molecules. Hydrogen bonds in the central helix zig-zag between host and guest oxygen atoms. Guest molecules may be DMSO (with the atom positions marked), ethanol (without the radial bond to the centre), or water

the water/ethanol crystal each guest provides one donor and one acceptor function, while in the water/DMSO crystal the guests alternate as double donor, double acceptor.

The helical columns are packed in hexagonal array, with slight interleaving of host molecules separated by $c/2 = 9$ Å. The cavities between the columns are too small to accommodate guest molecules.

This structure is unique, because the hydrogen bonding side of the DCA host faces the inside of the helix, and the outer surfaces of the columns are hydrophobic. Thus this structure is the inverse of the helical canal structure formed by the rigid diols, described in Section 3, where the hydrogen bonding spines are external to a hydrophobic canal. The DCA molecule is not unlike the diol host in being able to hydrogen bond at each end.

It does not appear to be likely that this structure type will occur with hydrogen bonding guests much larger than ethanol. The pitch of the guest helix is ca. 3.2 Å, which requires fairly tight packing of the guest tails, and is probably the reason for the co-inclusion of the smaller water guest molecules. The host helix cannot expand without disruption of the hydrogen bonds which maintain it, and reversion to the pleated bilayer inclusion structure.

Over the last few years the DCA and ACA compounds have been the subject of a number of interesting chemical developments including their use as templates in polymerisation [23,24], and as media for stereospecific chemical reactions [98–100]. This work has been carried out on the orthorhombic systems and is therefore outside the scope of this review.

6 Tri-o-thymotide

Unsolvated tri-o-thymotide (TOT) (*20*) crystallises in the orthorhombic space group $Pna2_1$ [101,102] but forms inclusion compounds with an extremely wide range of organic materials (> 100) in different arrangements [103].

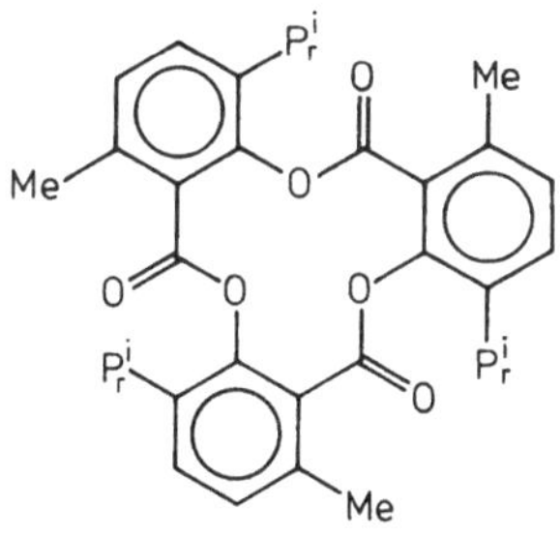

20: Tri-o-thymotide (TOT)

Both cage and canal structures are commonly produced, with the structure adopted almost depending entirely on the size of the guest molecule [5]. Guests less than 9.5 Å in length generally lead to formation of cage inclusion compounds with the trigonal structure $P3_121$. Recently reported crystal structures of this type include the inclusion compounds of TOT with 2-bromobutane [104], ethyl methyl sulfoxide [105], (*R*)-2-butanol [106], chlorocyclohexane, bromocyclohexane [8] and 2-chlorotetrahydropyran [107].

Guests over 9.5 Å in length give rise to canal structures, most commonly with space group $P6_1$ [5,102], but on occasion with $P6_2$ [5], $P3_1$ [5], or $P\bar{1}$ [108–110]. Recently inclusion compounds with other crystal structures (including $P2_1$, *Pbca*, *Pbcn* and *C2/c*) have been found but detailed structures have not yet been reported [74,111].

Early studies by Powell and Lawton [5] of the canal compounds formed by tri-o-thymotide with graded ranges of guest species illustrate the role of diffraction analysis for helical canal inclusions. TOT does not form canals in the absence of guest species, or with guest species less than 9.5 Å in length, and therefore the structure and symmetry of the TOT canal independent of the guest species cannot be determined directly. Diffraction photographs of canal type inclusion compounds reveal approximately equi-dimensional unit cells with composite space groups $P6_1$ ($P6_5$), $P6_2$ ($P6_4$), or $P3_1$ ($P3_2$). The symmetry of the TOT canal was cleverly determined to be $P6_1$ ($P6_5$) in an experiment in which the crystalline $HgEt_2$ inclusion compound, with symmetry $P6_2$, was observed by diffraction measurements to change to space group $P6_1$ as the guest molecules decomposed. The sixfold helical structure of TOT with cetyl alcohol guest has been determined [102].

Inclusion compounds of TOT with n-alkyl bromide and iodide guests extending from C_6 to C_{18} also display symmetry $P6_1$ in the sharp diffraction peaks, whereas it is impossible for the guests to adhere to this symmetry. These crystals, with relatively strong scattering from the guests, also manifest diffuse layer streaks on *c*-axis oscillation photographs, indicating that the crystal acts as a collection of one-dimensional diffraction gratings. The one-dimensional repeat distances measured from the layer streaks are very close to the calculated lengths of the guest molecules. Therefore the canals are filled with alkyl halide guests, without any registry of adjacent canals (consistent with definite separations of the canals by the host molecules). The host alone contributes to the sharp $P6_1$ diffraction, and the guests contribute to the diffuse diffraction.

The inclusion crystals with composite diffraction symmetry $P3_1$ occur for guests (such as hexane, n-$C_5H_{11}I$, $C_4H_9OCH_3$) where the guest length is close to one-third of the host canal repeat length. In these cases there is some degree of longitudinal ordering of guests along the canals, with three guest species head to tail for each complete turn of the host 6_1 helix. Powell and Lawton propose that large atoms such as Br or I in guest molecules may fit into the canal in a limited number of ways (i.e. lateral looseness or disorder is precluded or restricted), thereby reducing longitudinal disorder and anchoring a three-dimensional periodicity for the guest lattice. The composite host plus guest structure has $P3_1$ symmetry.

In common with the situation for DCA, the helical canal inclusion compounds of TOT have so far received little attention compared to the research activity on urea. Much interesting work has been carried out on TOT inclusion compounds with other crystal space groups including the study of monomeric carboxylic acid guests [111], the inclusion of thermodynamically disfavoured conformers of guests [74], and the photochemical reactivity of molecules [110,112]. These are discussed in a recent review [103]. One property which is relevant to this discussion of helical canals is that of enantiomer recognition. The 12-membered ring of TOT allows considerable conformational flexibility in solution. An enantiomeric pair of propeller conformations (all three carbonyl groups aligned on one side of the ring) and an enantiomeric pair of helical conformations (two carbonyl groups aligned on one side of the ring and the third on the other) equilibrate with each other at room temperature. The propeller conformer is the major one. For example in pentachloroethane at 68 °C the composition has been estimated as 86% propeller and 14% helix [113].

In the solid phase only the propeller conformation has so far been observed. Unsolvated TOT is racemic, but the cage $P3_121$ inclusion compounds of TOT with racemic solvents involve spontaneous resolution producing a conglomerate [9]. Any single crystal of the complex thus contains TOT propeller conformers of the one chirality, and these impose varying degrees of enantiomer recognition on the racemic mixture of potential guest molecules to an extent depending on the particular host-guest interaction. Typical results with $P3_121$ cage inclusion compounds have resulted in enantiomeric excesses of 2–47% [114,115]. If the solvent is optically pure then the $P3_121$ complex produced is also optically pure. The guest molecules are able to control the chirality of the TOT propeller since the left- and right-handed forms are in rapid equilibrium in solution.

Experiments on chiral discrimination using helical canal inclusion compounds of space group $P6_1$ have so far shown only low selectivity. The inclusion compounds of TOT with racemic 2-chlorooctane, 2-bromooctane, 3-bromooctane, 2-bromononane and 2-bromododecane gave enantiomeric excesses of 4–5% [114,115]. However it should be noted that if an enantiomerically enriched guest solvent is used then the enantiomer ratio in the complex increases markedly over the figure obtained when racemic solvent is used. Thus, for example, an enantiomer ratio in solution of 75:25 (50% optical purity of 2-bromooctane) gave an enantiomer ratio in the inclusion compound of 90:10 (80% optical purity) [98,115]. A possible explanation of this novel behaviour is that as the canal structure grows, guest-guest interactions are important in enantiomeric selection of potential guest molecules in the crystal. Such behaviour would then contrast with development of a cage structure where host-guest interactions would be the dominant factor and guest-guest interactions are not possible because the hosts are separators [115].

An alternative means of obtaining high guest optical purities is simply to add a powdered single crystal of the TOT inclusion compound to a saturated solution of TOT in the racemic solvent. Thus, use of the resolved $P6_1$ TOT/2-bromooctane inclusion compound as a seed gave polycrystalline material with an enantiomeric purity of 85% [115].

7 Polymeric Systems

7.1 DNA Intercalation

It has been realised for many years that planar aromatic molecules frequently have significant biological activity, some being mutagens and others being of value in chemotherapy. The genetic implications of the double helical structure of deoxyribonucleic acids (DNA) [116] quickly led to investigation of the nature of the interaction between these two disparate molecular structures, and the realisation that two major types of binding were involved. One was essentially due to electrostatic attractions while the other mode (with a much greater binding constant) gave rise to the biological effects. In 1961 Lerman [117] proposed that the available data were best explained by the intercalation of the planar molecule between adjacent base pairs [118,119].

The exploitation of this phenomenon has led to significant drug developments in several different pharmacological areas. Molecules such as ethidium bromide (*21*) and 9-methoxyellipticene (*22*) have antitrypanosomal activity, whereas quinine (*23*), quinacrine (*24*), and chloroquine (*25*) are antimalarials. All of these are known to intercalate with DNA. However the most exciting group of DNA intercalating drugs are those with antitumour activity, such as actinomycin D (*26*) [120], adriamycin (*27*) [121],

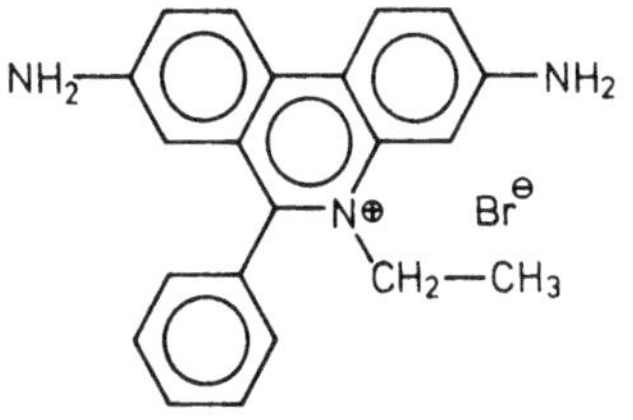

21: Ethidium bromide

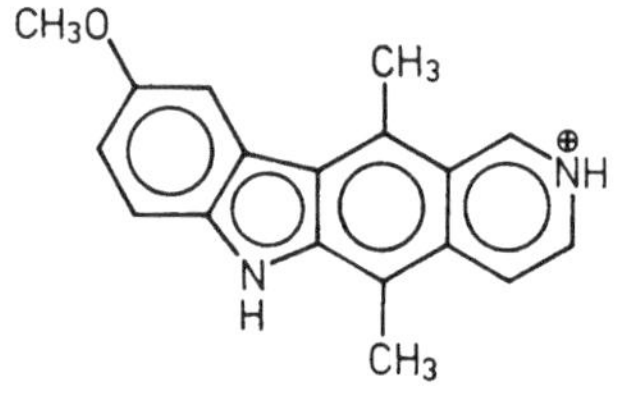

22: 9-Methoxyellipticine

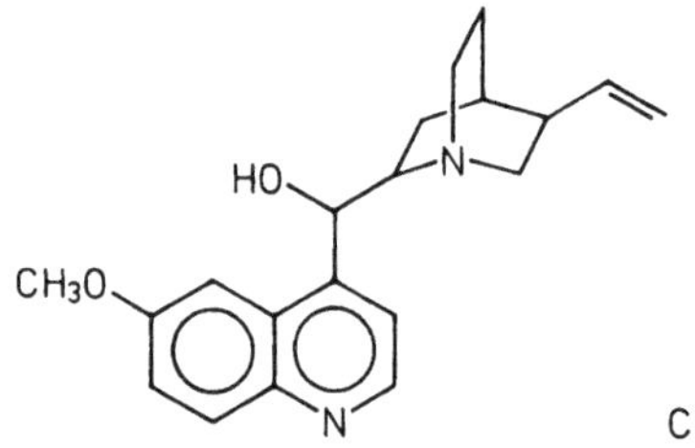

23: Quinine

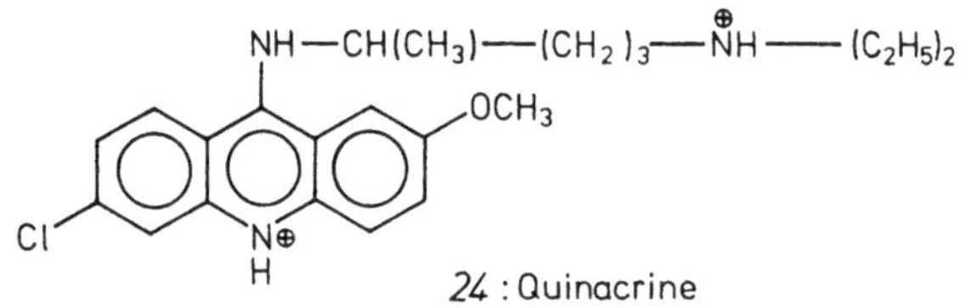

24: Quinacrine

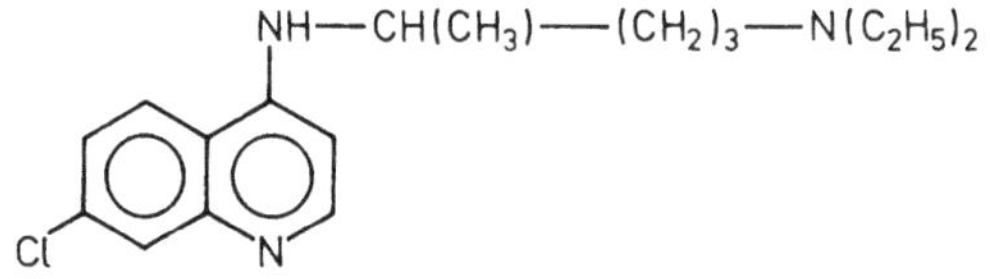

25: Chloroquine

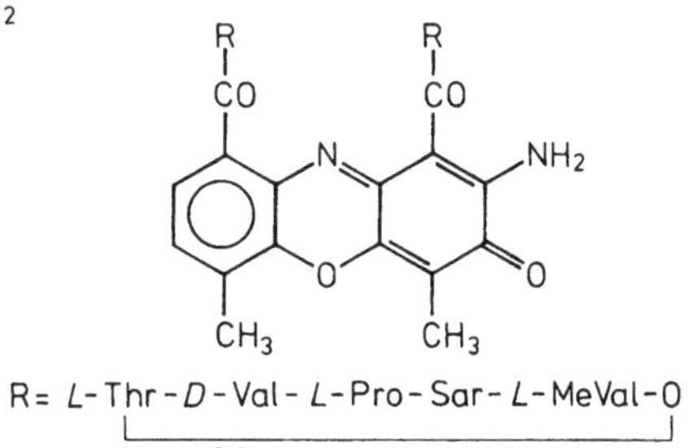

26: Actinomycin D

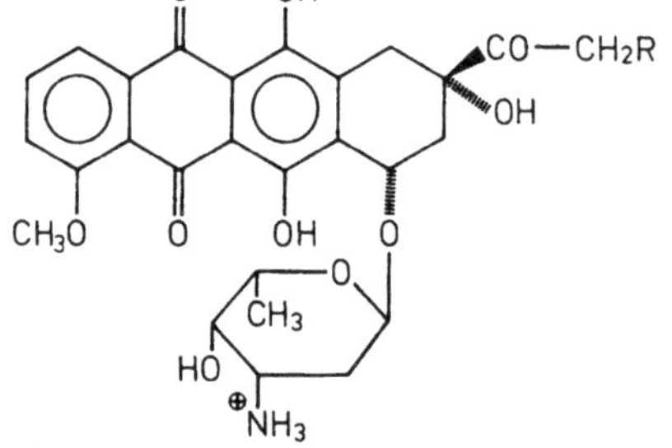

27: R = OH Adriamycin
28: R = H Daunomycin

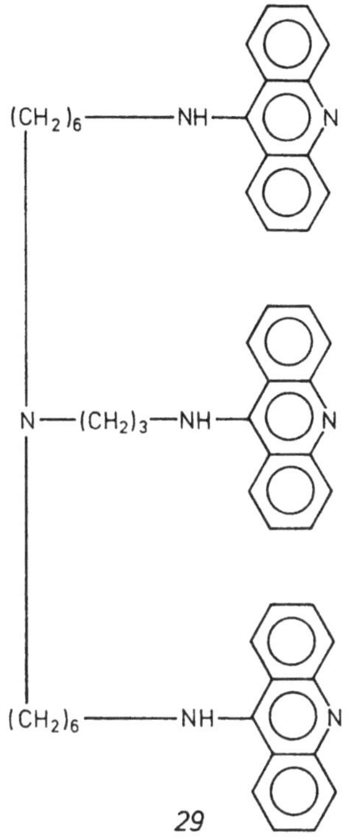

29

daunomycin (*28*) [122, 123] and bleomycin [124]. The importance of these areas may be gauged by the publication of several thousand papers.

Intercalation requires a certain degree of conformational change in the double helical host to produce a suitable cavity where the separation between adjacent base pairs has been increased from approximately 3.4 to 6.8 Å, and into which the aromatic guest molecule can slot lengthwise (Fig. 11). Generation of this cavity is achieved by a partial unwinding of the DNA double helix. The resultant lengthening of the DNA with increasing intercalation results in changes in properties such as sedimentation coefficient and viscosity which have been widely used to test predictions made on the basis of model building [119].

Intercalators with asymmetric substituents, such as the phenyl and ethyl groups of ethidium bromide (*21*), frequently cause a smaller increase in DNA length than expected from the simple model described above. In such cases these groups are inserted into the minor groove of the DNA helix with concomitant bending of the double helix towards the major groove. This alternative type of complexation is supported by X-ray studies on model systems [125].

Intercalation also occurs with the various classes of closed circular DNA, such as plasmids. Such complex formation has been widely used in studying the unwinding properties of DNA on complex formation. The unwinding angle varies according to the specific host-guest interactions and has been shown to range from 10° (anthracycline) to 26° (phenanthridine) with the most commonly observed value being 18°. Closed circular DNA is ideal for such studies since the natural right-handed super-

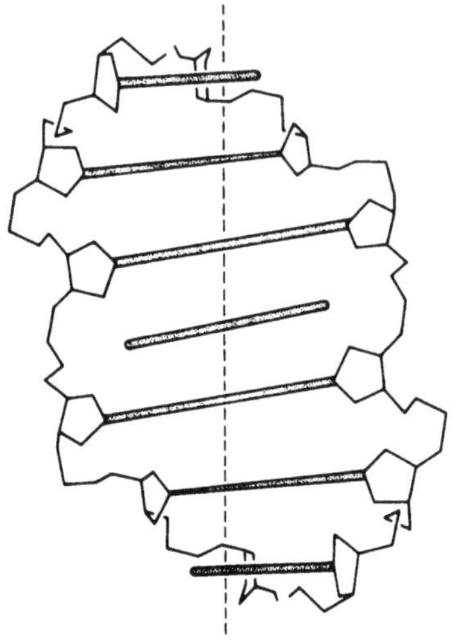

Fig. 11. Diagrammatic representation of a planar intercalating guest molecule complexed between adjacent base pairs of the double helical DNA host structure. The base pairs and intercalator are represented by stippled rods. Note the increased base pair separation caused by complexation with the guest.

helical coiling is gradually unwound as the extent of intercalation increases, with corresponding increase in viscosity to a maximum value when supercoiling has exactly been deleted. Further increases in intercalation beyond this point produces left-handed coiling with corresponding decrease in viscosity to a residual value higher than the original natural value [119]. It should be noted that intercalation does not occur at a higher ratio than one guest to two DNA base pairs. In other words although any two adjacent base pairs may participate as a binding site, once intercalation has taken place the immediately neighbouring sites become unavailable for further intercalation. This is illustrated in Fig. 12(a).

Efficient intercalation of a drug inhibits replication and/or transcription leading either to cell death or to more effective attack by the defences of the organism. Binding constants for drugs such as quinacrine (*24*) and daunomycin (*28*) are of the order of 10^5. One approach adopted to obtain high binding constants in synthetic guests has

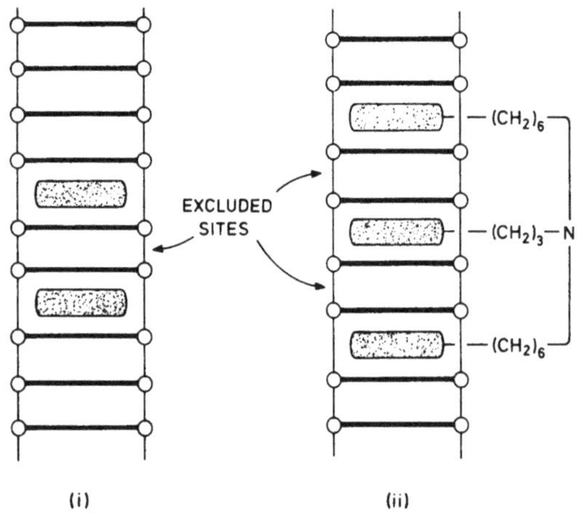

Fig. 12. Representation of the DNA-drug intercalation model showing locations of excluded sites which are not available for further intercalation, for: **(a)** two separate mono-intercalators such as daunomycin *28*; **(b)** one tris-intercalator such as compound *29*

been to design potential intercalators with two (or more) planar aromatic systems joined by flexible chains of appropriate length. In this manner bis- [126], tris- [127, 128] and tetrakis-intercalating guests [129] have been synthesised and studied. Figure 12(b) shows diagrammatically how a tris-intercalator such as (*29*) can bind to DNA.

Detailed investigation of the host-guest interactions between DNA and potential intercalating agents is a complex problem but one which is ideally suited to computer design approaches [130]. The design of new synthetic intercalating agents clearly offers enormous potential in the development of new and effective drugs, but at present the major drawback seems likely to be the classical problem of general toxicity versus specific uptake by the target cells.

7.2 Amylose Inclusion Compounds

The reaction between starch and iodine (or iodine-iodide mixtures) to form an inclusion compound was first reported in 1814 by Colin and de Claubry [131] and has since become familiar to all chemists through its applications in analytical chemistry. Its deep blue colour (λ_{max} 620 nm) has been known for years to result from a linear arrangement of "polyiodide" within a canal formed by a helical coil of amylose. The helical amylose structure will trap other molecules [132, 133] and other hosts will stabilise polyatomic iodide guests [134, 135].

The suggestion of a helical host molecule was originally put forward by Hanes [136] and then developed by Freudenberg and his colleagues [137]. Chemical [138–140] and X-ray diffraction studies by Rundle et al. [141–143] and by Bear [144, 145] demonstrated that these ideas were correct, and revealed that the helical structure had an outer diameter of 13.0 Å, an inner diameter of 5 Å, and a pitch of 8.0 Å with six glucose units per turn. The iodine atoms were arranged in a linear fashion with an average I-I separation of approximately 3.1 Å. These early results have been reviewed [146, 147]. They represent the first confirmed example of helical structure for a biopolymer.

This helical arrangement of amylose, known as the V-form, may be precipitated from certain solutions (e.g. in butanol or DMSO) of amylose. Either hydrated (V_h-) or anhydrous (V_a-) amylose absorbs I_2 vapour to produce the blue compound with the necessary I^- being produced in situ. Alternatively the compound may be formed from iodine-iodide mixtures in solution which allows the V-form to be produced and stabilised as the polyiodide compound [141]. The compound was reported [142] to have the orthorhombic space group $P2_12_12_1$.

More recent crystallographic work has been directed at inclusion compounds of V-amylose (including more detailed examination of the polyiodide compound) and at the structures of V_a-amylose and V_h-amylose. Despite the terminology used, both forms have significant water content and may therefore be classed as inclusion compounds in their own right.

The structures of V_a- and V_h-amylose are similar. In both cases the amylose chains form left-handed helices packed in an antiparallel manner in the space group $P2_12_12_1$. V_a-amylose [148–150] was originally believed to have a sixfold helical conformation [148, 149] but further refinement [150] has shown only twofold symmetry due to different rotational positions of glucose hydroxymethyl groups. The structure contains significant amounts of water. This refinement locates these guest molecules in interstitial sites between neighbouring helices of amylose [150].

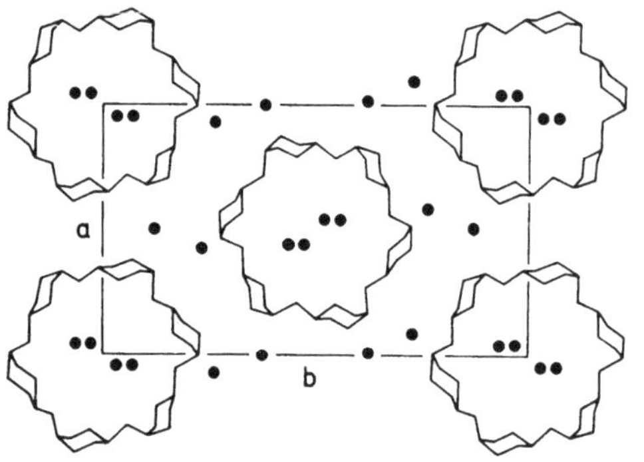

Fig. 13. Projection view of V_h-amylose on the *a, b* plane. The amylose chains are packed in an antiparallel manner in space group $P2_12_12_1$. Guest water molecules are represented as filled circles in the helical canals and in interstitial sites between the helices

V_h-amylose shows an increase in water molecules from 4 to 16 per unit cell [151]. Eight of these guests occupy interstitial sites and eight are present within the helical canals (Fig. 13). All unit cell dimensions are increased over the V_a-form, thereby increasing the unit cell volume from 2304 to 2604 Å^3. The sixfold helical arrangement has been confirmed for V_h-amylose.

X-ray crystallographic studies have also been carried out on V-amylose alcohol inclusion compounds [152], the DMSO inclusion compound [153] and on the V_h-amylose polyiodide inclusion compound [154]. The DMSO compound contains guest DMSO molecules within the helical canals, while the interstitial sites are occupied by both DMSO and water molecules. Although the space group is still $P2_12_12_1$ the compound has a pseudotetragonal unit cell in contrast to the pseudohexagonal cell more typical of V-amyloses. The latter is restored on removal of DMSO during hydration indicating that this arrangement is due to hydrogen bonding between the helical chains and the interstitial DMSO.

It is now realised that amylose-polyiodide compounds can also involved V_a- and V_h-forms [155]. Recent studies on the V_h-compound [154] have shown that the early structural ideas described above are essentially correct but that rather than $P2_12_12_1$ the monoclinic space group $P2_1$ is applicable. An almost linear arrangement of iodine atoms was present in the helical canals but the length of the iodide chain could not be determined. In addition to this guest, eight water molecules were present in the unit cell and once again these were present in the interstitial sites (see Fig. 14).

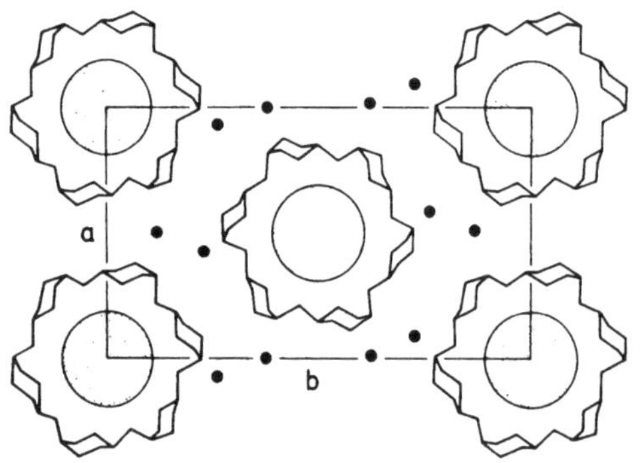

Fig. 14. The V_h-amylose polyiodide inclusion compound shown as a simplified projection on the *a, b* plane. Guest water molecules are shown as filled circles in the interstitial sites. The guest polyiodide chains are situated inside the helical amylose chains and their cross-sections are shown here as stippled circles

Thus, despite all the work carried out on starch-iodine, the exact nature of the guest iodine atoms is still not totally resolved. Teitelbaum, Ruby and Marks [156] have examined the compound using Raman and ^{129}I Mössbauer spectroscopy and concluded that the pentaiodide ion I_5^- was the major chromophore present. However work based on iodine compounds of cyclodextrins [3, 134, 135] has shown that a variety of polyiodide species is possible, and the starch-iodine inclusion compound could conceivably involve $I_2 \cdot I_3^-$, $I_2 \cdot I^- \cdot I_2$, or I_5 species [157]. The anhydrous amylose-iodine compound has recently been found to exhibit semiconductor behaviour [158].

As mentioned earlier, amylose has been known to complex with a wide range of small organic molecules for many years [132, 133, 145, 159]. Further examples investigated recently have included monoglyceride derivatives [160, 161], fatty acids [162, 163], sodium palmitate [164], surfactant stilbenes [165] and others [166]. In many of these cases X-ray diffraction patterns have indicated that the materials are genuine inclusion compounds involving V-amylose. This has not yet been confirmed by detailed structural analysis and, if correct, it is not known whether the guest molecules actually occupy the helical canals or occupy interstitial sites.

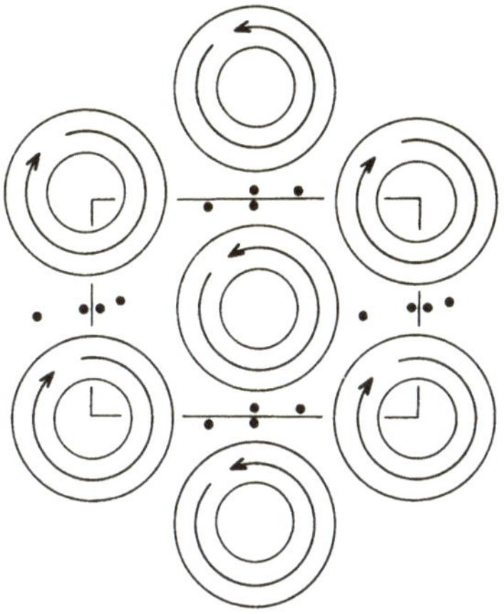

Fig. 15. Projection view of A-amylose on the *a*, *b* plane showing the guest water molecules (black dots) situated in the interstitial sites between helical double strands of amylose, which are represented diagrammatically as rings. The arrows illustrate the anti-parallel nature of the packing

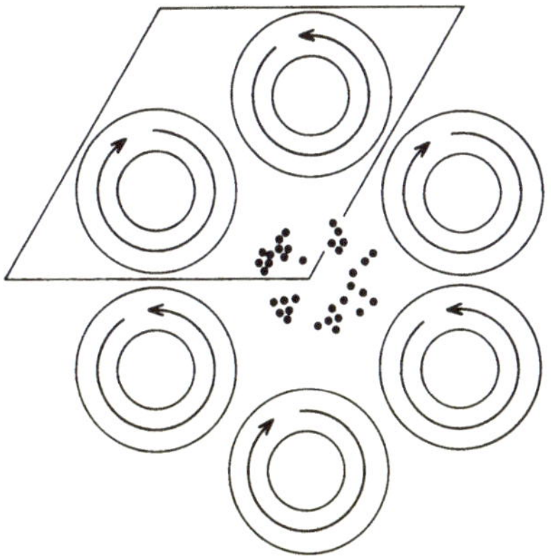

Fig. 16. Projection view of B-amylose showing the large central canal occupied by guest water molecules (black dots). The helical double strands of amylose are repesented as rings. The arrows illustrate the antiparallel nature of the packing

Recent work has, however, revealed the structure of other starch polymorphs not involved in inclusion behaviour of the above type, but which are inclusion compounds in their own right by virtue of their co-crystallisation with water.

The crystal structure of A-amylose is made up of parallel-stranded, *right*-handed *double* helices which pack in an antiparallel manner to give an orthorhombic unit cell [167]. About eight water molecules are present per unit cell and these guests are present in the interstitial sites (Fig. 15).

The B-polymorph is also based on right-handed double helices, packed in an antiparallel manner but this time to give a hexagonal unit cell (see Fig. 16). A hexagonal array of double helices surrounds a large open canal, approximate diameter 8 Å, containing large amounts of water (up to 3.5 per glucose residue) [168]. C-amylose has been found to be a mixture of the A- and B- unit cells [169].

7.3 Inclusion Compounds of Other Helical Polymers

The linear arrangement of iodine atoms in the amylose inclusion compound has generated much interest right from the early days with respect to its spectroscopic and optical properties [138–140]. It has also been known for many years that polyvinyl alcohol (PVA) behaves similarly, and this was applied by Polaroid Corporation for the manufacture of polarising plastic using stretched sheets of iodine-stained polyvinyl alcohol [170].

Unlike the starch case, a mixture of iodine and boric acid is used to generate the PVA-polyiodide compound. This method may be used to determine PVA [171]. While amylose is a stereoregular polymer, PVA has a more irregular configuration resulting in a more flexible polymer chain. The boric acid is therefore required to act as a bridge linkage to lock the PVA chain into its helical arrangement. At full iodine saturation of PVA it is believed that twelve vinyl alcohol residues constitute one turn of the helix surrounding an iodine atom of the polyiodide chain [172]. A representation of the proposed structure is shown in Fig. 17.

Substituted derivatives of amylose such as diethylaminoethylamylose and carboxymethylamylose also form polyiodide compounds [173]. Triethylamylose crystallises in the space group $P2_12_12_1$ involving antiparallel left-handed fourfold (4_3) helices. This polymer also forms.inclusion compounds with small organic guest molecules such as nitromethane, dichloromethane and chloroform [174]. These compounds have the same pseudotetragonal unit cell and also have space group $P2_12_12_1$. In these structures the helical canals have insufficient diameter to accommodate guest molecules. Neither are these situated in interstitial sites, but rather they form a solvation shell around the helices with individual molecules being located in the grooves of the helices.

Another polymorph of triethylamylose with dichloromethane or chloroform has also been studied [175]. Here the number of guest molecules decreases from 8 to 4 per unit cell and the solvent molecules are now located in interstitial sites. The orthorhombic cell has space group $P2_12_12_1$.

The trimethylamylose-dichloromethane inclusion compound [176] has a pseudotetragonal unit cell and space group $P2_12_12_1$, but this time all 8 guest molecules are in interstitial sites.

Hui et al. [177] have investigated the inclusion compounds between carboxymethylamylose and the two aromatic keto-acid salts (*30*) and (*31*). The binding constants

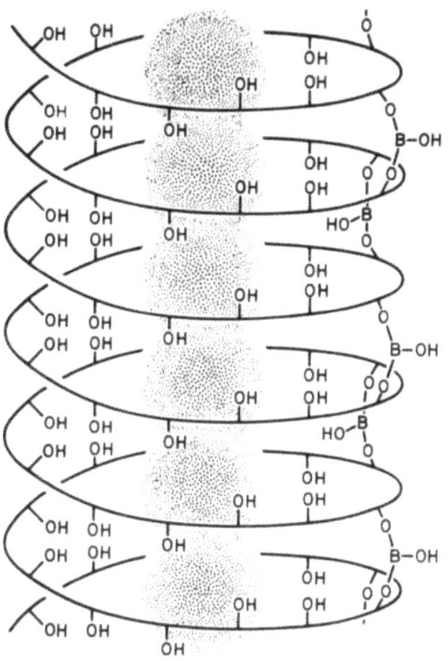

Fig. 17. Proposed structure of the PVA-polyiodide inclusion compound. The polymer chain is shown as a helical line with appended hydroxyl groups and bridging borate groups. The stippled core represents the polyiodide guest chain

were found to increase with chain length. Quantum yields for the Norrish type-II photoelimination process were low indicating conformational restriction of the guests in the helical canals.

$Ph{-}CO{-}(CH_2)_3{-}CO_2^-Na^+$ (*30*)

$Ph{-}CO{-}(CH_2)_{10}{-}CO_2^-Na^+$ (*31*)

Crystalline samples of syndiotactic poly(methyl methacrylate) (st-PMMA) may be obtained from chloroacetone [178]. This guest could be completely replaced by a variety of other guest molecules such as acetone, 1,3-dichloroacetone, bromoacetone, pinacolone, cyclohexanone, acetophenone and benzene. The X-ray diffraction patterns for these inclusion compounds were similar. These data indicate that the st-PMMA chains adopt a helical conformation of radius about 8 Å and pitch 8.85 Å. The guest molecules are located both inside the helical canals and in interhelix interstitial sites.

Isotactic poly(methyl methacrylate) (it-PMMA) can form a stereocomplex with st-PMMA. Recent X-ray studies [179] of this material indicate that the two polymer chains probably interact to form a double helical structure. The it-PMMA chain forms the inner helix and is surrounded by the st-PMMA helical chain which winds around it. If subsequent work confirms this model, this material would constitute a most unusual inclusion compound involving only one monomeric substance.

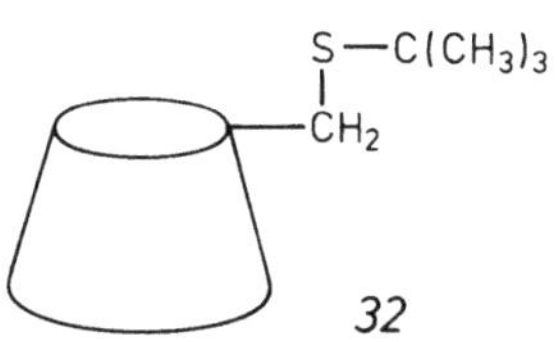

32

Another novel synthetic helical polymer [180] has been obtained from the cyclodextrin derivative 6-deoxy-6(t-butylthio)-β-cyclodextrin, represented as *32*. It is well known that cyclodextrins can form canal-type inclusion compounds as well as cage structures. Description of these and the requirements for canal formation have been reviewed by Saenger [3,4]. These canals, formed by the hydrogen-bonded stacking of cyclodextrin molecules in columns, are not helical and therefore lie outside the scope of this article. In contrast, the cyclodextrin *32* forms an inclusion compound of empirical formula *32* · 22 H_2O in which the t-butylthio group of one molecule is included by the hydrophobic cavity of the next such that a helical polymer is built up around a 2_1 screw axis. Although this is described as a "polymer" it should be noted that the monomeric units are linked by host-guest inclusion rather than by a covalent bond. The crystal space group is $P2_12_12_1$. This unusual structure is the first polymeric inclusion compound formed from a single molecule acting simultaneously as host and guest. The water molecules are chiefly situated in the helical grooves of the polymer (see Fig. 18).

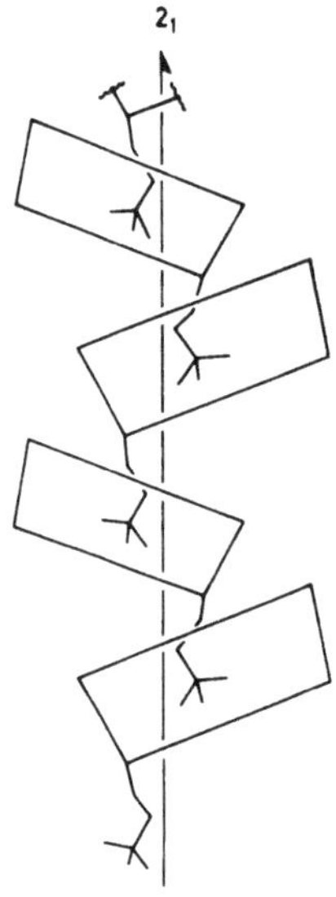

Fig. 18. Diagrammatic representation of the helical polymeric form of the cyclodextrin *32* around the 2_1 screw axis

Inclusion compounds have been reported from derivatives of chitosan [181] and from collagen [182] but structural information on these is unavailable at present.

In reviewing these helical structures based on polymeric systems it is worth closing by emphasising that although X-ray diffraction methods provide the most definitive results, the difficulties involved are considerable. Nearly all the structural work discussed has been reported with large values of the crystallographic structure factor residual R which would be unacceptable for work on small molecules. Even if the material is obtained in a crystalline state, the polymer sample frequently will contain regions that are imperfectly crystalline. If the polymeric material is not stereoregular, is not entirely linear, or contains non-repeating units, then the problems mount rapidly.

On the other hand, as biological molecules become larger their tendency to be associated with water molecules, metal ions, and other materials increases. Crystalline proteins, for example, routinely contain 27–65% of the solvent used for their crystallisation [183]. Such associated materials may be difficult to locate by crystallography and it may become a question of terminology whether such molecules should be regarded as inclusion complexes, non-specific aggregates, or merely contaminated biomolecules.

Putting aside such considerations, the reader is encouraged to examine the sections of Klug's Nobel Lecture [184] dealing with the structure and the growth of Tobacco Mosaic Virus to see how helical structures and concepts of inclusion phenomena can relate to molecular biology.

8 Transmembrane Ion Channels

8.1 Ionophore Behaviour

Although the transport of ions and neutral molecules across biological membranes has been widely investigated for many years, the means by which these vital processes can be accomplished at a molecular level have proved difficult to determine.

It is now recognised that a wide range of organic molecules, collectively termed ionophores [185, 186] or complexones [187], are able to facilitate ion (usually cation) transport. Two major mechanisms have been revealed for this process, namely the involvement of transmembrane ion carriers and transmembrane pores or channels (see Fig. 19). The majority of ionophores studied to date are natural antibiotics and their synthetic analogues which are, on a biological scale, comparatively small molecules lending themselves to study outside the biological system. In contrast far less is known about the molecular structures involved in normal transport processes. Such molecules are likely to be more complex or present in small amounts and may require

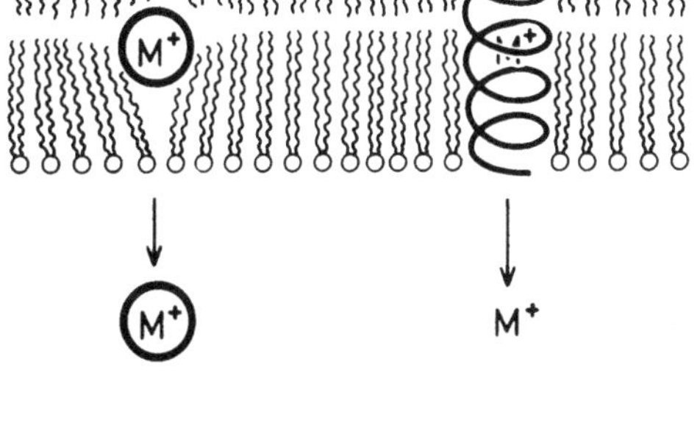

Fig. 19. Representation of the movement of metal ions (M^+) across a bilayer involving an ion carrier and a transmembrane channel

study in situ to reveal their genuine natural function and properties. The systems described in this section do not involve helicity but the concepts introduced are necessary for descriptions of ion channels involving helical structures in subsequent sections.

The ion carrier antibiotics fall into several structural classes but have the common feature that they present a hydrophilic interior to complex the cation and a lipophilic exterior to assist passage through the membrane. In addition a further requirement is that the complex should be only moderately stable — a trade-off between rapid ion trapping and the need for rapid ion release after traversal of the membrane. Antibiotics such as nigericin and monensin achieve this by having an open polyoxygenated structure which wraps around the cation to form a pseudocyclic complex [188]. Other major classes of ion carriers are cyclic molecules including the macrotetrolide antibiotics such as nonactin and the depsipeptide group such as valinomycin and enniatin. These diverse groups of compounds may be regarded as the natural forerunners of the crown ethers and subsequent families of synthetic ion carriers. The behaviour of these fascinating compounds lies outside the scope of this article and readers are directed to the excellent reviews available on this subject [185–194].

Turning to channel forming antibiotics, one major class contains the macrocyclic polyene lactones produced by *Streptomycetes* bacteria [195]. These antifungal agents have a polyhydroxylated lactone ring of 23–37 atoms containing a segment of 4–7 conjugated olefinic bonds and often also have a carboxylic acid and an amino sugar group attached to the ring. Typical examples of this family are nystatin (*33*), amphotericin B (*34*) and filipin (*35*).

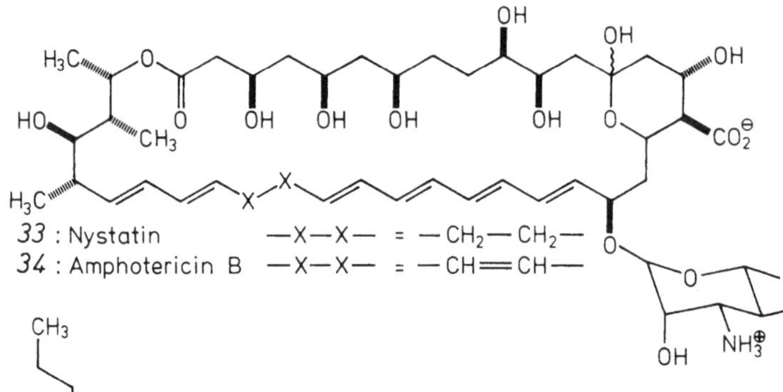

33 : Nystatin —X—X— = —CH₂—CH₂—
34 : Amphotericin B —X—X— = —CH═CH—

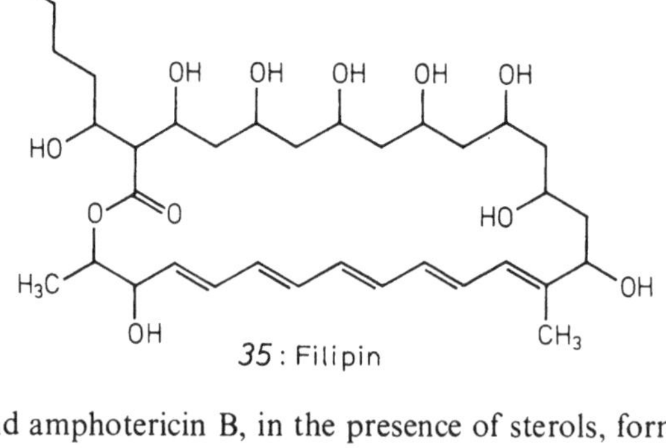

35 : Filipin

Nystatin and amphotericin B, in the presence of sterols, form temporary channels across lipid bilayers. Neutral molecules under the size of glucose can traverse the bilayer via these pores indicating a channel diameter of about 8 Å [196]. The obvious supposition would be that the macrocyclic rings stack horizontally on top of each other to produce the channel, but this can be ruled out since the smaller-ring molecule filipin produces larger channels. It is believed instead that several of the antibiotic molecules stack together vertically like a ring of columns across the membrane presenting a hydrophilic internal face (the oxygen functional groups) and a lipophilic outer face (the polyene chains). The membrane sterol molecules are associated with the lipophilic face [197].

For the aggregate of amphotericin B and cholesterol, De Kruijff and Demel [198] have suggested a 16-column circular array of alternating amphotericin B and cholesterol molecules. However, Finkelstein [199] favours a pore structure where the amphotericin B molecules form a channel with a turbine-like cross section and the sterol molecules fit into the wedge shaped cavities of the "turbine blades" of the outer lipophilic face [199]. Recent studies on nystatin and sodium iodide in methanol solution using ^{1}H and ^{23}Na NMR [200] have revealed weak association between the alkali metal ions and a specific region of the macrocycle. This finding provides an alternative model for ion transport which does not require such a highly organised channel structure.

It is known that nystatin and amphotericin B can form "single length" (ca. 22 Å) and "double length" (ca. 44 Å) channels (see Fig. 20). The latter are believed to result from tail to tail dimerisation of the "single length" channel via hydrogen bonding of the single hydroxyl group of the non-polar segment of the molecule with similar others [197,198].

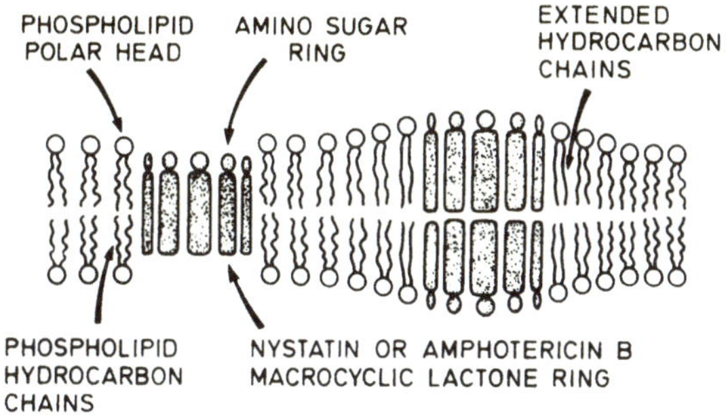

Fig. 20. Diagram showing a "single-length" channel and a "double-length" channel formed across a phospholipid bilayer by a circular cluster of nystatin or amphotericin B aggregates

Nystatin applied to both sides of a bilayer produces channels which are anion selective, whereas application to one side gives cation selectivity. This remarkable difference is in accord with the double channel having identical head ends, whereas the single channel has dissimilar ends [199].

8.2 Peptaibol Antibiotic Channels

The peptaibol antibiotics are a group of linear oligopeptides (of 15 to 24 residues) characterised by the presence of a high content of α-amino isobutyric acid (Aib) residues and an amino alcohol group at the C-terminus. These substances are frequently obtained as a complex mixture of closely related compounds. For example

ALAMETHICIN I:
Ac—Aib—Pro—Aib—Ala—Aib—Ala—Gln—Aib—Val—Aib—Gly—Leu—Aib—Pro—Val—Aib—Aib—Glu—Gln—Phol

ALAMETHICIN II:
Ac—Aib—Pro—Aib—Ala—Aib—Aib—Gln—Aib—Val—Aib—Gly—Leu—Aib—Pro—Val—Aib—Aib—Glu—Gln—Phol

EMERIMICIN III:
Ac—Phe—Aib—Aib—Aib—Val—Gly—Leu—Aib—Aib—Hyp—Gln—Iva—Hyp—Ala—Phol

TRICHOTOXIN A-40:
Ac—Aib—Gly—Aib—Ala—Aib—Glu—Aib—Aib—Aib—Ala—Aib—Aib—Pro—Leu—Aib—Iva—Gln—Valol

ZERVAMICIN IIA:
Ac—Trp—Ile—Gln—Aib—Ile—Thr—Aib—Leu—Aib—Hyp—Glu—Aib—Hyp—Aib—Pro—Phol

Fig. 21. Structures of some peptaibol antibiotics

alamethicin isolated from *Trichoderma viride* comprises a mixture of at least twelve oligopeptides. The structures of the two major compounds alamethicin I and II, and some of the other peptaibol antibiotics are given in Fig. 21.

Many of these materials form voltage-gated transmembrane ion channels, i.e. they exhibit non-linear current voltage curves, unlike the polyene antibiotics and gramicidin A which have linear current voltage curves. Of the peptaibol antibiotics the most thoroughly investigated substance to date is alamethicin [201-203].

Fox and Richards [204] have determined the crystal structure of alamethicin in the form of its inclusion compound from methanol/acetonitrile. The crystals were of space group $P2_1$ comprising three molecules of alamethicin per asymmetric unit and about 30% organic solvent. Each of the peptides exists essentially in an α-helical conformation but with small (and different) deviations near the C-terminus due to local 3_{10} helical structure over regions of one or two residues. Consequently the helices are bent slightly around Pro-14. Solution investigations involving NMR [205] confirm variable conformation in the region of the C-terminus. It would therefore seem reasonable that these gross features will also be present when alamethicin is inserted into a bilayer, but this is not totally certain at present (see Fig. 22).

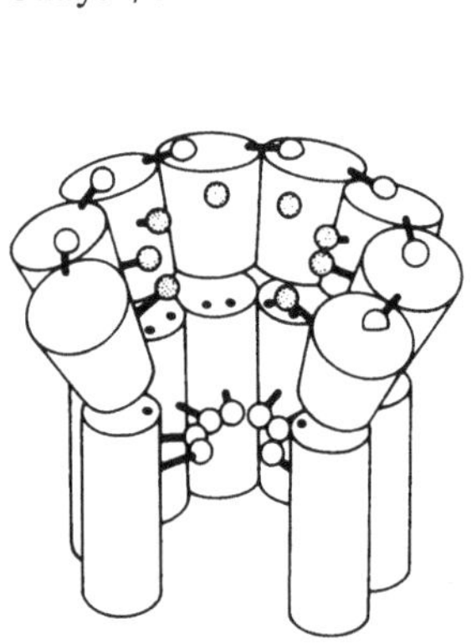

Fig. 22. The Fox and Richards model of the alamethicin transmembrane channel shown diagrammatically. Interruption of the α-helical hydrogen bonding by the Pro-14 residue is signified by representation of each monomer as two cylindrical sections. The stippled spheres at the mouth of the channel represent the Glu-18, the spheres at the centre the Gln-7, and the spheres at the top the Gln-19 residues. Fox, R. O., Richards, F. M.: Reprinted by permission from *Nature 300*, 325 (1982). Copyright Macmillan Journals Limited

It is known that the helical conformation of alamethicin would provide a channel too small to allow passage of ions, since it has been calculated that the interior of an α-helix of poly-L-alanine has an energy barrier of approximately 840 kJ mol^{-1} for passage of protons [206]. The alamethicin pores arise from circular aggregates of parallel peptide helices across the bilayer, the number of monomers varying with alamethicin concentration and the ionic strength of the medium. Channel diameters may range from 4 to 20 Å, involving four to eleven monomers [203].

A variety of hypothetical models have been proposed for the alamethicin pores. These generally invoke the bent helix monomer conformation, hydrogen bonding between adjacent helical monomers, and structural features compatible with the voltage dependence of the channels [201-204]. Unfortunately experimental evidence is insufficient to distinguish between these models at present.

8.3 Gramicidin

Naturally occurring gramicidin is a mixture of at least six linear pentadecapeptide antibiotics of closely related structure produced by *Bacillus brevis* [207]. The amino acid sequence of the major component [Val]gramicidin A was reported by Sarges and Witkop [208] in 1965:
$OHC\text{-L-Val}_1\text{-Gly}_2\text{-L-Ala}_3\text{-D-Leu}_4\text{-L-Ala}_5\text{-D-Val}_6\text{-L-Val}_7\text{-D-Val}_8\text{-L-Trp}_9\text{-D-Leu}_{10}\text{-}$
$\text{L-Trp}_{11}\text{-D-Leu}_{12}\text{-L-Trp}_{13}\text{-D-Leu}_{14}\text{-L-Trp}_{15}\ NHCH_2CH_2OH$.
Gramicidin B and C differ only at position 11 where the L-tryptophan is replaced by L-phenylalanine and L-tyrosine, respectively. In addition, some 5–20% of the natural material has an L-Ile residue at position 1 in place of L-Val [209]. The biological function of gramicidin in *B. brevis* is believed to be inhibition of RNA polymerase resulting in spore production [210], but the material has sprung to prominence through the recognition of its novel ionophoric properties. Most of these studies have been carried out on the natural mixture or on [Val]gramicidin A and the current state of research is the subject of several recent reviews [210-212].

It has been known for some years that gramicidin forms transmembrane ion channels in lipid bilayers and biological membranes and that these channels are assembled from two molecules of the polypeptide [213]. The channels are permeable specifically to small monovalent cations [such as H^+, Na^+, K^+, Rb^+, Cs^+, Tl^+, NH_4^+, $CH_3NH_3^+$, but not $(CH_3)_2NH_2^+$] and small neutral molecules (such as water, but not urea). They do not allow passage of anions or multivalent cations [211].

Four helical models have been proposed for the conformation of the gramicidin channel in order to accommodate these and other data (see Fig. 23):

(i) The single left-handed (β) helical channel formed by head (formyl end) to head dimerisation [214, 215].
(ii) The corresponding tail (ethanolamine end) to tail dimer [216].
(iii) The antiparallel β-double-helix [217].
(iv) The corresponding parallel β-double-helix [217].

All four of these helical arrangements could, in principle, provide a channel long enough to span a lipid bilayer and wide enough to permit ion permeability.

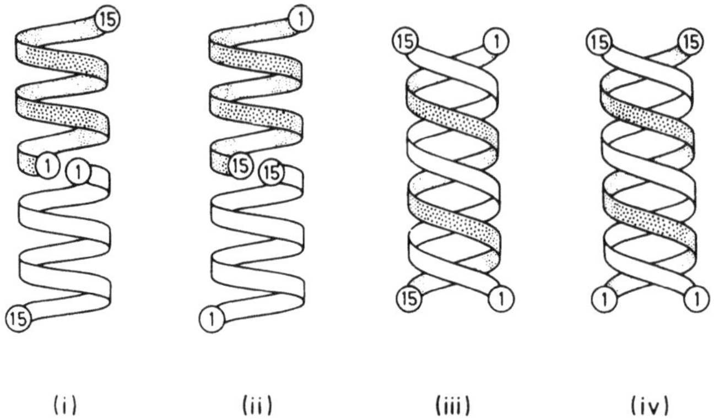

Fig. 23. Possible helical models for the gramicidin A ion channel. The polypeptides are represented as helical strips, one molecule being stippled for clarity. Numbers refer to the substituted terminal amino acid residues. Model (i), proposed originally by Urry, is the one now generally accepted

X-ray diffraction studies on gramicidin commenced as early as 1949 [218,219] and this early work pointed to a helical structure [220]. Recent work by Koeppe et al. [221] on gramicidin A crystallised from methanol ($P2_1$) and ethanol ($P2_12_12_1$) has shown that the helical channel has a diameter of about 5 Å and a length of about 32 Å in both cases. The inclusion complexes of gramicidin A with CsSCN and KSCN ($P2_12_12_1$) have channels that are wider (6–8 Å) and shorter (26 Å) than the uncomplexed dimer [221, 222]. Furthermore there are two cation binding sites per channel situated either 2.5 Å from either end of the channel or 2.5 Å on each side of its centre [222] Unfortunately these data do not permit a choice to be made from the helical models (i)–(iv) and it is not certain if the helical canals studied are the same as those involved in membrane ion transport.

Major effort has therefore been applied to study of the helical channels by non-crystallographic methods in order to discriminate unambigously between the likely models. This has resulted in the gramicidin ion channel not only being the first clearly identified example but also the most characterised in structural terms. The bulk of

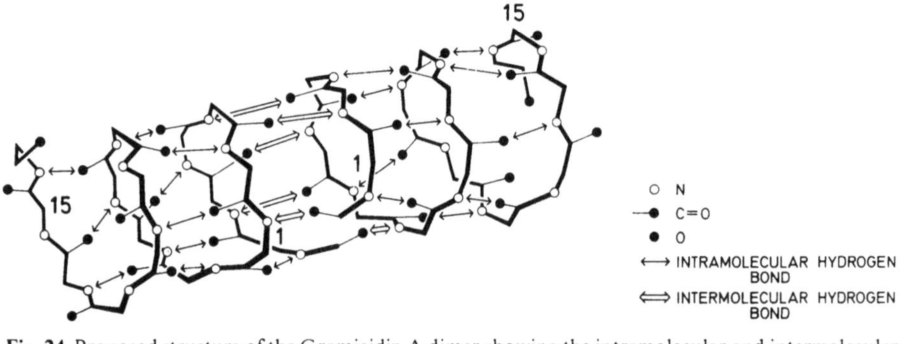

Fig. 24. Proposed structure of the Gramicidin A dimer showing the intramolecular and intermolecular hydrogen bonding arrangement. Numbers refer to the amino acid residues; Peptide side groups are omitted for clarity.

this evidence [210-212] supports the model (i) which was the original proposition of Urry [214], and these findings have now gained almost unanimous acceptance [223].

The structure of gramicidin is unusual in having alternating L and D configurations for its amino acid residues, for the numerous hydrophobic residues, and for the absence of ionizable side chains or end groups. Urry [214,215] has described a series of π_{LD}-helices which, in principle, could be constructed from a molecule of gramicidin A. As a consequence of the alternating amino acid configurations the carbonyl groups of the L-residues are all oriented towards the ethanolamine terminus and those of the D-residues and glycine are oriented towards the formyl terminus of the chain. The most reasonable of these was the helix with 6.3 amino acid residues per turn, designated

Fig. 25. Top view of a space-filling model of the proposed Gramicidin A dimer structure, showing the transmembrane channel. [Photograph courtesy of D. A. Haydon]

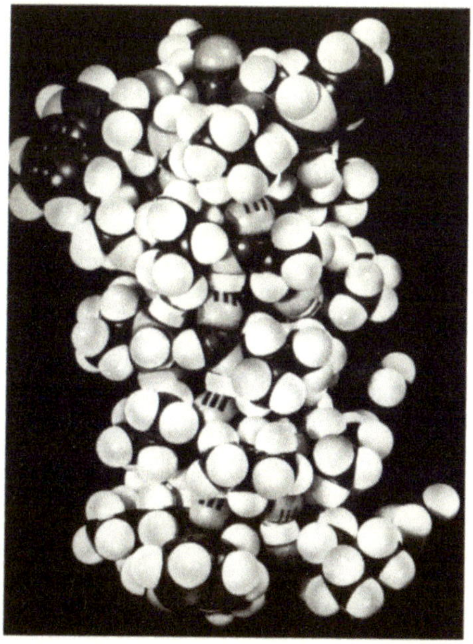

Fig. 26. Side view of the proposed Gramicidin A dimer structure. [Photograph courtesy of D. A. Haydon]

the $\pi_{LD}^{6.3}$-helix, which was held in this conformation by ten intramolecular hydrogen bonds. Six intermolecular hydrogen bonds allowed linkage of two such helices in a head (formyl end) to head arrangement producing an ion channel with an internal diameter of about 4 Å and a length of approximately 28 Å (see Fig. 24, 25, and 26). This should be compared with about 27 Å for the hydrocarbon thickness of a typical lipid bilayer. Unlike the crystal structure results it has been determined that the width and length of the channel are not greatly affected by the inclusion of cations when the channel is bridging a lipid membrane [224].

It should be noted that in forming this dimeric channel structure all the hydrogen bonds are parallel to the channel axis and that the inner surface is lined with the polar polypeptide groups. In addition the various lipophilic side chains coat the outer wall of the structure and are thus in contact with the lipid hydrocarbon chains. The resulting gramicidin A channel is a most efficient means of ion transport with approximately 10^7 sodium ions traversing the channel per second, under conditions of 1 M NaCl, 100 mV applied potential and a temperature of 25 °C [225]. The detailed mechanism by which this can be achieved is under active study [226].

8.4 Bacteriorhodopsin and Rhodopsin

Bacteriorhodopsin is a transmembrane protein found in the bacterium *Halobacterium halobium* which lives in environments rich in sodium chloride [227]. Stoeckenius et al. discovered that when oxygen supplies are limited the organism synthesises a specialised region of cell membrane known as "the purple membrane" which utilises light energy to pump hydrogen ions across the membrane out of the cells with concomitant synthesis of intracellular ATP as an energy source [228, 229]. The purple colour is due to a molecule of retinal covalently bonded to a protein (MW ≈ 26,000) which makes up about 75% of the mass of the purple membrane, along with 25% of lipid.

Electron microscopy was utilised by Henderson and Unwin [230] to determine the arrangement of protein in the purple membrane. It was found to comprise seven, closely packed α-helical arrays roughly perpendicular to the plane of the membrane with bilayer regions occupying the remaining space. The seven helices were all 10–12 Å apart and approximately 35–40 Å long, the membrane itself being 45 Å thick. These helices made up about 75% of the polypeptide but the connectivity between them was not resolved.

Ovchinnikov et al. [231] later determined the sequence of the polypeptide showing it to be a linear protein containing 247 amino acid residues, and proposing its arrangement as seven helices spanning the membrane. Thus bacteriorhodopsin was the first membrane protein to have its structure determined. A more involved analysis of the protein conformation was performed in 1980 by Engelman et al. [232] using the Russian amino acid sequence and proteolytic data, plus the electron density data of Henderson and Unwin [230]. Since it was known which membrane face was which they had to examine 5040 possible models (7!) of fitting the seven helices with the seven regions of helical density. Fortunately a preferred model could be proposed for the conformation of the polypeptide (see Fig. 27).

Khorana et al. [233] also carried out a sequence of bacteriorhodopsin, which was largely in agreement with the earlier study but which found another tryptophan residue

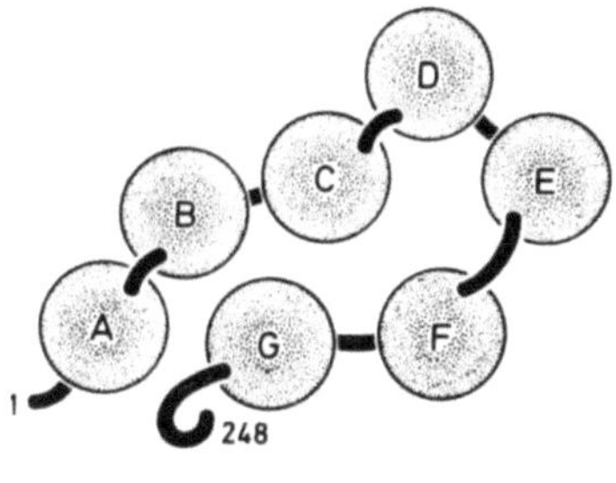

Fig. 27. A schematic representation of the seven transmembrane helical peptide chains (A–G) viewed from inside the cell. The numbering denotes the first and last amino acid residues. The proton channel is believed to be the volume between helices C, D, F and G

(position 137), bringing the total to 248 amino acid residues. This is now accepted by both research groups [234].

Ovchinnikov et al. [231] had commented on the high content of hydrophobic amino acid residues present ($\approx 66\%$) and their presence in definite regions of the polypeptide chain. Using neutron scattering techniques Engelman and Zaccai [235] were able to show that the polypeptide was an "inside-out" protein in the sense that most of the charged and polar groups were situated towards the interior of the helical arrays where they might participate in the hydrogen ion pumping mechanism, while the non-polar groups were directed outward in contact with the membrane lipids. This interpretation is shown in more detail in Fig. 27, where the region between helices C, D, F and G is believed to be the site of the proton channel.

Quite apart from the molecular structure of the channel it must also allow proton movement in only one direction and a pumping mechanism. Stoeckenius [236] has proposed an ingenious means by which the Schiff base linkage of the protein and retinal performs both these functions.

Ovchinnikov [234, 237] has shown that bovine rhodopsin, although quite different in amino acid sequence (348 residues), also forms seven transmembrane helices. This structural similarity between bacterial and mammalian light activated membrane proteins is remarkable. Since the two amino acid sequences have little in common it would appear that the necessary requirement is seven transmembrane helices to form a channel which is specific for proton migration. For example it has been suggested that a similar arrangement and function is performed by the lactose permease of *E. coli* [237].

It appears probable that other transmembrane proteins function by providing related channels between helical arrays of polypeptide chains [238–240].

8.5 Synthetic Approaches

An exciting area in inclusion chemistry is the design and synthesis of molecules which could behave as ion channels. Future developments in this field offer the potential for developing new synthetic antibiotic molecules, model systems for investigating transport across membranes, and ion channels specific for particular ions. Such studies are so far only in their infancy.

One approach is to synthesise polymeric molecules containing crown ether rings which stack on top of each other to produce linear canals through which small cations could migrate. The poly(iminomethylene) crown ether *36* of van Beijnen et al. [241]

illustrates this approach. In this polymeric structure the central poly(iminomethylene) chain constitutes a tight helix with four repeating units per turn. As a result the peripheral benzo crown ether groups are stacked to form four channels parallel to the polymer axis while the outer surface is lipophilic and therefore suitable for insertion into membranes.

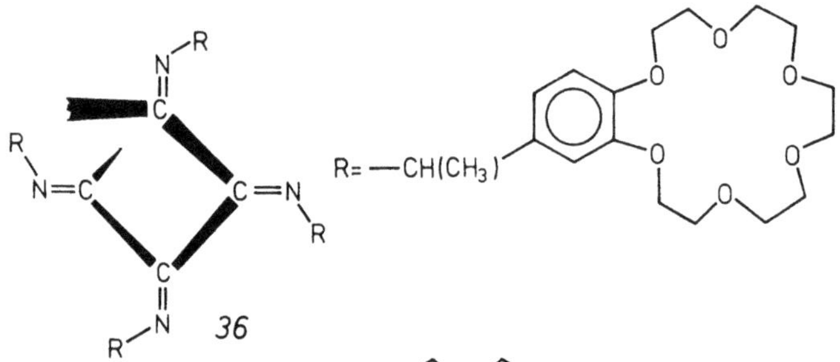

36

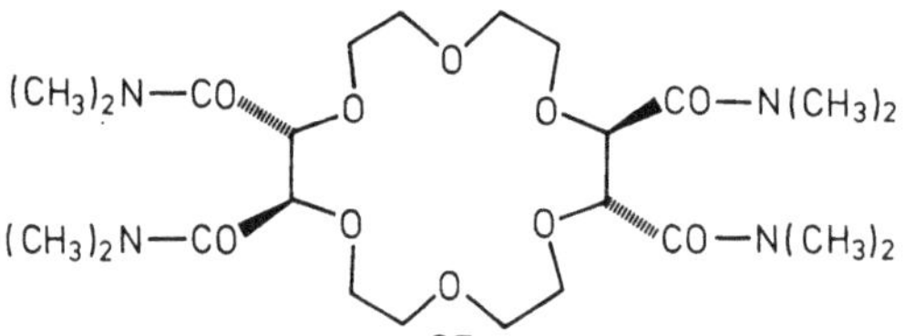

37

Lehn [242,243] has described a solid phase model of a K^+ channel based on the crown ether *37*. The crystal structure of this inclusion complex reveals stacking of the crown ethers into vertical columns, empirical formula $[2 \cdot 37, 2\,K, 3\,H_2O]^{2+}$, linked by water and potassium ions. The counter ions, empirical formula $[K, 3\,Br, 4\,H_2O]^{2-}$, comprise a polymeric chain running parallel to the columns.

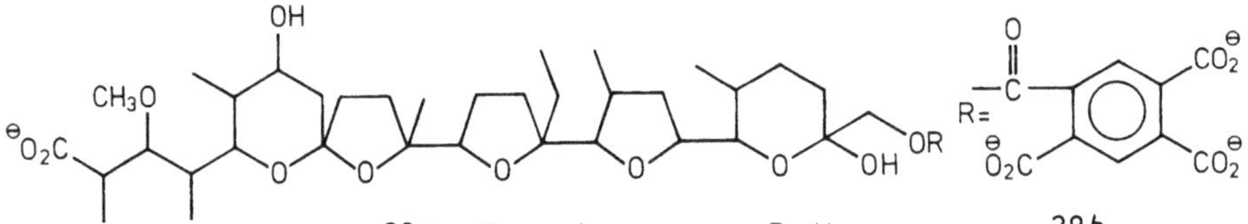

38a: Monensin R = H *38b*

An elegant alternative approach [244] has been to convert the ionophore monensin (*38a*) (see Section 8.1) into its monopyromellitic ester (*38b*), thereby producing a molecule of length ca. 24 Å with a hydrophilic group at *both* ends, but retaining its original hydrophilic and lipophilic faces. Instead of forming the pseudocyclic structure favoured by *38a*, the modified compound *38b* can now bridge a 20 Å diameter model membrane. Aggregation of bridging *38b* molecules apparently produces an ion channel of roughly circular cross-section similar to the "single length" ion channels of nystatin or amphotericin B (Fig. 20, Section 8.1). Incorporation of such channels into a monolayer vesicle makes the vesicle membrane permeable to lithium ions. Addition of appropriate organic cations can irreversibly or partially reversibly seal the ion channels [245]. This represents a first step towards development of a switching mechanism for artificial ion channel systems.

The third current approach is synthesis of peptide chains as models for the helical peptaibol (Section 8.2) and gramicidin (Section 8.3) ion channels. Considerable work has been carried out in the former area, involving synthesis of Aib-containing small peptides, in order to obtain conformational data applicable to the more complex oligopeptide antibiotics. By working with such fragments it has been possible to obtain valuable X-ray crystal structure information on the helical conformation of alamethin [246], emerimicin [247], suzukacillin [248], and other members of the peptaibol series.

9 Acknowledgements

We gratefully acknowledge the valuable synthetic and structural contributions and intellectual input from Stephen Hawkins and Dr. Marcia Scudder in our helical tubuland work, which has been supported financially through the Australian Research Grants Scheme. We also wish to thank Professor D. A. Haydon for kindly supplying photographs of the gramicidin channel model, and Martin Dudman for the preparation of diagrams.

10 References

1. For a general review of helical molecules in organic chemistry see Meurer, K. P., Vögtle, F.: Top. Curr. Chem. *127*, 1 (1985)
2. Weber, E., Josel, H.-P.: J. Incl. Phenom. *1*, 79 (1983)
3. Saenger, W.: in "Inclusion Compounds" (Atwood, J. L., Davies, J. E. D., MacNicol, D. D., eds.), Academic Press, London 1984, vol. 2, p. 231
4. Saenger, W.: J. Incl. Phenom. *2*, 445 (1984)
5. Lawton, D., Powell, H. M.: J. Chem. Soc. *1958*, 2339
6. Bishop, R., Dance, I. G.: J. Chem. Soc., Chem. Commun. *1979*, 992
7. Dance, I. G., Bishop, R., Hawkins, S. C., Lipari, T., Scudder, M. L., Craig, D. C.: J. Chem. Soc., Perkin 2, 1299 (1986)
8. Dance, I. G., Bishop, R., Scudder, M. L.: ibid. 1309 (1986)
9. Jacques, J., Collet, A., Wilen, S. H.: "Enantiomers, Racemates and Resolutions", J. Wiley and Sons, New York 1981, ch. 2.2
10. The convention adopted for designation of the stereochemistry of the hydroxy groups in this and subsequent tricyclic diols is *syn* or *anti* with respect to the larger of the bridges across the molecular twofold or pseudo-twofold axis
11. Bishop, R. Choudhury, S., Dance, I. G.: J. Chem. Soc., Perkin Trans. 2 *1982*, 1159
12. Landa, S., Kiefmann, J.: Coll. Czech. Chem. Commun. *35*, 1005 (1970)
13. Bishop, R., Dance, I. G., Hawkins, S. C.: J. Chem. Soc., Chem. Commun. *1983*, 889
14. Bishop, R., Craig, D. C., Dance, I. G., Hawkins, S. C., Scudder, M. L.: in preparation
15. Bengen, F.: German Patent Application, 869,070 (I.G. Farbenindustrie, March 18, 1940)
16. Schlenk, W.: Liebigs Ann. Chem. *565*, 204 (1949)
17. Smith, A. E.: J. Chem. Phys. *18*, 150 (1950)
18. Smith, A. E.: Acta Crystallogr. *5*, 224 (1952)
19. Fetterly, L. C.: in "Non Stoichiometric Compounds" (Mandelcorn, L., ed.), Academic Press, New York 1964, p. 491
20. Frank, S. G.: J. Pharm. Sci. *64*, 1585 (1975)
21. Takemoto, K., Sonoda, N.: in "Inclusion Compounds" (Atwood, J. L., Davies, J. E. D., MacNicol, D. D., eds.), Academic Press, London 1984, vol. 2, p. 47

22. Sybilska, D., Smolkova-Keulemansova, E.: in "Inclusion Compounds" (Atwood, J. L., Davies, J. E. D., MacNicol, D. D., eds.), Academic Press, London 1984, vol. 3, p. 173
23. Takemoto, K., Miyata, M.: J. Macromol. Sci., Rev. Macromol. Chem. *C18*, 83 (1980)
24. Farina, M.: in "Inclusion Compounds" (Atwood, J. L., Davies, J. E. D., MacNicol, D. D., eds.), Academic Press, London 1984, vol. 3, p. 297
25. Parsonage, N. G., Staveley, L. A. K.: ibid p. 1
26. Davies, J. E. D.: ibid p. 37
27. Swaminathan, S., Craven, B. M., Spackman, M. A., Stewart, R. F.: Acta Crystallogr. B, *B40*, 398 (1984)
28. Vaughan, P., Donohue, J.: Acta Crystallogr. *5*, 530 (1952)
29. Caron, A., Donohue, J.: Acta Crystallogr. B, *B25*, 404 (1969)
30. Wyckoff, R. W. G., Corey, R. B.: Z. Kristallogr. *89*, 102 (1934)
31. Swaminathan, S., Craven, B. M., McMullan, R. K.: Acta Crystallogr. B, *B40*, 300 (1984)
32. Pryor, A., Sanger, P. L.: Acta Crystallogr. A, *A26*, 543 (1970)
33. Guth, H., Heger, G., Klein, S., Treutmann, W., Scheringer, C.: Z. Kristallogr. *153*, 237 (1980)
34. Truter, M. R.: Acta Crystallogr. *22*, 556 (1967)
35. Rutherford, J. S., Calvo, C.: Z. Kristallogr. *128*, 229 (1969)
36. Sas, T. M., Suvorov, V. V., Efremov, V. A., Rudnev, V. V., Chukurov, P. M.: Zhur. Obshch. Khim. *54*, 587 (1984)
37. van Bekkum, H., Remijnse, J. D., Wepster, B. M.: J. Chem. Soc., Chem. Commun. *1969*, 67
38. Hadicke, E., Schlenk, W.: Liebigs Ann. Chem. *764*, 103 (1972)
39. Zimmerschied, W. J., Dinerstein, R. A., Weitkamp, A. W., Marschner, R. F.: Ind. Eng. Chem. *42*, 1300 (1950)
40. McAdie, H. G.: Can. J. Chem. *41*, 2144 (1963)
41. Logvinenko, V. A., Gegola, O. V., Chekhova, G. N., Dyadin, Yu. A.: Tezisy Dokl. Soveshch. Kinet. Mekh. Khim. Reacts. Tverd Tele, 7th 150 (1977); see Chem. Abstr. *88*, 198423y (1978)
42. Chatani, Y., Taki, Y., Tadokoro, H.: Acta Crystallogr. B, *B33*, 309 (1977)
43. Chatani, Y., Aranku, H., Taki, Y.: Mol. Cryst. Liq. Cryst. *48*, 219 (1978)
44. Yokoyama, F., Monobe, K.: J. Polymer Sci., Polymer Lett. Ed. *19*, 91 (1981)
45. Yokoyama, F., Monobe, K.: Polymer *24*, 149 (1983)
46. Chatani, Y., Kuwata, S.: Macromolecules *8*, 12 (1975)
47. Lenne, H. U.: Z. Kristallogr. *118*, 454 (1963)
48. Lenne, H. U., Mez, H. C., Schlenk, W.: Chem. Ber. *101*, 2435 (1968)
49. Chatani, Y., Yoshimori, K., Tatsuta, Y.: Polymer Prep., Am. Chem. Soc., Div. Polymer Chem. *19*, 132 (1978); see Chem. Abstr., *93*, 205097d (1980)
50. Clement, R., Mazieres, C., Guibe, L.: J. Solid State Chem. *5*, 436 (1972)
51. Otto, J.: Acta Crystallogr. B, *B28*, 543 (1972)
52. Casal, H. L.: J. Phys. Chem. *89*, 4799 (1985)
53. Le Brumant, J., Jaffrain, M., Lacrampe, G.: ibid *88*, 1548 (1984)
54. Hori, Y., Aoyuma, S., Kashiwabara, H.: J. Chem., Phys. *75*, 1582 (1981)
55. Ichikawa, T.: J. Phys. Chem. *81*, 2132 (1977)
56. Suryanarayana, D., Chamulitrat, W., Kevan, L.: ibid *86*, 4822 (1982)
57. Clement, R., Gourdji, M., Guibe, L.: J. Magn. Res. *20*, 345 (1975)
58. Casal, H. L., Cameron, D. G., Kelusky, E. C.: J. Chem. Phys. *80*, 1407 (1984)
59. Ripmeester, J. A.: Chem. Phys. Letters *74*, 536 (1980)
60. Kacher, M. L.: Eur. Pat. Appl. EP73,541 (Procter and Gamble Co); see Chem. Abstr., *98*, P217655a (1983)
61. Leadbetter, G.: U.S. Pat. 4,170,601; see Chem. Abstr., *92*, P6037p (1980)
62. Fujiwara, H., Sato, Y., Nishi, K.: (Takeda Chem. Ind. Ltd.); Takeda Kenkyusho Ho, *35*, 217 (1976); see Chem. Abstr., *87*, 1114y (1977)
63. Casal, H. L., de Mayo, P., Miranda, J. F., Scaiano, J. C.: J. Am. Chem. Soc. *105*, 5155 (1983)
64. Schiessler, R. W., Flitter, D.: ibid *74*, 1720 (1952)
65. Nishikawa, M.: Chem. Pharm. Bull. *11*, 977 (1963)
66. Fukushima, K.: J. Mol. Struct. *34*, 67 (1976)
67. Fukushima, K., Sugiura, K.: ibid *41*, 41 (1977)
68. Gustavsen, J. E., Klaeboe, P., Kvila, H.: Acta Chem. Scand. A, *A32*, 25 (1978)
69. Allen, A., Fawcett, V., Long, D. A.: J. Raman Spec. *4*, 285 (1976)

70. Christian, S. D., Grundnes, J., Klaeboe, P., Toerneng, E., Woldbaek, T.: Acta Chem. Scand. A, *A34*, 391 (1980)
71. Fukushima, K., Ohata, A.: Nippon Kagaku Kaishi 317 (1983); see Chem. Abstr., *98*, 214904v (1983)
72. Woldbaek, T., Klaeboe, P.: J. Mol. Struct. *63*, 195 (1980)
73. Fukushima, K., Takeda, S.: ibid *49*, 259 (1978)
74. Gerdil, R., Frew, A.: J. Incl. Phenom. *3*, 335 (1985)
75. Lenne, H. U.: Acta Crystallogr. *7*, 1 (1954)
76. Clement, R., Mazieres, C., Gourdji, M., Guibe, L.: J. Chem. Phys. *67*, 5381 (1977)
77. Meirovitch, E., Krant, T., Vega, S.: J. Phys. Chem. *87*, 1390 (1983)
78. Hough, E., Nicholson, D. G.: J. Chem. Soc., Dalton Trans. *1978*, 15
79. Luckhurst, G. R., Setaka, M.: J. Magn. Res. *25*, 539 (1977)
80. Hori, Y., Ohno, H., Shimada, S., Kashiwabara, H.: J. Phys. Chem. *89*, 3 (1985)
81. Lee, T. D., Birrell, G. B., Bjorkman, P. J., Keana, J. F. W.: Biochem. Biophys. Acta *550*, 369 (1979)
82. Bayle, J. P., Courtieu, J., Jullien, J., Kan, S. K.: Org. Magn. Res. *16*, 85 (1981)
83. Jpn. Kokai Tokkyu Koho, 80, 85,538 (Hisamitsu Pharmaceutical Co.); see Chem. Abstr., *93*, P220419d (1980)
84. Jpn. Kokai Tokkyu Koho 82, 75,950 (Suwa Seikosha Co.); see Chem. Abstr., *97*, P127288h (1982)
85. Ijam, M. J., Al-Zaid, K. A. H.: Ind. Eng. Chem., Prod. Res. Dev. *16*, 78 (1977)
86. Polyakova, A. A., Sergienko, S. R., Aidogdyev, A. A., Niyazov, B. G., Kogan, L. O.: Izv. Akad. Nauk. Turkm. SSR, Ser. Fiz.-Tekh. Khim. Geol. Nauk 83 (1978); see Chem. Abstr., *89*, 199923u (1978)
87. Serra Masia, A., Martinez Moreno, J. M.: Grasas Aceites (Seville) *32*, 313 (1981); see Chem. Abstr., *96*, 124913b (1982)
88. Chatani, Y., Nakatani, S.: Z. Kristallogr. *144*, 175 (1977)
89. Ichikawa, T., Nakao, O.: Asahi Garasu Kogyo Gijutsu Shoreikai Kenkyu Hokoku *30*, 1 (1977); see Chem. Abstr., *89*, 110492g (1978)
90. Sergeev, G. B., Komarov, V. S., Zvonov, A. V.: Zhur. Obshch. Khim. *53*, 2496 (1983)
91. Sergeev, G. B., Komarov, V. S., Zvonov, A. V.: ibid *54*, 985 (1984)
92. For a recent review see Giglio, E.: in "Inclusion Compounds" (Atwood, J. L., Davies, J. E. D., MacNicol, D. D., eds.), Academic Press, London 1984, vol. 2, p. 207
93. Cram, D. J., Cram, J. M.: Acc. Chem. Res. *11*, 8 (1978)
94. Coiro, V. M., Giglio, E., Mazza, F., Pavel, N. V.: J. Incl. Phenom. *1*, 329 (1984)
95. Coiro, V. M., Mazza, F., Pochetti, G., Pavel, N. V.: Acta Crystallogr. C, *C41*, 229 (1985)
96. De Sanctis, S. C., Coiro, V. M., Giglio, E., Pagliuca, S., Giglio, E., Pavel, N. V., Quagliata, C.: Acta Crystallogr. B, *B34*, 1928 (1978)
97. De Sanctis, S. C., Giglio, E., Petri, F., Quagliata, C.: ibid *B35*, 226 (1979)
98. Arad-Yellin, R., Green, B. S., Krossow, M., Tsoucaris, G.: in "Inclusion Compounds" (Atwood, J. L., Davies, J. E. D., MacNicol, D. D., eds.), Academic Press, London 1984, vol. 3 p. 263
99. Popovitz-Biro, R., Tang, C. P., Chang, H. C., Lahav, M., Leiserowitz, L.: J. Am. Chem. Soc. *107*, 4043 (1985)
100. Tang, C. P., Chang, H. C., Popovitz-Biro, R., Frolow, F., Lahav, M., Leiserowitz, L., McMullan, R. K.: ibid *107*, 4058 (1985)
101. Brunie, S., Tsoucaris, G.: Cryst. Struct. Commun. *3*, 481 (1974)
102. Williams, D. J., Lawton, D.: Tetrahedron Lett. 111 (1975)
103. Gerdil, R.: Top. Curr. Chem. *140*, 71 (1987)
104. Allemand, J., Gerdil, R.: Acta Crystallogr. B, *B38*, 1473 (1982)
105. Allemand, J., Gerdil, R.: ibid *B38*, 2312 (1982)
106. Allemand, J., Gerdil, R.: Cryst. Struct. Commun. *10*, 33 (1981)
107. Gerdil, R., Frew, A.: J. Incl. Phenom. *3*, 335 (1985)
108. Allemand, J., Gerdil, R.: Acta Crystallogr. C, *C39*, 260 (1983)
109. Brunie, S., Navaza, A., Tsoucaris, G., Declerq, J. P., Germain, G.: Acta Crystallogr. B, *B33*, 2645 (1977)

110. Arad-Yellin, R., Brunie, S., Green, B. S., Knossow, M., Tsoucaris, G.: J. Am. Chem. Soc. *101*, 7529 (1979)
111. Arad-Yellin, R., Green, B. S., Knossow, M., Rysanek, N., Tsoucaris, G.: J. Incl. Phenom. *3*, 317 (1985)
112. Gerdil, R., Barchietto, G., Jefford, C. W.: J. Am. Chem. Soc. *106*, 8004 (1984)
113. Downing, A. P., Ollis, W. D., Sutherland, I. O.: J. Chem. Soc. B *1970*, 24
114. Arad-Yellin, R., Green, B. S., Knossow, M.: J. Am. Chem. Soc. *102*, 1157 (1980)
115. Arad-Yellin, R., Green, B. S., Knossow, M., Tsoucaris, G.: ibid *105*, 4561 (1983)
116. Watson, J. D., Crick, F. H. C.: Nature *171*, 737 (1953)
117. Lerman, L. S.: J. Mol. Biol. *3*, 18 (1961)
118. Patel, D. J.: Acc. Chem. Res. *12*, 118 (1979)
119. Wilson, W. D., Jones, R. L.: Adv. Pharmacol. Chemother. *18*, 177 (1981)
120. Sobell, H. M.: "The Stereochemistry of Actinomycin Binding to DNA and its Implications in Molecular Biology" in: Progress in Nucleic Acid Research and Molecular Biology (Davidson, J. N., Cohn, W. E., eds.), Academic Press, New York 1973, p. 153
121. Angeloni, L., Smulevich, G., Marzocchi, M. P.: Spectrochim. Acta *38A*, 213 (1982)
122. Pigram, W. J., Fuller, W., Hamilton, L. D.: Nature (New Biology) *235*, 17 (1972)
123. Quigley, G. J., Wang, A. H.-J., Ughetto, G., van der Marel, G., van Boom, J. H., Rich, A.: Proc. Natl. Acad. Sci. (USA) *77*, 7204 (1980)
124. Shields, H., McGlumphy, C., Hamrick, P. J.: Biochem. Biophys. Acta *697*, 113 (1982)
125. Sobell, H. M., Reddy, B. S., Bhandary, K. K., Jain, S. C., Sakore, J. D., Seshadri, T. P.: Cold Spring Harbor Symp. Quant. Biol. *42*, 87 (1977)
126. Shafer, R. H.: Biopolymers *19*, 419 (1980)
127. Atwell, G. J., Leupin, W., Twigden, S. J., Denny, W. A.: J. Am. Chem. Soc. *105* 2913 (1983)
128. Hansen, J. B., Koch, T., Buchardt, O., Nielsen, P. E., Norden, B., Wirth, M.: J. Chem. Soc., Chem. Commun. *1984*, 509
129. Hansen, J. B., Thomsen, T., Buchardt, O.: J. Chem. Soc., Chem. Commun. *1983*, 1015
130. Miller, K. J., Newlin, D. D.: Biopolymers *21*, 633 (1982)
131. Colin, J. J., de Claubry, H. G.: Ann. Chim. *90*, 87 (1814)
132. Schoch, T., Williams, C.: J. Am. Chem. Soc. *66*, 1232 (1944)
133. Mikus, F. F., Hixon, R. M., Rundle, R. E.: ibid *68*, 1115 (1946)
134. Cramer, F., Bergmann, U., Manor, P. C., Noltemeyer, M., Saenger, W.: Liebigs Ann. Chem. *1976*, 1169
135. Noltemeyer, M., Saenger, W.: J. Am. Chem. Soc. *102*, 2710 (1980)
136. Hanes, C. S.: New Phytologist *36*, 101 and 189 (1937)
137. Freudenberg, K., Schaaf, E., Dumpert, G., Ploetz, T.: Naturwissenschaften *27*, 850 (1939)
138. Rundle, R. E., Baldwin, R. R.: J. Am. Chem. Soc. *65*, 554 (1943)
139. Rundle, R. E., French, D.: ibid *65*, 558 (1943)
140. Rundle, R. E., Foster, J. F., Baldwin, R. R.: ibid *66*, 2116 (1944)
141. Rundle, R. E., French, D.: ibid *65*, 1707 (1943)
142. Rundle, R. E., Edwards, F. C.: ibid *65*, 2200 (1943)
143. Rundle, R. E.: ibid *69*, 1769 (1947)
144. Bear, R. S.: ibid *64*, 1388 (1942)
145. Bear, R. S.: ibid *66*, 2122 (1944)
146. Cramer, F.: Rev. Pure Appl. Chem. *5*, 143 (1955)
147. Senti, F. R., Erlander, S. R.: in "Non-Stoichiometric Compounds" (Mandelcorn, L., ed.), Academic Press, New York 1964, p. 568
148. Winter, W. T., Sarko, A.: Biopolymers *13*, 1447 (1974)
149. Murphy, V. G., Zaslow, B., French, A. D.: ibid *14*, 1487 (1975)
150. Zugenmaier, P., Sarko, A.: ibid *15*, 2121 (1976)
151. Rappenecker, G., Zugenmaier, P.: Carbohydr. Res. *89*, 11 (1981)
152. Valletta, R. M., Germino, F. J., Lang, R. E., Moshy, R. J.: J. Polymer Sci. *A2*, 1085 (1964)
153. Winter, W. T., Sarko, A.: Biopolymers *13*, 1461 (1974)
154. Bluhm, T. L., Zugenmaier, P.: Carbohydr. Res. *89*, 1 (1981)
155. Zaslow, B., Miller, R. L.: J. Am. Chem. Soc. *83*, 4378 (1961)
156. Teitelbaum, R. C., Ruby, S. L., Marks, T. J.: ibid *102*, 3322 (1980)
157. Saenger, W.: Naturwissenschaften *71*, 31 (1984)

158. Rietman, E. A.: J. Mater. Sci. Letter *3*, 1043 (1984)
159. Whistler, R., Hilbert, G.: J. Am. Chem. Soc. *67*, 1161 (1945)
160. Carlson, T. L.-G., Larsson, K., Dinh-Nguyen, N., Krog, N.: Starch *31*, 222 (1979)
161. Hoover, R., Hadziyev, D.: ibid *33*, 290 (1981)
162. Takeo, K., Tokumura, A., Kuge, T.: ibid *25*, 357 (1975)
163. Davies, T., Miller, D. C., Procter, A. A.: ibid *32*, 149 (1980)
164. Bulpin, P. V., Welsh, E. J., Morris, E. R.: ibid *34*, 335 (1982)
165. Hui, Y., Russell, J. C., Whitten, D. G.: J. Am. Chem. Soc. *105*, 1374 (1983)
166. Szejtli, J., Banky-Elod, E.: Starch *30*, 85 (1978)
167. Wu, H. C. H., Sarko, A.: Carbohydr. Res. *61*, 27 (1978)
168. Wu, H. C. H., Sarko, A.: ibid *61*, 7 (1978)
169. Sarko, A., Wu, H. C. H.: Starch *30*, 73 (1978)
170. Land, E. H.: U.S. Pat. 2,237,567 (Polaroid Corp.)
171. Joshi, D. P., Lan-Chun-Fung, Y. L., Pritchard, J. G.: Anal. Chim. Acta *104*, 153 (1979)
172. Zwick, M. M.: J. Appl. Polymer Sci. *9*, 2393 (1965)
173. Cesaro, A., Brant, D. A.: Biopolymers *16*, 983 (1977)
174. Bluhm, T. L., Zugenmaier, P.: Prog. Colloid and Polymer Sci. *64*, 132 (1978)
175. Bluhm, T. L., Zugenmaier, P.: Polymer *20*, 23 (1979)
176. Buchele, C., Zugenmaier, P.: Colloid and Polymer Sci. *258*, 768 (1980)
177. Hui, Y., Winkle, J. R., Whitten, D. G.: J. Phys. Chem. *87*, 23 (1983)
178. Kusuyama, H., Miyamoto, N., Chatani, Y., Tadokoro, H.: Polymer Commun. *24*, 119 (1983)
179. Bosscher, F., ten Brinke, G., Challa, G.: Macromolecules *15*, 1442 (1982)
180. Hirotsu, K., Higuchi, T., Fujita, K., Ueda, T., Shinoda, A., Imoto, T., Tabushi, I.: J. Org. Chem. *47*, 1143 (1982)
181. Yalpani, M., Hall, L. D.: Macromolecules *17*, 272 (1984)
182. Newesely, H., Hosemann, R., Uther, B.: Z. Naturforsch. *35C*, 177 (1980)
183. Lindquist, Y., Branden, C.-I.: J. Mol. Biol. *143*, 201 (1980)
184. Klug, A.: Angew. Chem. *95*, 59 (1983); Angew. Chem., Int. Ed. Engl. *22*, 565 (1983)
185. Pressman, B. C.: Ann. Rev. Biochem. *45*, 501 (1976)
186. Dobler, M.: "Ionophores and Their Structure", Wiley, New York 1981
187. Ovchinnikov, Yu. A., Ivanov, V. T., Shkrob, A. M.: "Membrane-Active Complex... B.B.A. Library Vol. 12, Elsevier, Amsterdam 1974
188. Vögtle, F., Weber, E.: Angew. Chem. *91*, 813 (1979); Angew. Chem., Int. Ed. Engl. *18*, 753 (1979)
189. Westley, J. W.: Ann. Rep. Med. Chem. *10*, 246 (1975)
190. Hilgenfeld, R., Saenger, W.: Top. Curr. Chem. *101*, 1 (1982)
191. Painter, G. R., Pressman, B. C.: ibid *101*, 83 (1982)
192. Shinkai, S., Manabe, O.: ibid *121*, 67 (1984)
193. Shanzer, A., Libman, J., Frolow, F.: Acc. Chem. Res. *16*, 60 (1983)
194. Racker, E.: ibid. *12*, 338 (1979)
195. Oroshnik, W., Mebane, A. D.: "The Polyene Antifungal Antibiotics" in: Fortschritte der Chemie Organischer Naturstoffe (Zechmeister, L., ed.), Springer Verlag, Wien 1963, vol. 21, p. 17
196. Holz, R., Finkelstein, A.: J. Gen. Physiol. *56*, 125 (1970)
197. Cass, A., Finkelstein, A., Krespi, V.: ibid *56*, 100 (1970)
198. De Kruijff, B., Demel, R. A.: Biochim. Biophys. Acta *339*, 57 (1974)
199. Marty, A., Finkelstein, A.: J. Gen. Physiol. *65*, 515 (1975)
200. Brown, J. M., Derome, A. E., Kimber, S. J.: Tetrahedron Lett. *26*, 253 (1985)
201. Nagaraj, R., Balaram, P.: Acc. Chem. Res. *14*, 356 (1981)
202. Mathew, M. K., Balaram, P.: Molec. Cell Biochem. *50*, 47 (1983)
203. Hall, J. E., Vodyanoy, I., Balasubramanian, T. M., Marshall, G. R.: Biophys. J. *45*, 233 (1984)
204. Fox, R. O., Richards, F. M.: Nature *300*, 325 (1982)
205. Bannerjee, H., Tsui, F. P., Balasubramanian, T. M., Marshall, G. R., Chan, S. I.: J. Mol. Biol. *165*, 757 (1983)
206. Van Duijnen, P. Th., Thole, B. T.: Chem. Phys. Lett. *83*, 129 (1981)
207. Dubos, R. J.: J. Exp. Med. *70*, 1 (1939)
208. Sarges, R., Witkop, B.: J. Am. Chem. Soc. *87*, 2011 (1965)
209. Sarges, R., Witkop, B.: Biochem. *4*, 2491 (1965)

210. Ivanov, V. T., Sychev, S. V.: "The Gramicidin A Story" in: "Structure of Complexes Between Biopolymers and Low Molecular Weight Molecules" (Snatzke, G., Bartmann, W., eds.), Wiley, New York 1982, p. 107
211. Hladky, S. B., Haydon, D. A.: "Ion Movements in Gramicidin Channels" in: "Current Topics in Membranes and Transport" (Bronner, F., Stein, W. D., eds.), Academic Press, New York 1984, vol. 21, p. 327
212. Andersen, O. S.: Ann. Rev. Physiol. *46*, 531 (1984)
213. Tosteson, D. C., Andreoli, T. E., Tieffenberg, M., Cook, P.: J. Gen. Physiol. *51*, 373s (1968)
214. Urry, D. W.: Proc. Natl. Acad. Sci. (USA) *68*, 672 (1971)
215. Urry, D. W., Goodall, M. C., Glickson, J. D., Mayers, D. F.: ibid *68*, 1907 (1971)
216. Bradley, R. J., Urry, D. W., Okamoto, K., Rapaka, R.: Science *200*, 435 (1978)
217. Veatch, W. R., Fossel, E. T., Blout, E. R.: Biochem. *13*, 5249 (1974)
218. Hodgkin, D. C.: Cold Spring Harbor Symp. Quant. Biol. *14*, 65 (1949)
219. Synge, R. L. M.: ibid *14*, 191 (1949)
220. Cowan, P. M., Hodkin, D. C.: Proc. Roy. Soc. Ser. B *141*, 89 (1953)
221. Koeppe, R. E., Hodgson, K. O., Stryer, L.: J. Mol. Biol. *121*, 41 (1978)
222. Koeppe, R. E., Berg, J. M., Hodgson, K. O., Stryer, L.: Nature *279*, 723 (1979)
223. Weinstein, S., Durkin, J. T., Veatch, W. R., Blout, E. R.: Biochem. *24*, 4374 (1985)
224. Wallace, B. A., Veatch, W. R., Blout, E. R.: ibid *20*, 5754 (1981)
225. Henze, R., Neher, E., Trapane, T. L., Urry, D. W.: J. Membrane Biol. *64*, 233 (1982)
226. Urry, D. W., Alonso-Romanowski, S., Venkatachalam, C. M., Bradley, R. J., Harris, R. D.: ibid *81*, 205 (1984)
227. Larsen, H.: Adv. Microbial Physiol. *1*, 97 (1967)
228. Oesterhelt, D., Stoeckenius, W.: Nature, New Biol. *233*, 149 (1971)
229. Stoeckenius, W., Lozier, R., Bogomolni, R. A.: Biochim. Biophys. Acta *505*, 215 (1979)
230. Henderson, R., Unwin, P. N. T.: Nature *257*, 28 (1975)
231. Ovchinnikov, Yu. A., Abdulaev, N. G., Feigina, M. Yu., Kiselev, A. V., Lobanov, N. A.: FEBS Letters *100*, 219 (1979)
232. Engelman, D. M., Henderson, R., McLachlan, A. D., Wallace, B. A.: Proc. Natl. Acad. Sci. (USA) *77*, 2023 (1980)
233. Khorana, H. G., Gerber, G. E., Herlihy, W. C., Gray, C. P., Anderegg, R. J., Nihei, K., Bieman, K.: ibid *76*, 5046 (1979)
234. Ovchinnikov, Yu. A.: FEBS Letters *148*, 179 (1982)
235. Engelman, D. M., Zaccai, G.: Proc. Natl. Acad. Sci. (USA) *77*, 5894 (1980)
236. Stoeckenius, W.: Acc. Chem. Res. *13*, 337 (1980)
237. Ovchinnikov, Yu. A.: in "Physical Chemistry of Transmembrane Ion Motions" (Spach, G., ed.), Studies in Physical and Theoretical Chemistry, Elsevier, Amsterdam 1983, vol. 24, p. 437
238. Klingenberg, M.: Nature *290*, 449 (1981)
239. Kyte, J.: ibid *292*, 201 (1981)
240. Deisenhofer, J., Epp, O., Miki, K., Huber, R., Michel, H.: ibid *318*, 618 (1985)
241. van Beijnen, A. J. M., Nolte, R. J. M., Zwikker, J. W., Drenth, W.: Rec. Trav. Chim. Pays-Bas *101*, 409 (1982)
242. Behr, J.-P., Lehn, J.-M., Dock, A.-C., Moras, D.: Nature, *295* 526 (1982)
243. Lehn, J.-M.: in "Physical Chemistry of Transmembrane Ion Motions" (Spach, G., ed.), Studies in Physical and Theoretical Chemistry, Elsevier, Amsterdam 1983, vol. 24, p. 181
244. Fuhrhop, J.-H., Liman, U.: J. Amer. Chem. Soc. *106*, 4643 (1984)
245. Fuhrhop, J.-H., Liman, U., David, H. H.: Angew. Chem. *97*, 337 (1985); Angew. Chem., Int. Ed. Engl. *24*, 339 (1985)
246. Butters, T., Hutter, P., Jung, G., Pauls, N., Schmitt, H., Sheldrick, G. M., Winter, W.: Angew. Chem. *93*, 904 (1981); Angew. Chem., Int. Ed. Engl. *20*, 889 (1981)
247. Toniolo, C., Bonora, G. M., Benedetti, E., Bavoso, A., Di Blasio, B., Pavone, V., Pedone, C.: Biopolymers *22*, 1335 (1983)
248. Francis, A. K., Iqbal, M., Balaram, P., Vijayan, M.: ibid *22*, 1499 (1983)

Recent Progress in Molecular Recognition

Julius Rebek, Jr.

Department of Chemistry, University of Pittsburgh, Pittsburgh, PA 15260, USA

Table of Contents

Molecular recognition is the most recent term for chemical phenomena as old as Fischer's lock-and-key notion of enzyme action. Over the years, as model systems became available a number of other terms have been used: inclusion complexes, weak molecular complexes and host-guest chemistry, but molecular recognition has the advantage of being appreciated by a wider audience. Synthetic organic chemists, for example, understand it as reagent selectivity whereas mechanistic chemists can interpret it in terms of stereoelectronic effects. The recent, gradual drift of organic chemists into more biorelevant research has resulted in a rich array of sophisticated model systems. Such models are intended to imitate the specific recognition of substrates by their enzymes, antigens by antibodies or messengers by their receptors. A number of recent reviews [1] and monographs [2] on this subject have appeared, and even an entire journal is devoted to inclusion phenomena [3]. These vehicles provide momentum to a rapidly growing field. In the present review, we explore some of the more recent issues and discoveries.

1 Molecular Architecture

In most model systems for enzymes the intent is to form a complex, then perform chemical operations between the components of the complex (Scheme 1). The recognition step involves the matching of shapes, sizes and functionality between the two reacting partners and most of the literature deals with the energetics and dynamics of this process. A striking feature of structures used has been their macrocyclic shape. Cyclodextrins, crown ethers and, more recently, cyclophanes (*1–3*) have been the most frequently used systems in the literature of bioorganic chemistry. Crown ethers and especially their 3-dimensional cryptand or spherand-type derivatives have been developed to an extraordinary degree of refinement for the selective binding of spherical metal ions[4]. Such ions, after all, present complementary outer surfaces to the inner surfaces of structures which are lined with Lewis bases. At the molecular level, most research has dealt with the primary ammonium group, a positively charged "knob" that protrudes from larger structures[5], and some success has also been scored in recognition of slightly larger cations such as the guanidinium ions[6].

Scheme 1

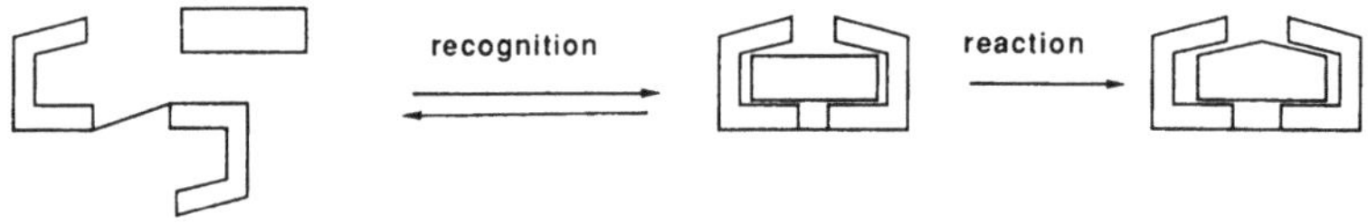

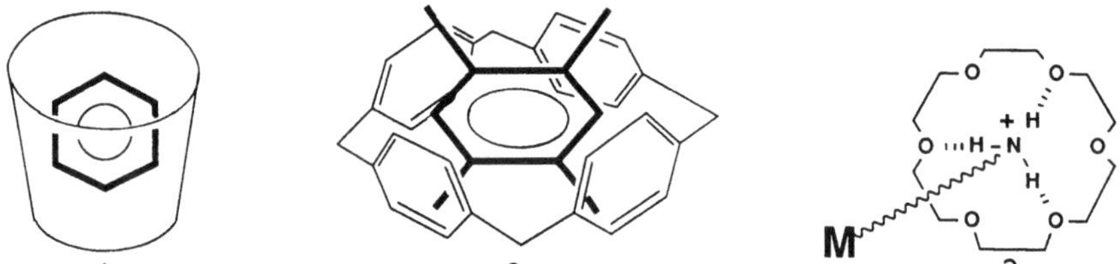

1 Cyclodextrin - benzene complex

2 Cyclophane - durene complex

3 Crown ether - ammonium complex

It has also been possible to develop macrocyclic structures for the recognition of anions[7] and Lewis basic sites. There are relatively few of these, however, and their numbers reflect the generally more difficult task of arranging Lewis acids in a convergent sense vs. arranging Lewis bases in such a manner. Indeed, the special problems posed by converging carboxylic acids led our own research, and this is an account of our recent adventures in this area.

The geometric nature of the problem is posed by the very structure of a carboxylic acid *4*. If two, or even more, of such functions are to converge on a single basic site, the least one can anticipate is that a U-turn must be engineered into the system. This arises from the direction of the carbonyl-C_α bond which points in a direction roughly opposite to the direction indicated by the OH bond. From this perspective, the tricarboxylic acid *5a* first described by Kemp [8] offers an unusually favorable architecture. In this, three methyls force the smaller carboxyl groups to assume a triaxial arrangement and cause a U-shaped relationship between any two carboxyl functions. This conformation was originally deduced by Kemp [8] from basis of NMR evidence and it was confirmed by crystallographic analysis [9] of either the triacid or the trimethyl ester *5a*, each of which feature the triaxial orientation.

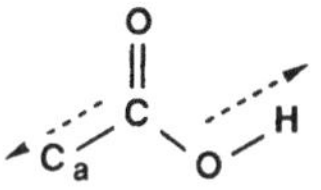

4

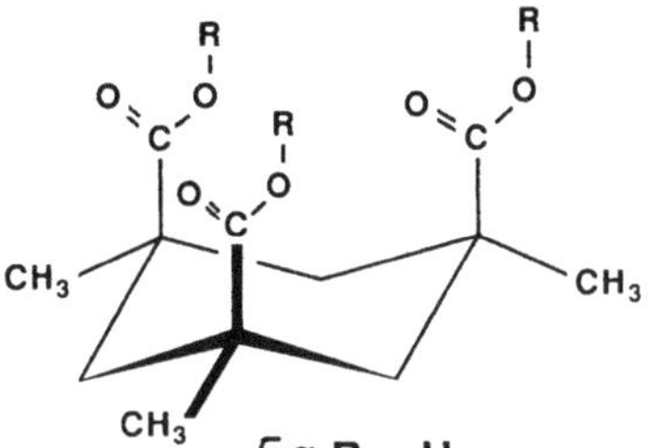

5a R = H
5b R = Me

It was our decision to combine two diacid units with a diamine as shown in Eq. (1) that gave the solution to the problem. The diamine acts as a spacer group that converts two U-shaped subunits into a C-shaped molecule *6* which now involves a near perfect focus of two carboxyl groups. Enforcing this shape, rather than the S-shape *7* which is also possible can be accomplished by further remote structural elements.

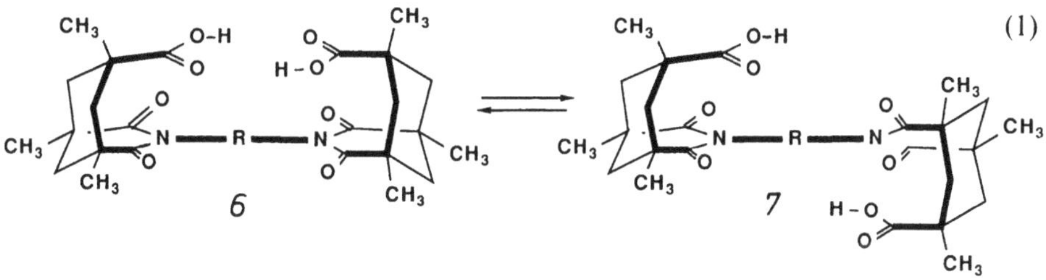

(1)

Condensation of the triacid with aniline gave an imide *8* for which the barrier to rotation about the C_{aryl}-N_{imide} bond indicated was about 13 kcal/mole[9]. However, ortho substituents stop this rotation and fix the conformation in such a way that the alkyl (or practically any other group) is directed away from the carboxyl function as in *9*. The seemingly passive methyl groups of this structure thereby limit the internal rotations and enforce a fairly rigid structure.

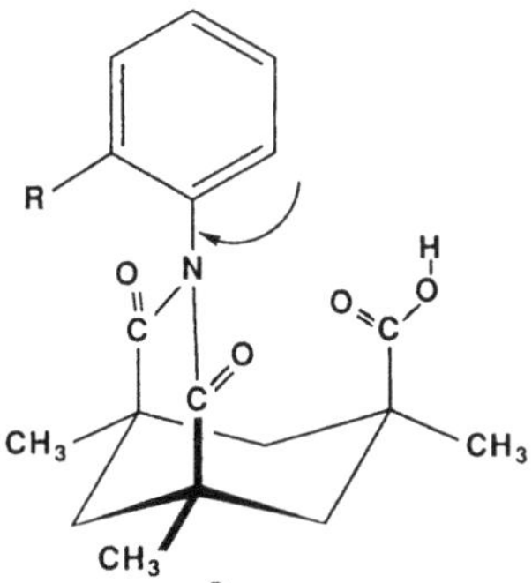

8 R = H
9 R = CH_3

The simplest case for complete enforcement of the convergent conformation was provided by meta xylidine diamine *10*. Condensation of two triacids with this amine in the melt gave excellent yields of the dicarboxylic acid *11* (Eq. (2)). The architectural cliche is easily repeated in the naphthalene series to provide *12* and in the acridine series to provide *13*. The latter is derived from the commercially available acridine yellow at very low cost.

(2)

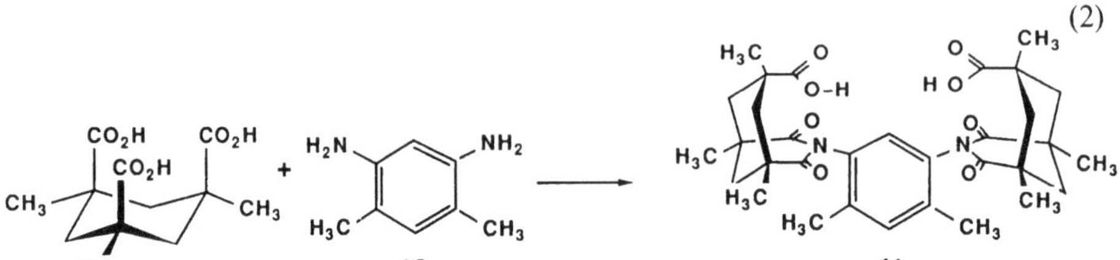

10 11

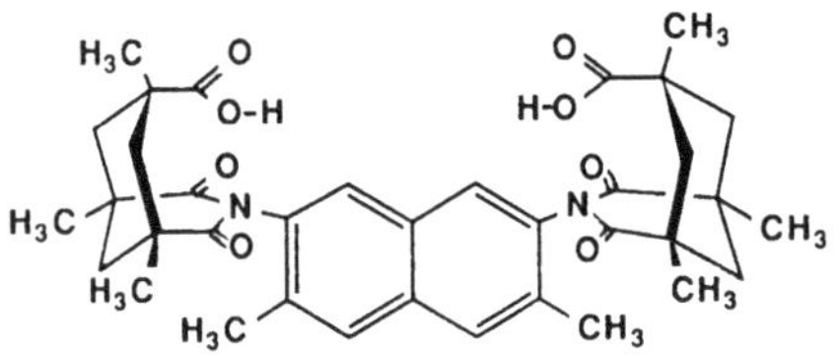

12

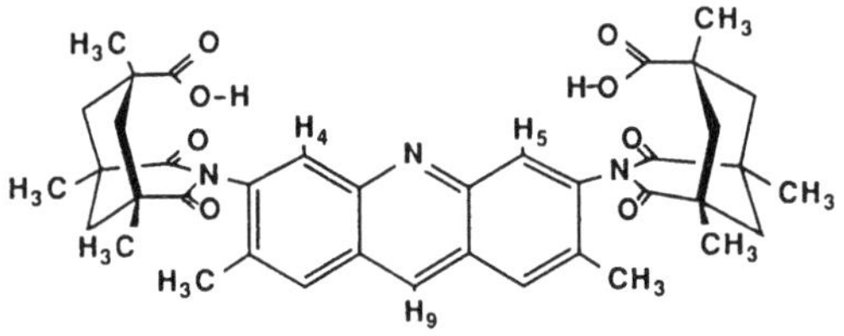

13

These three structures represent a simple series of molecules which feature a molecular cleft formed by the convergence of two carboxyl groups. In principle, any primary amines are capable of generating the corresponding acids on condensation and we have converted a series of α,ω-diamino-alkanes to the diacids *14* using this technology. The reaction is not limited to diamines. For example, the triamine TREN also condenses with three equivalents of triacid to give a molecule *15* capable of having three convergent carboxylic acids.

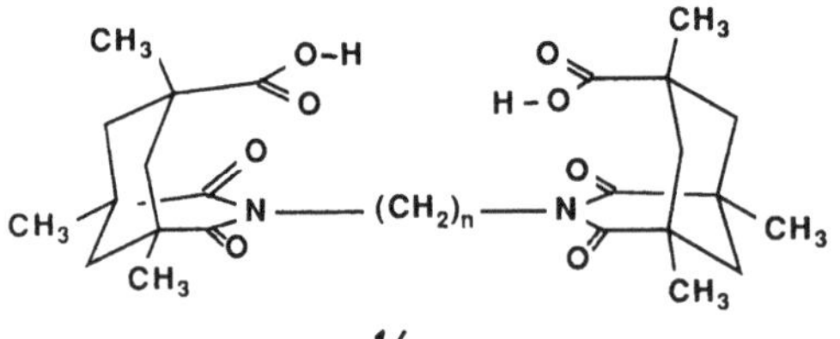

14

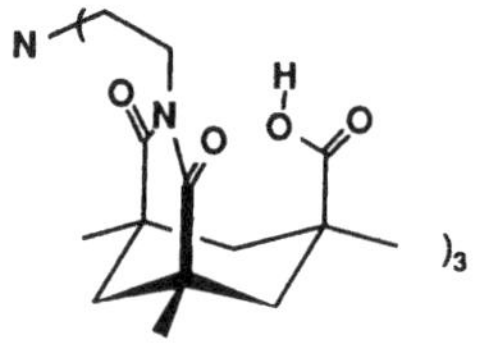

15

1.1 Rigidity vs. Flexibility

Of course, with these alkane spacers rotation is possible about any number of σ-bonds and the convergent conformation is only one of many available to the molecule. This flexibility permits variable distances between opposing carboxyl functions. The price paid for this adjustability is the relatively low probability of any single conformation. The advantage of fine-tuning then, is somewhat offset by the entropic price of selecting one conformation from among many. This situation has its counterpart in biological systems[10]. For example, the high specificity that antibodies show their antigens suggests a rigid binding site, capable of little flexibility in attaining an exquisite fit. Enzymes however, need to conform to more dynamic situations that involve substrates, transition states, intermediates and products. Their active sites must enjoy some flexibility to accommodate this range of structures. Between these two functions, receptors might represent intermediate flexibility. Some conformational changes must take place on binding the messenger because information could be transmitted to the rest of the system through such motion. Conformational changes are also involved in allosteric effects [11] that give rise to regulation in subunit enzyme systems.

The optimal degree of rigidity or flexibility in model systems is a matter of some discussion. With molecules as small as these, finding a perfect fit between substrate and receptor would be fortunate. Some structurally rigid cavities do show tight binding to metal ions [4], a feature attributed to preorganization. However, the enthalpic destabilization that such systems suffer in the uncomplexed state also must contribute to the observed binding tenacity. With larger cavities and substrates it may be possible to preorganize without such destabilization. The importance of rigidity in binding could then be properly assessed.

1.2 Template Effects

The condensation reactions described above are unique in yet another sense. The conversion of an amine, a basic residue, to a neutral imide occurs with the simultaneous creation of a carboxylic acid nearby. In one synthetic event, an amine acts as the template and is converted into a structure that is the complement of an amine in size, shape and functionality. In this manner the triacid *15* shows high selectivity toward the parent triamine in binding experiments. Complementarity in binding is self-evident. Cyclodextrins for example, provide a hydrophobic inner surface complementary to structures such as benzenes, adamantanes and ferrocenes having appropriate shapes and sizes [12] (cf. *1*). Complementary functionality has been harder to arrange in macrocycles; the lone pairs of the oxygens of crown ethers and the π-surfaces of the cyclophanes are relatively inert [13]. Catalytically useful functionality such as carboxylic acids and their derivatives are available for the first time within these new molecular clefts.

2 Properties of the New Diacid Systems

2.1 Physical Properties

The carboxyl functions in the new structures are buried within the clefts in a manner that discourages the formation of the intermolecular hydrogen bonded dimers which are usually observed in the solution and solid phases. Unusual acid-base behavior is one consequence. In the smallest system *11* (represented by the benzene spacer) a tremendous difference in pK_a's (6 units) can be observed for the two ionizations [14].

The dianion *16* derived from this structure has such an intensity of negative charge compressed into a small space that an ideal microenvironment is created for divalent metals ions *17* [Eq. (3)]. The molecule enjoys a vise-like shape and a grip that holds calcium or magnesium ions tightly [15].

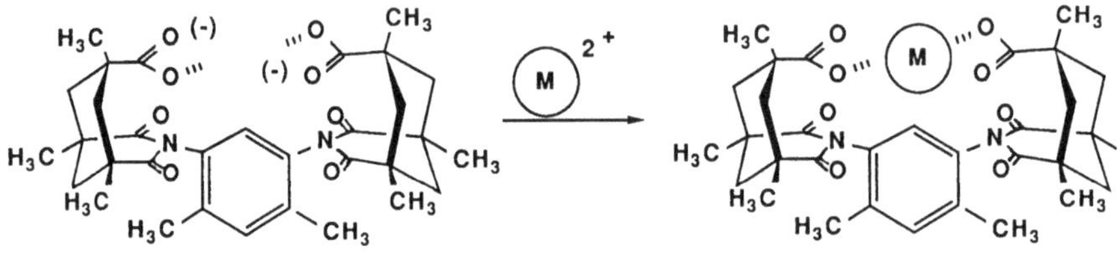

16 17 (3)

This affinity for metals results not only from the structural organization of the new diacids but from stereoelectronic effects at carboxyl oxygen as well. The in-plane lone pairs of a carboxylate *18* differ in basicity by several orders of magnitude [16]. Conventional chelating agents [17] derived from carboxylic acids such as EDTA, *19a* are constrained by their shape to involve the less basic anti lone pairs [Eq. (4)]. The new diacids are permitted the use of the more basic syn lone pairs in contact with the metal *19b*. These systems represent a new type of chelate for highly selective recognition of divalent ions.

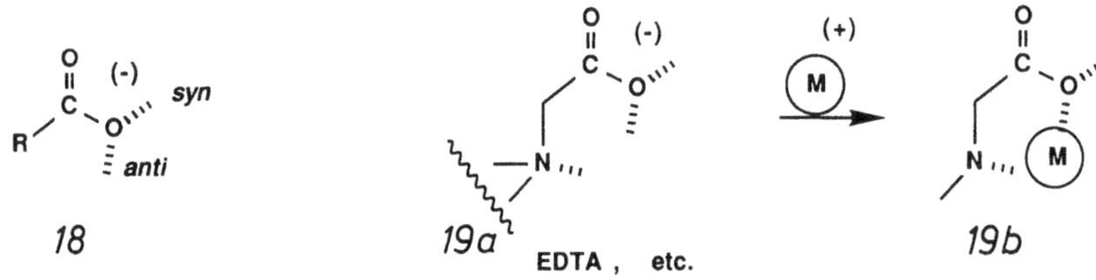

18 19a 19b (4)

2.2 Complexation of Diamines

The distance of 8–9 Å between opposing carboxyl functions in the acridine derived *13* suggested that selective binding to diamines would result since these structures could provide complementary functionality to the convergent dicarboxylic acids. A number of diamines have been tested in this context, using NMR techniques in $CDCl_3$, and other solvents in which proton transfer is not expected to contribute much to the binding. After all, it would be unfortunate to overwhelm the subtleties of molecular recognition with mere acid-base chemistry. The results are listed in Table 1[18,19].

The high selectivity that the system shows to pyrazine *20* compared to the stronger base pyridine, indicates that the diamine is chelated between the carboxylic acid functions as in *21*. Spectroscopic evidence in the form of upfield shifts in the NMR spectra of the complexes supports such structures. Not only aromatic diamines are accommodated but also aliphatics such as 1,4-diazabicyclo[2.2.2]octane (DABCO) in complex *22*. Typically, exchange rates into and out of these complexes are such that they appear fast on the NMR time scale at ambient temperature, but exchange can be frozen out at low temperatures[20]. For DABCO, an activation barrier of 10.5 kcal M^{-1} was observed at $T_c = 208$ °K.

Table 1. Association constants for complexes of *13*, $CDCl_3$, 25°

Base + *13* $\xrightarrow{K_a}$ Complex

Base	K_a (M^{-1} L)	pK_a (BH^+)
1,4-Diazabicyclo-[2.2.2]octane (DABCO)	1.6×10^5	8.2, 4.2
Pyrazine	1.4×10^3	0.65
Quinoxaline	23×10^3	0.56
Phenazine	2.2×10^3	1.2
Pyrimidine	7×10^3	1.3
Quinazoline	1.6×10^3	1.9[a]
Imidazole	$K_1 = 1.0 \times 10^6$, $K_2{}^b = 5.5 \times 10^4$	6.9
Benzmidazole	$K_1 = 1.5 \times 10^4$, $K_2{}^b = 7.5 \times 10^3$	5.5
Purine[c]	$\sim 8 \times 10^3$	2.4
Pyridine	$K_1 = 1.2 \times 10^2$, $K_2{}^b = < 1$	5.2

[a] For the unhydrated form: "The Chemistry of Heterocyclic Compounds", D. J. Brown, Ed., Interscience, NY, 1967, Vol. 24, Part 2 (Quinalzolines), Ch. 11.
[b] (M^{-2} L^2).
[c] Low solubility of this base required such high dilution that some uncertainty was encountered in determining nmr shifts.

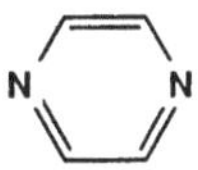

20

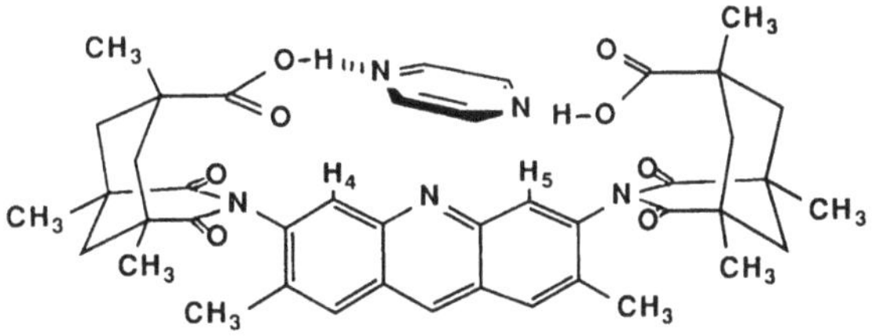

21

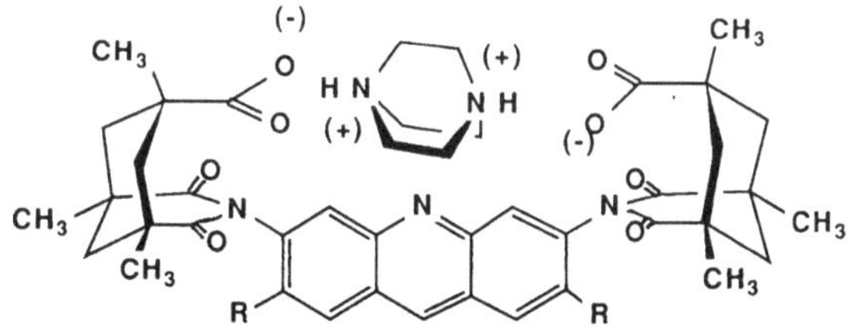

22

Experiments with quinoxaline complex *23* and phenazine complex *24* established that additional binding interactions were available in the form of aryl-aryl stacking between aromatic subunits in the components. In the case of quinoxaline this accounts for about 1.6 kcal or a factor of 15 in K_a. In *24*, these attractive forces are partially offset by steric effects introduced by the remote ring as shown.

Imidazole and its derivatives revealed yet another binding modality. In imidazole both a hydrogen bond donor and an acceptor are present on the opposite sides of the molecule, and this relatively strong base can elicit some proton transfer with the

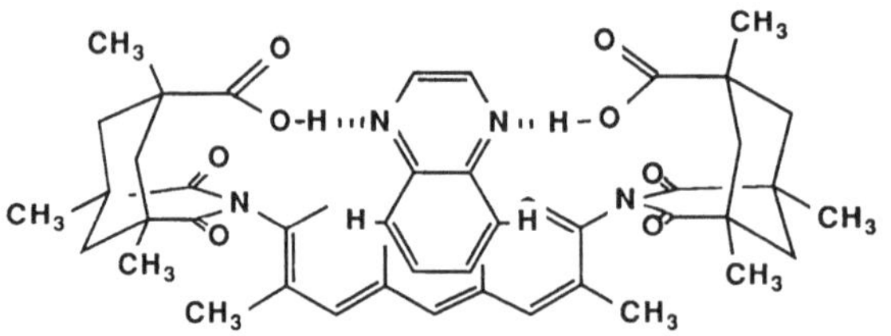

23

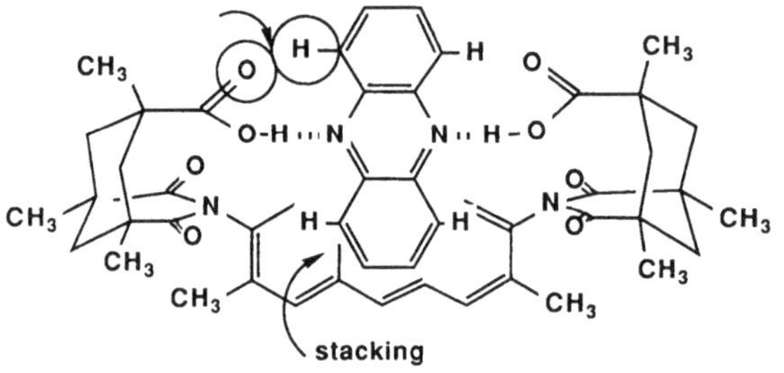

24

diacids. Titration experiments indicated 2:1 stoichiometry, as may be expected from the complexation in a complementary sense as shown in Eq. (5). It is surprising that the two association constants involved in the formation of the termolecular complex *25b* are comparable in size, since some degree of electrostatic repulsion is expected when the second imidazolium ion/carboxylate pair is created. Perhaps a cooperative effect is operational on hydrogen bonding to both nitrogens of an imidazole and both oxygens of a carboxylate in the same way that hydroxyl groups can be involved in cooperative hydrogen bonds[21].

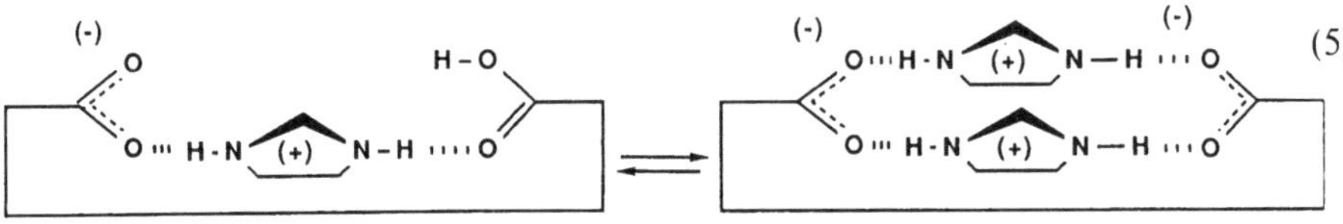

(5)

25a 25b

A measure of the promiscuity of these diacids was attempted with pyrimidine *26* vs pyrazine *20*. The two heterocycles differ in size, shape and also basicity; in addition, stacking interactions can be observed in the pyrimidine complex but not with pyrazine. These four variables make interpretation with confidence somewhat futile since it is difficult to attribute the binding differences to any particular feature.

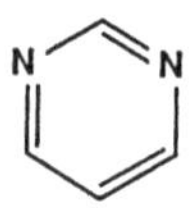

26

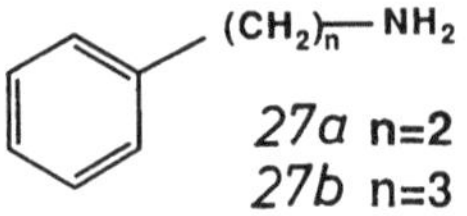

27a n=2
27b n=3
27c n=4

2.3 Aryl Amine Complexes

A number of aliphatic amines bearing aromatic nuclei revealed an unusual range of stoichiometries in titration experiments with the acridine-derived diacid[22a]. Figure 1 shows a plot of the chemical shifts as a function of stoichiometry for β-aryl-ethylamines *27a*. Identical behavior can be seen with phenylalanine derivatives such as the alcohol *28a* and the methyl ester *28b*, or tryptamine *29*. In contrast, γ-phenyl-propylamine *27b*, or δ-phenyl-butylamine *27c* show conventional aromatic signals throughout the titration protocols. Direct competition experiments showed these latter bases fail to interfere with the tight binding to the β-aryl-ethylamines; a very special relationship between the acridine diacid and these amines is apparent. The inescapable conclusion from the titration experiments is that ternary complexes having two molecules of the receptor are involved. An initially formed carboxylate ammonium ion pair is further stabilized by hydrogen bonding and, most likely, stacking interactions involving a second receptor. For all of the amines that form 2:1 complexes, the association constants are quite large, on the order of 10^7 M^{-2} L^2.

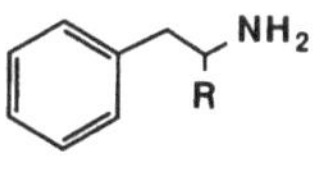

28a R = CH_2OH
28b R = CO_2CH_3

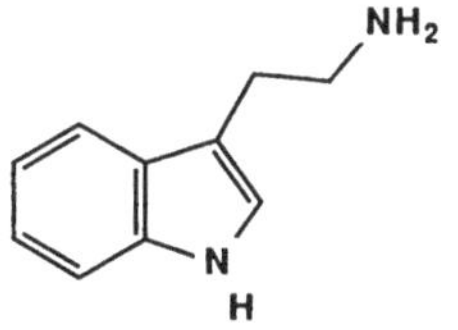

29

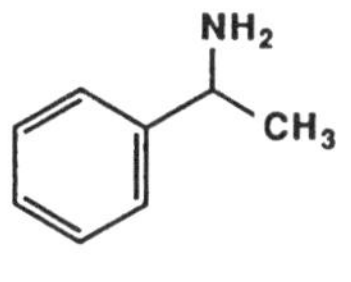

30

The formation of such complexes apparently involves a delicate balance of binding forces, since α-phenyl-ethylamine *30* shows only modest tendencies to form 2:1 complexes and its stacking efficiency is reduced. The structural details of these complexes are not known, but intermolecular NOE experiments favor structures such as shown in *31*. The distance between the aromatic and amine recognition sites in the

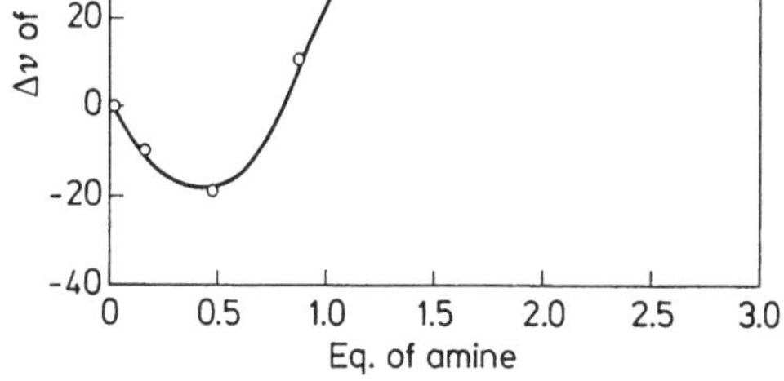

Fig. 1. Changes in the chemical shift of protons lining the cleft of *13* (H_4, H_5) as β-phenyl-ethylamine is added

acridine-derived diacid is quite similar to that proposed in a recent model for the receptors of the central nervous system [22b)].

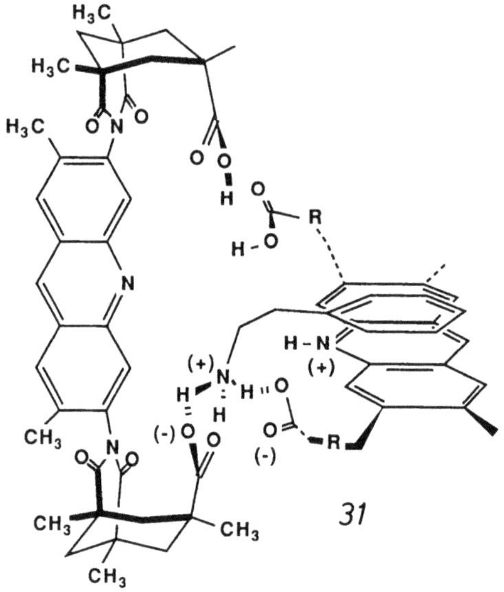

31

2.4 Recognition of Dicarboxylic Acids

The aromatic spacer group of the model receptors prevent the formation of intramolecular hydrogen bonds between the opposing carboxyls yet these functions are ideally positioned for intermolecular hydrogen bonds of the sort indicated in *32*. The acridine derivatives do indeed form stoichiometric complexes with oxalic, malonic (and C-substituted malonic acids) as well as maleic and phthalic acids. Fumaric, succinic or glutaric acids did not form such complexes. Though protonation appears to be a necessary element in the recognition of these diacids, the receptor has more to

32

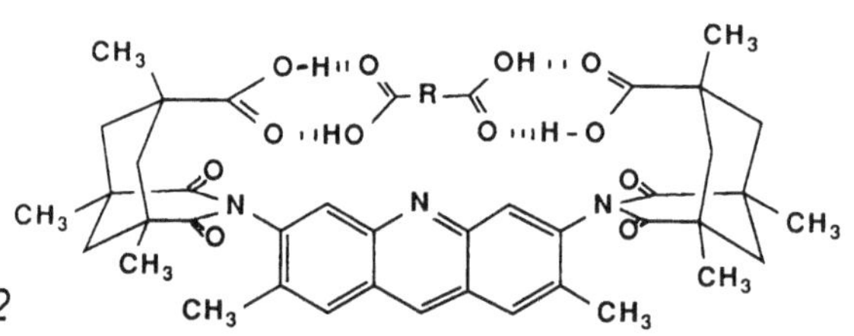

33

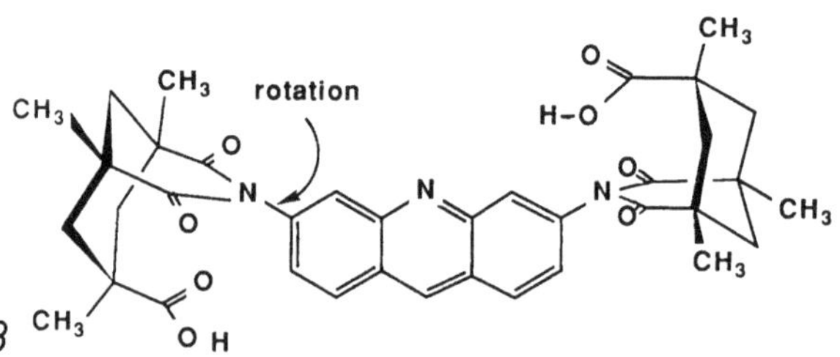

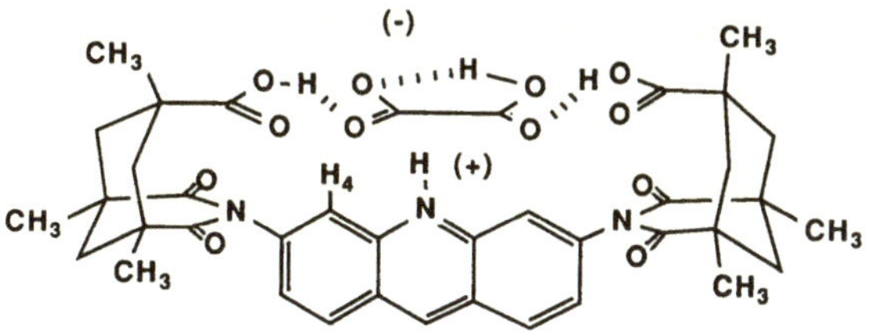

34

offer than simple basicity. The picrate of *13* also dissolves oxalic and malonic acid in a process that results in the release of free picric acid.

The protonated acridine must therefore provide special stabilization to the conjugate bases of small dicarboxylic acids. Evidence for the nature of this special stabilization was provided by some of the receptors which are not constrained to convergent conformations. Figure 2 shows the ambient temperature spectrum of *33* and its simple salts such as the picrate. At low temperature, complex spectra are observed as interconversion between the three possible conformations become slow. In the presence of appropriate diacids such as oxalic acid the spectra are sharpened and are no longer temperature dependent [22c)].

The binding of dicarboxylic acids of suitable size, shape and pK_a restricts receptor molecules to a single conformation, most likely the one featuring the convergent carboxyl functions. Only these conformations can provide the specific stabilization of the substrates by involving both carboxyls of the receptor in hydrogen bonding. Heteronuclear NOE experiments involving H_4 (resp. H_5) of the receptor and the ^{13}C of the oxalate suggest these two elements are close in the complex as suggested in *32*. Structures similar to that suggested, with possible contributions of tautomers such as

c

b

a

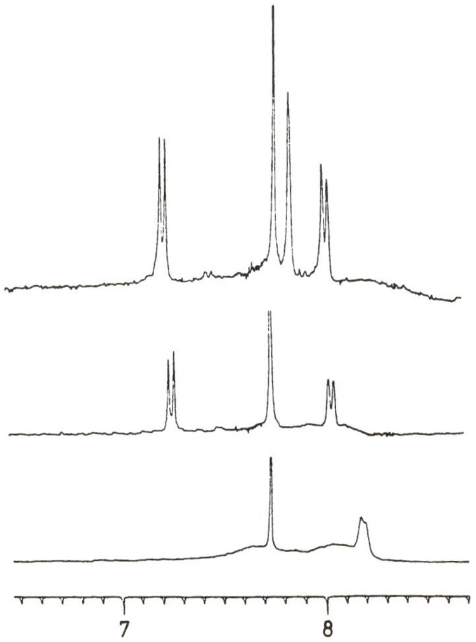

Fig. 2a. Ambient temp. (297 °K) 300 MHz pmr spectrum of *33* broadened by rotation; **b**. spectrum after the addition of 2 equiv. glutaric acid at 297 °K; **c** as in **b**, but at 210 °K

34 are also likely for the complexes of malonate and maleate and other anions that can be chelated between the convergent carboxyls in these models.

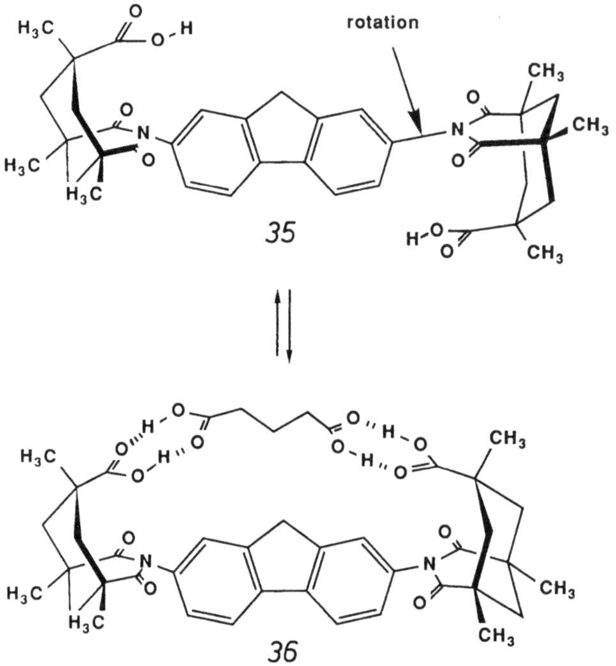

35

36

(6)

In addition, the fluorene derived diacid *35* shows selective binding to glutaric acid [Eq. (6)] and camphoric acid but not the smaller oxalic or succinic acids. Structure *36* is likely for the glutaric acid complex. Again, highly specific means of stabilization can be observed with substrates bearing suitably placed aromatic functions. The stacking interactions between the pendant aryl nucleus and the large π-surface can be seen, for example, in the complex with benzyl malonic acid *37*. Intermolecular NOE is also seen in *37* between the benzylic protons and those lining the cleft of the receptor. Specific stabilization of conjugate bases, leading to changes in apparent pK_a's has

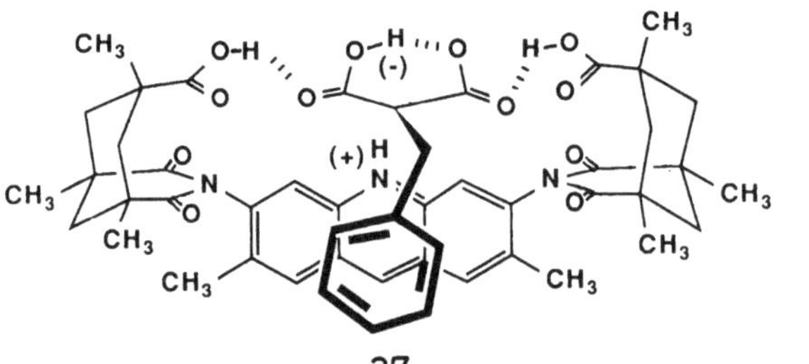

37

also been observed by Kimura [23] in the chemistry of carboxylic acids in contact with macrocyclic polyamines.

Amino Acids

In polar solvents, the structure of the acridine *13* involves some zwitterionic character *13a* [Eq. (7)] and the interior of the cleft becomes an intensely polar microenvironment. On the periphery of the molecule a heavy lipophilic coating is provided by the hydrocarbon skeleton and methyl groups. A third domain, the large, flat aromatic surface is exposed by the acridine spacer unit. This unusual combination of ionic, hydrophobic and stacking opportunities endows these molecules with the ability to interact with the zwitterionic forms of amino acids which exist at neutral pH [24]. For example, the acridine diacids can extract zwitterionic phenylalanine from water into chloroform, and NMR evidence indicates the formation of 2 : 1 complexes *39* such as were previously described for other β-phenyl-ethylammonium salts. Similar behavior is seen with tryptophan *40* and tyrosine methyl ether *41*. The structures lacking well-placed aromatics such as leucine or methionine are not extracted to measureable degrees under these conditions.

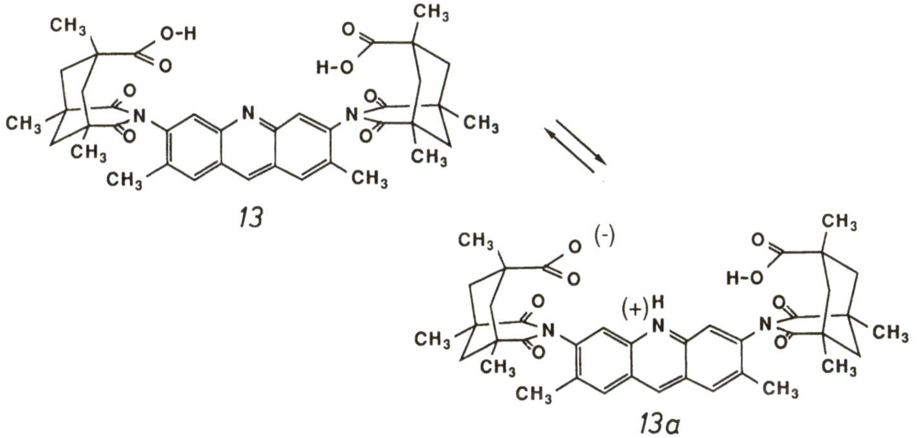

13 ⇌ *13a* (7)

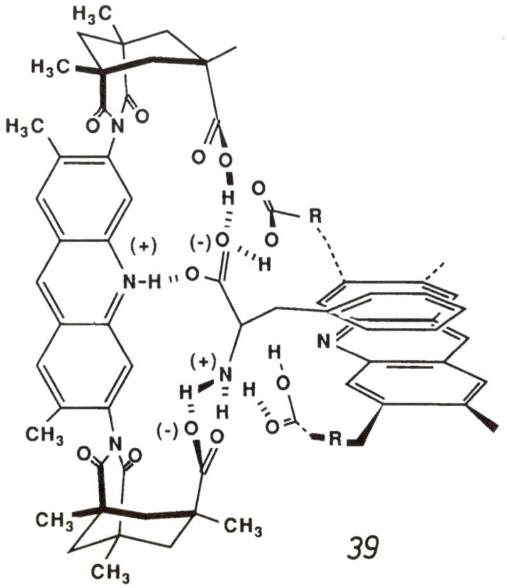

39

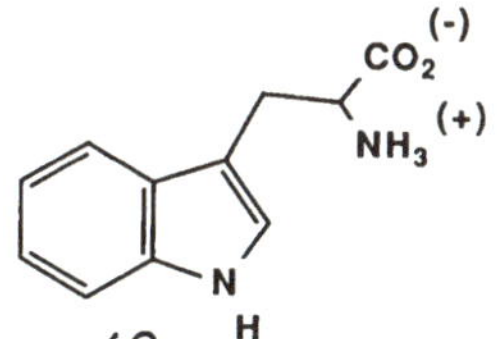

40

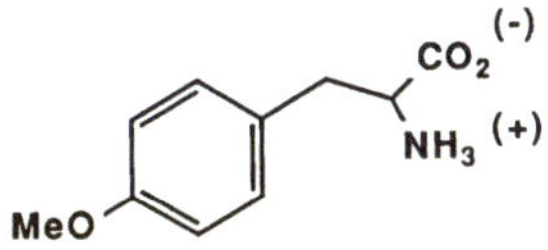

41

Table 2. Transport of amino acids

Transport rates: Millimoles transported[a], hr^{-1}, cm^{-2}, $[carrier]^{-1}$

Carrier	Phenyl-alanine	Trypto-phane (*40*)	Leucine	Tyrosinemethyl ether (*41*)	γ-phenyl-butyrine (*43*)
13	8.5	42	<0.1	18	0.2
Aliquat 366[b]	7	6.7	3.5	—	—

[a] Initial conc. of amino acids in the source phase were 0.03 *M*; carrier was 2×10^{-3} *M*; stirring rate is 500 rpm. Analysis by ninhydrin assay.
[b] Methyltricapryl ammonium chloride in toluene; source phase at pH = 13, data is from Ref. [25b)]
[c] γ-Phenyl-α-amino-butyric acid.

Transport of amino acids across a chloroform liquid membrane with these carriers also revealed a high specificity (Scheme 2). For efficient transport, an aromatic side chain must be present and the distance between the aryl and ammonium functions is optimal in the β-aryl systems. Neither α-phenyl-glycine *42* nor γ-phenyl-butyrine *43* are transported to significant extents [25a)]. These results are shown in Table 2. The selectivity with *13* contrasts sharply for that observed with typical detergents wherein side chain hydrophobicity determines the relative transport rates.

Scheme 2

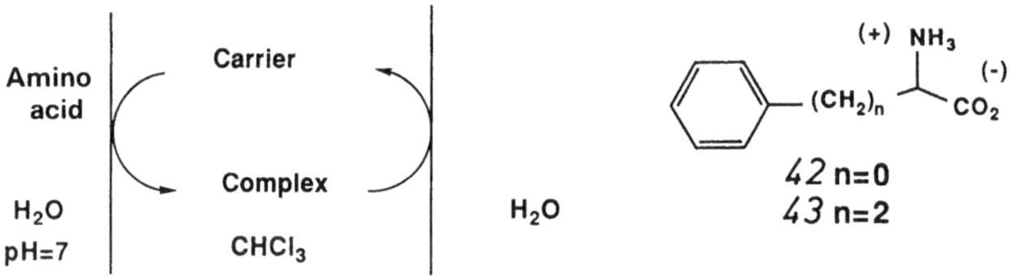

2.5 Nucleic Acid Complexes

The classical form of molecular recognition is the base pairing of purines and pyrimidines in nucleic acids formulated by Watson and Crick[26)]. The complementary hydrogen bonding surfaces of adenine and thymine [Eq. (8)] provide a vehicle for information transfer, while stacking interactions between adjacent base pairs provide additional stability for the double-stranded structures. The hydrogen bonding aspects have been examined in detail[27)] while the stacking of individual bases in water has also been observed[28)].

We have now adjusted our molecular systems to provide a model in which both forces can operate simultaneously. The U-shaped relationship that exists between the imide function and amides of aryl amines creates a hydrogen bonding edge and a planar stacking surface that converge from perpendicular directions as in *44* to provide a microenvironment complementary to nucleic acid components. A large number of aromatic rings can be functionalized with this simple scaffold, and spacers (R) can also be incorporated. The imide function is a mimic of the thymine residues.

(8)

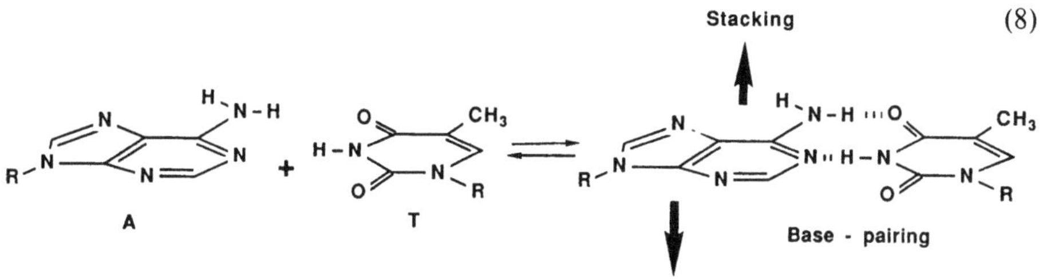

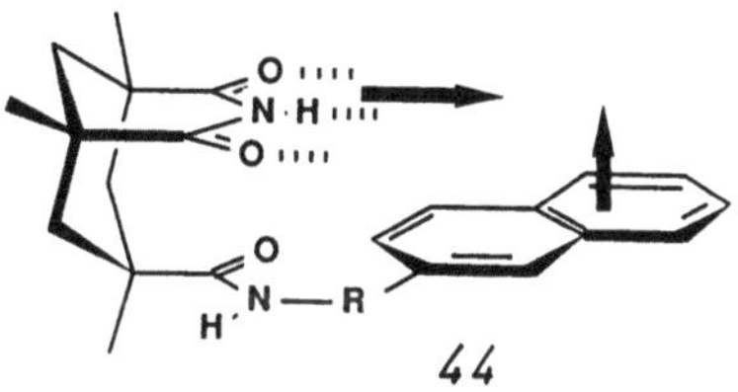

44

The binding affinity of adenine to several aromatic stacking surfaces has been determined [29]. The naphthalene derivative *45* provided a useful model for the shape of an intact base pair in $CDCl_3$ [Eq. (9)] even though the additional binding provided by naphthalene stacking was marginal. Both the Watson-Crick *46* and Hoogsteen geometries (not shown) are observed in base pairing.

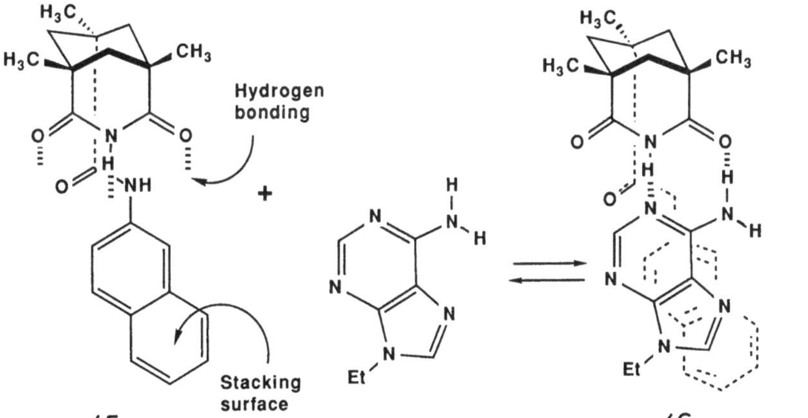

(9)

45 *46*

One of the goals of this type of research is the synthesis of reagents capable of sequence-specific binding to double-stranded nucleic acids. To do so without disturbing the base-pairing requires that the agent recognize the acid and base functions exposed in the major or minor grooves (Scheme 3). For the A–T base pair, the minor groove presents only lone pairs of electrons, a feature which accounts for the affinity of distamycin and other hydrogen donor structures to regions with high frequency of A–T pairs [30]. The major groove of the G–C pair is also unique and offers a number of acid and base sites. A remarkable similarity exists between the major groove side of an A–T base pair and the minor groove side of a G–C base pair. The array of hydrogen bond donors and acceptors are nearly identical! The only structural difference is the position of N_7(A) vs. N_3(G), but functionally these are expected to behave in the same way, the stereoelectronics of binding to N_7(A) are such as is observed in Hoogsten base-pairing of adenine derivatives [26b].

Scheme 3

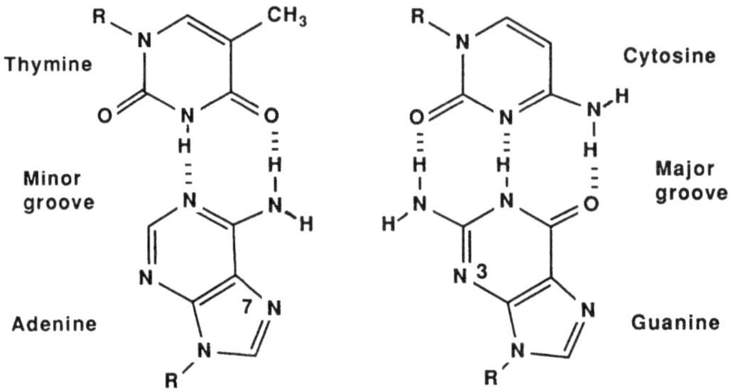

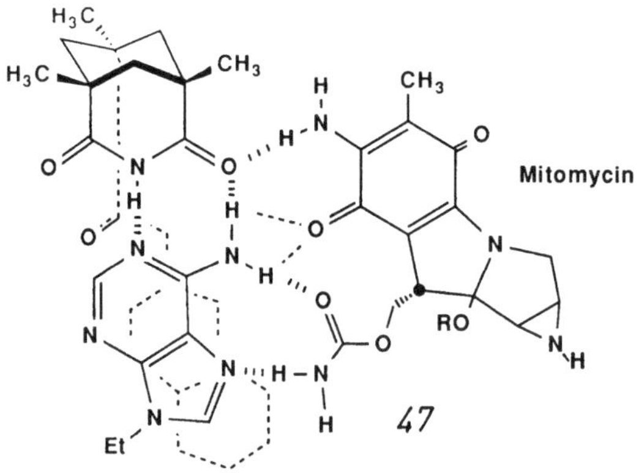

47

The feasibility of identifying these "edges" of water base pairs has been supported by our studies of mitomycin C interacting with the model system for A–T base pairs [29]. Interactions of either component with mitomycin C are not observed but a complex is formed when all three components are present. Chemical shift changes observed in the NMR spectra support the structure *47* for the termolecular complex. The broader implication is that mitomycin C will likewise recognize the minor groove side of a G–C pair (it is known to alkylate the guanidine on this side) [31].

2.6 Neutral Substrate Complexes

With acid or basic substrates protein transfer may contribute to recognition by the model receptor, but binding to lactams involves hydrogen bonds exclusively. Recent research[32] has concluded that hydrogen bonds are not so often created as they are merely exchanged in aqueous solutions, and the overall enthalpic changes are modest.

For binding within the clefts the amide components must be either primary amides or lactams, functional groups which show high affinity for water, particularly at their oxygen atoms [33].

Not surprisingly, the diacid *13* and its diamide are "waterlogged" with 2–4 molecules of H_2O from which they are difficult to liberate. Binding experiments in $CHCl_3$, a non-competing solvent, revealed that stoichiometric complexes, e.g. *48* were formed with diketopiperazines [40] ($K_a \sim 10^4$) and amides such as malonamide. With structures of inadequate hydrogen bonding capacity, such as sarcosine anhydride, complexation does not occur.

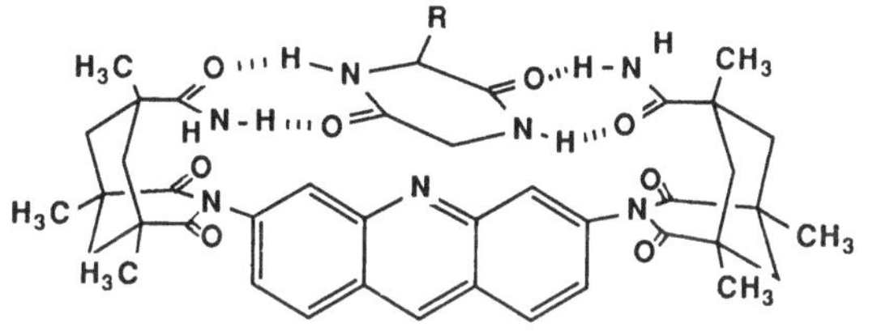

48

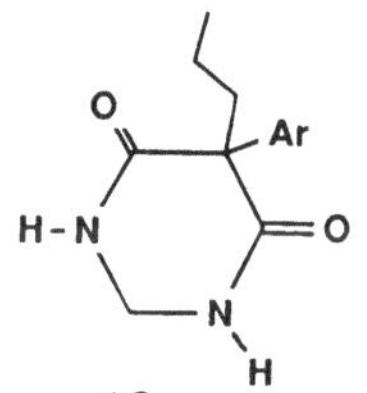

49

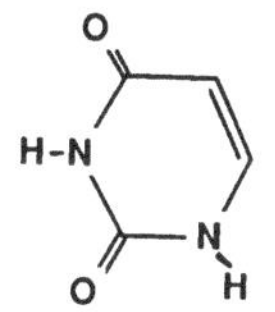

50

Slight changes of geometry in the substrate are also accommodated by *13*. For example, the pyrimidine dione *49* formed 1 : 1 complexes but only traces of uracil *50* were bound. This behavior provides another measure of the promiscuity of these receptors. The carboxyls of *13* have sufficient in-plane flexibility to accommodate diketopiperazines, or Y, but binding to uracil requires that the opposing carboxyl oxygens must come within ~ 6.5 Å of each other, at least if idealized hydrogen bonds are involved. This distance is prevented by the rigidity of the acridine spacer group. Another system capable of this level of discrimination is the cucurbituril recently described by Mock [34].

3 Asymmetric Recognition

One of the more fashionable pursuits in modern synthetic organic chemistry is the development of reagents capable of high asymmetric induction. The success of a number of reagents and catalysts that incorporate C_2 symmetry has led to a growing sense of confidence concerning the use of this symmetry element in reagent design[35]. Sharpless's tartrates [36], Masamune's boranes [37], and Noyori's biaryls [38] are some of the more spectacular examples [39]. This shape seems to be well suited for distinguishing between the top and bottom face of a π-system and we are incorporating it into the molecular clefts in the form of the oxazoline shown in *51*. Here, "almost" the same asymmetric environment is exposed to the carboxylic acid regardless of rotation about the bond indicated. Such a molecule should be able to make distinctions between the three coordinates: up and down, left and right and back and front. We intend to test it within the context of oxidizing olefins to epoxides with the corresponding peracids.

Relatively little progress has been achieved in the identification and recognition of single asymmetric centers. For this purpose, it is not obvious that C_2 symmetry offers any advantages or is even appropriate. Chemical intuition suggests that 3 domains

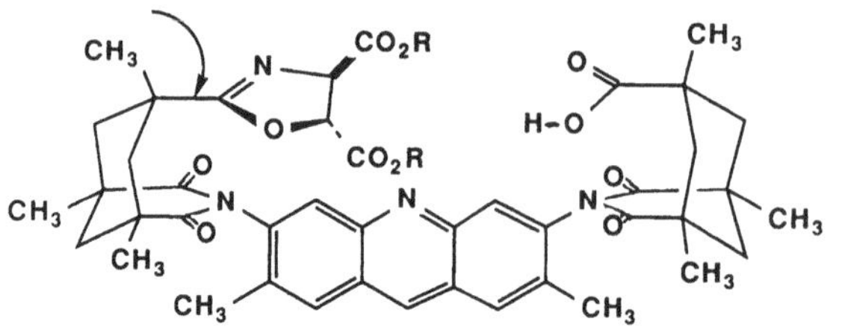

51

would offer a better fit between substrate and receptor than the two inherent structures featuring C_2 symmetry. Accordingly, we have launched a program to test these notions by constructing molecules in which 3 groups appear on a single surface. Specifically, the triacid derivative *52* has been converted to *53* which features large, medium and small groups protruding in the same direction [Eq. (10)]. This indeed shows the formation of diastereomeric complexes in NMR experiments using racemic amines as substrates [40].

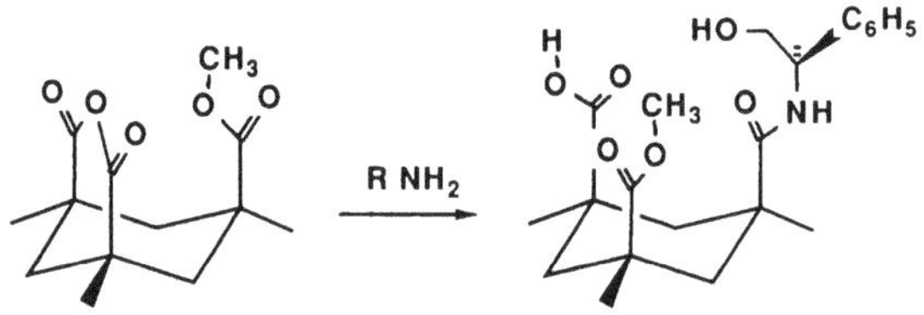

(10)

52 53

Within the context of the molecular clefts, we have also made progress in chiral recognition. Structures in the class of *54* have already shown use as chiral solvating agents for racemic amines and alcohols [40]. More refined versions such as the p-NO_2-Phenyl benzyl ester *55* exhibit highly selective recognition of aromatic amino acids. Again stacking interactions permit multipoint binding and greater discrimination. Since amino acids are excellent sources of both functional groups and asymmetric centers, lining the molecular cleft with a given amino acid derivative can provide structures for recognition of antipodes of specific amino acids.

The molecular scaffolds also provide an unusual opportunity for the identification and recognition of meso structures. For example, successive acylation of L and D phenylalanine with the acridine diacid gives the structure *56* and we anticipate that

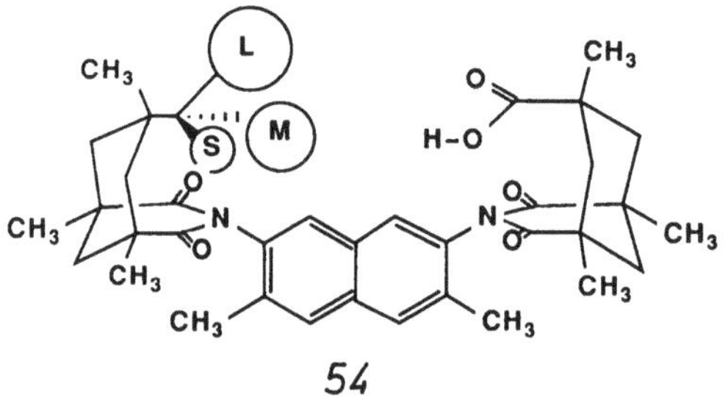

54

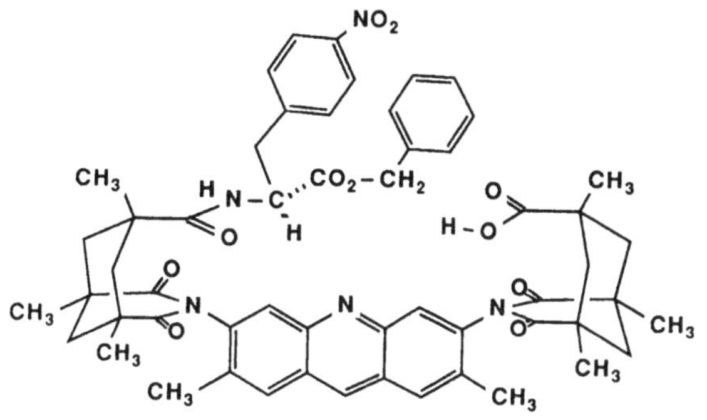

55

it will favor complexation with meso structures. The isomer having C_2 symmetry *57* is more receptive to chiral structures of complementary shape[41]. The recognition of meso structures has been largely ignored but offers all of the same chemical challenges as those involved in recognition of single asymmetric centers.

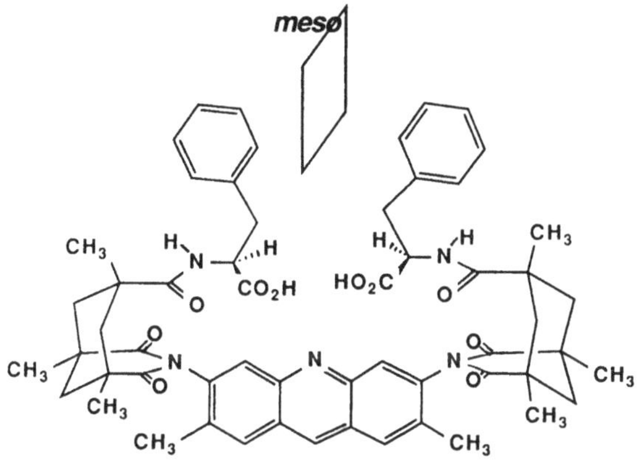

56

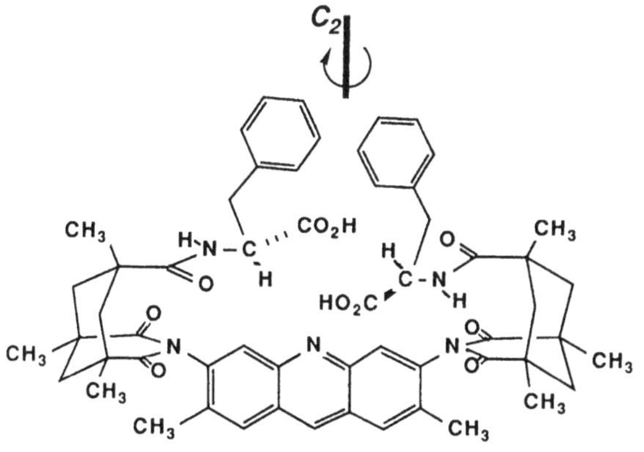

57

4 Enzyme Models and Catalysis

The general skeletons of these molecular clefts resemble two enzymes which feature converging carboxylic acids: lysozyme (a), which is involved in glycolysis and the aspartic proteinases (Scheme 4) such as renin and pepsin (b). Both structures have carboxyl groups at a distance of 4–6 Å. Gandour [16] has suggested that this geometry permits the more basic *syn* lone pairs to converge on the substrates, and the opposite carboxylic acids are in their stable, *syn* conformation as well. Surprisingly, the model systems for lysozyme to date involve exclusively the less basic anti lone pairs [42]. While rate enhancements can be seen in such systems, their relevance to the enzymatic question is shrouded by the question of stereoelectronics.

Scheme 4

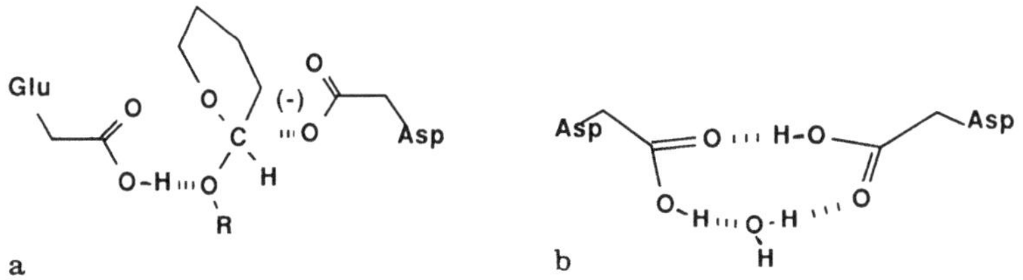

a b

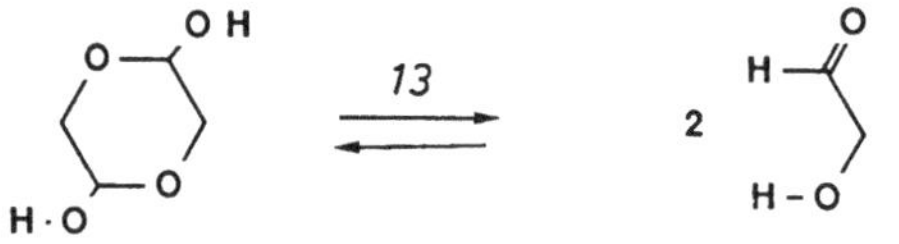

(11)

58 59

We are now in a position to test some of these effects at carboxyl oxygen and our preliminary results with hemiacetal cleavage are encouraging. For example, treatment of the dimer *58* of hydroxy acetaldehyde *59* with catalytic amounts of the acridine diacid *13* results in rapid dissociation at room temperature in a matter of minutes [Eq. (11)]. The use of comparable acids and bases (e.g. mixtures of acetic acid and pyridine) even in large excess results in slow cleavage of the dimer over a period of several days [41]. These observations are in accord with the importance of lone pair orientation in catalysis. In general, the original notions[43] of "push-pull" or concerted general acid-base catalysis can be tested by these molecular clefts, as they permit the arrangement of nucleophile and electrophile on opposite sides of the substrate. We shall report on these developments in due course.

5 Acknowledgements

It is a pleasure to acknowledge the experimental skill of my coworkers on this project: Mr. Ben Askew, Dr. Pablo Ballester, Dr. Chris Buhr, Dr. Timothy Costello, Professor Ana Costero, Dr. Maria Doa, Mr. Robert Duff, Professor Pilar Gil, Professor William Gordon, Dr. Nafisa Islam, Dr. Sharon Jones, Dr. Mary Killoran, Professor Santi Luis, Dr. Luann Marshall, Dr. James McManis, Mr. David Nemeth, Mr. Kevin Parris,

Mr. Glenn Russo, Mr. Tjama Tjivikua, Mr. Kevin Williams and Mr. Raymond Wolak. Financial support from the National Institutes of Health, the National Science Foundation and the J. S. Guggenheim Foundation is acknowledged with gratitude.

6 References

1. Breslow, R.: Science *218*, 532 (1982); Hayward, R. C.: Chem. Soc. Rev., *1983*, 285; Vögtle, F., Löhr, H.-G., Franke, J., Worsch, D.: Angew. Chem. Int. Ed. Engl. *24*, 727 (1985); Cram, D. J.: Science *219*, 1177 (1983); Lehn, J. M.: Science *227*, 849 (1985); Rebek, J., Jr.: Acc. Chem. Res. *17*, 258 (1984); Kellogg, R. M.: Top. Curr. Chem. *101*, 111 (1982)
2. Top. Curr. Chem. vols. *98* (1981), *101* (1982), *121* (1984), *128* (1985), *132* (1986), *136* (1986), *140* (1987), Atwood, J. L., Davies, J. E. D., MacNicol, D. D.: Inclusion Compounds vol. 1, 2, 3, Academic Press, London 1984
3. Journal of Inclusion Phenomena (Atwood, J. L., Davies, J. E. D., eds., Reidel, Dordrecht, Holland)
4. Cram, D. J., Ho, S. P.: J. Am. Chem. Soc. *108*, 2998 (1986). For a recent review see: Izatt, R. M., Bradshaw, J. S., Nielsen, S. A., Lamb, J. D., Christensen, J.: Chem. Rev. *85*, 271 (1985)
5. Behr, J.-P., Lehn, J.-M., Vierling, P.: Helv. Chim. Acta *182*, 1853 (1982); Cram, D. J., Lam, P. Y.-S., Ho, S. P.: J. Am. Chem. Soc. *108*, 839 (1986)
6. Uiterwijk, J. W. H. M., Harkema, S., Geevers, J., Reinhoudt, D. N.: J. Chem. Soc. Chem. Comm. 1982, 200
7. Schmidtchen, F. P.; Muller, G. J.: Chem. Soc. Chem. Commun. 1984, 1115; Schmidtchen, F. P.: Top. Curr. Chem. *132*, 101 (1986); Vögtle, F., Sieger, H., Müller, W. M.: ibid. *98*, 107 (1981), and references therein
8. Kemp, D. S., Petrakis, K. S.: J. Org. Chem. *46*, 514 (1981)
9. Rebek, J., Jr., Marshall, L., Wolak, R., Parris, K., Killoran, M., Askew, B., Nemeth, D., Islam, N.: J. Am. Chem. Soc. *107*, 7476 (1985)
10. For a discussion see: Rebek, J.: Science, *235*, 1437 (1987)
11. Rebek, J., Jr., Costello, T., Marshall, L., Wattley, R., Gadwood, R. C., Onan, K.: J. Am. Chem. Soc. *107*, 7481 (1985); Tabushi, I., Kugimiya, S., Kinnaird, M., Sasaki, T.: J. Am. Chem. Soc. *107*, 4192 (1985)
12. Bender, M. L., Komiyama, M.: Cyclodextrin Chemistry, Springer, Berlin, Heidelberg, New York, 1978; Breslow, R., Czarniecki, M. F., Emert, J., Hamaguchi, H. J.: Am. Chem. Soc. *102*, 762 (1980); Trainor, G., Breslow, R. ibid. *103*, 154 (1981)
13. Stetter, H., Roos, E.-E.: Chem. Ber. *88*, 1390, 1395, (1955); Odashima, K., Itai, A., Iitaka, Y., Koga, K.: J. Am. Chem. Soc. *102*, 2504 (1980); Miller, S. P., Whitlock, H. W., Jr.: ibid. *106*, 1492 (1984); Winkler, J., Coutouli-Argyropopoulou, E., Leppkes, R., Breslow, R.: ibid. *105*, 7198 (1983); Diederich, F., Griebel, D.: ibid. *106*, 8037 (1984); Gutsche, C. D.: Acc. Chem. Res. *16*, 161 (1983)
14. Rebek, J., Jr., Duff, R. J., Gordon, W. E., Parris, K.: J. Am. Chem. Soc. *108*, 6068 (1986)
15. S. Luis, unpublished observations
16. Gandour, R.: Bioorg. Chem. *10*, 169 (1981)
17. Mehrotra, R. C., Bohra, R.: Metal Carboxylates, Academic Press, London, 1983
18. Rebek, J. Jr., Nemeth, D.: J. Am. Chem. Soc. *108*, 5637 (1986)
19. Rebek, J., Jr., Askew, B., Killoran, M., Nemeth, D., Lin, F.-T.: J. Am. Chem. Soc. *109*, 2426 (1987)
20. Rebek, J., Jr., Askew, B., Islam, N., Killoran, M., Nemeth, D., Wolak, R.: J. Am. Chem. Soc. *107*, 6736 (1985)
21. Jeffrey, G. A., Takasi, S.: Acc. Chem. Res. *11*, 264 (1978); cf. Czugler, M., Ángýáu, J. G., Náraý-Szabó, G., Weber, E.: J. Am. Chem. Soc. *108*, 1275 (1986)
22. a) Askew, B., Lin, F. T., Ballester, P., Askew, B., Costero, A.: J. Am. Chem. Soc. in press.
    b) Lloyd, E. J., Andrews, P. J.: J. Med. Chem. *29*, 453 (1986)
    c) Rebek, J. Jr., Nemeth, D., Ballester, P., Lin, F.-T.: J. Am. Chem. Soc. *109*, 3474 (1987)
23. Kimura, E., Sakonaka, A.: J. Am. Chem. Soc. *104*, 4984 (1982). For other studies of selective

binding of carboxylic acids see: Breslow, R., Rajagopalan, R., Schwartz, J.: J. Am. Chem. Soc. *103*, 2905 (1981); Kimura, E., Sakonaka, A., Yatsunami, T., Kodama, M.: ibid *103*, 3041 (1981); Hosseini, M. W., Lehn, J. M.: ibid. *104*, 3525 (1982)
24. Rebek, J., Jr., Nemeth, D.: J. Am. Chem. Soc. *107*, 6738 (1985)
25. a) Rebek, J., Jr., Askew, B., Nemeth, D., Parris, K.: J. Am. Chem. Soc. *109*, 2432 (1987)
b) For studies involving transport under acidic or basic conditions see: Behr, J.-P., Lehn, J. M.: J. Am. Chem. Soc. *95*, 6108 (1973)
c) Tsakube, H.: Tetrahedron Lett. *22*, 3981 (1981)
26. a) Watson, J. D., Crick, F. H. C.: Nature *171*, 737 (1953)
b) Saenger, W.: Principles of Nucleic Acid Structure, Springer, Berlin, Heidelberg, New York 1984, Ch. 6
27. Kyogoku, Y., Lord, R. G., Rich, A.: Proc. Nat. Acad. Sci. *57*, 250 (1967)
28. Chan, S. I., Schweitzer, M. P., Tso, P. O. P., Helmkamp, G. K.: J. Am. Chem. Soc. *86*, 4182 (1964); Schweitzer, M. P., Chan, S. I., Ts'o, P. O. P.: ibid. *87*, 5241 (1965)
29. Rebek, J. Jr., Askew, B., Buhr, C., Jones, S., Nemeth, D., Williams, K.: J. Am. Chem. Soc. in press.
30. For a recent review see: Dervan, P.: Science *232*, 464 (1986)
31. For leading references see Tomasz, M., Chowdary, D., Lipman, R., Shimotakahara, S., Viero, D., Walker, V., Verdine, G.: Proc. wat. Acad. Sci. *83*, 6702 (1986)
32. For a discussion see Fersht, A. R., Shi, J.-P., Knill-Jones, J., Lowe, D. M., Wilkinson, A. J., Blow, D. M., Brick, P., Carter, P., Waye, M. M. Y., Winter, G.: Nature *314*, 235 (1985); Stahl, N., Jencks, W. P.: J. Am. Chem. Soc. *108*, 4196 (1986)
33. Wolfenden, R., Anderson, P. M., Cullis, P. M., Southgate, C. C. B.: Biochemistry *20*, 849 (1981)
34. Mock, W. L., Shih, N.-Y.: J. Org. Chem. *51*, 4440 (1986)
35. Kagan, H. B., Dang, T. P.: J. Am. Chem. Soc. *94*, 6429 (1972); Kagan, H. B., in: Asymmetric Synthesis, Vol. 5, (ed.) Morrison, J. D., Academic Press, New York, 1985, Ch. 1
36. Finn, M. G., Sharpless, K. B.: ibid, Ch. 8
37. Masamune, S., Kim, B. M., Petersen, J. S., Sato, T., Veensta, S. J.: J. Am. Chem. Soc. *107*, 4549 (1985)
38. Noyori, R. et al.: J. Am. Chem. Soc. *102*, 7932 (1980); Noyori, R.: Pure Appl. Chem. *53*, 2315 (1981)
39. For asymmetric recognition in chromatography and its origins in multisite binding see: Pirkle, W. O., Pochapsky, J. C.: J. Am. Chem. Soc. *108*, 5627 (1986)
40. Rebek, J., Jr., Askew, B., Islam, N., Killoran, M., Nemeth, D., Wolak, R.: J. Am. Chem. Soc. *107*, 6736 (1985); Rebek, J. Jr., Askew, B.; Ballester, P., Doa, M.: J. Am. Chem. Soc. *109*, 4119 (1987)
41. Askew, B., Nemeth, D.: unpublished observations
42. Fife, T. H., Przystas, T. J.: J. Am. Chem. Soc. *102*, 292 (1980); Loudon, G. M., Ryono, D. E.: ibid. *98*, 1900 (1976)
43. Swain, C. G., Brown, J. F., Jr.: J. Am. Chem. Soc. *74*, 2538 (1952)

Reaction Control of Guest Compounds in Host-Guest Inclusion Complexes

Fumio Toda

Department of Industrial Chemistry, Faculty of Engineering, Ehime University, Matsuyama 790, Japan

Table of Contents

1 Introduction

When guest molecules are arranged together in the channel of a host-guest inclusion complex, intermolecular reactions of the guest compound may proceed stereoselectively and efficiently. An enantioselective reaction is expected when optically active host compounds are used.

We carried out some of these reaction controls mostly on the photoreaction of guest compounds in their host-guest complexes. We also succeeded in preparing chiral crystals of an achiral oxoamide compound which upon irradiation gives an optically active β-lactam. The host-guest complexation method can also be used to freeze an equilibrium of a guest compound in one isomer. Corresponding examples are described here. In addition, reaction control of the pinacol rearrangement in a host-guest complex and reaction acceleration of the m-chloroperbenzoic acid oxidation of ketones in the solid state are also demonstrated.

The chapter covers studies which have been carried out by our research group since 1987.

2 Host Compounds

All host compounds except *8–10* have been described in vol. 140 of this series, "Molecular Inclusion and Molecular Recognition-Clathrates I" [1]. The new host compound (−)-*8a* was prepared by reaction of PhMgBr and (−)-trans-4,5-bis(ethoxycarbonyl)-2,2-dimethyl-1,3-dioxolane. The latter was obtained from (+)-diethyl tartrate with 2,2-dimethoxypropane [2]. Analogously, (+)-*8b* was synthesized from (−)-diethyl tartrate [2]. The new host compound *8* was found to be effective for the optical resolution of various bicyclic enones through complexation of them [2]. The optically active amide derivatives *9*, *10*, also obtained from tartaric acid [3], are useful for the optical resolution of 2,2′-dihydroxy-1,1′-binaphthyl (*6*) and 10,10′-dihydroxy-9,9′-biphenanthryl (*7*), respectively, by complex formation of them [3].

3 Control of Regio- and Stereoselective Photoreactions

The photodimerization of chalcone *11a* is not easy both in solution and in the solid state. For example, irradiation of *11a* in solution gives a mixture of *11a* and its cis isomer [4] or a polymer [5]. Irradiation of *11a* in the solid state results in a complex mixture of all possible stereoisomeric photodimers in low yield [5]. X-Ray crystal structural studies of two dimorphs of *11a* show that the distances between the double bond centers are 5.2 Å in one dimorph and greater than 4.8 Å in another dimorph although the molecules are packed in parallel manner in both cases [6]. Those distances are longer than the limit of an intermolecular reaction (4.2 Å) and therefore the two dimorphs are photoinactive.

By way of contrast, irradiation for 6 h in the solid state of the 1:2 inclusion complex *12a* composed of *1* and *11a* which has been prepared by recrystallization of the two components from a solvent gave the syn-head-to-tail dimer *13a* selectively in 90% yield [7,8]. Similar irradiation of the 1:2 inclusion complexes *12b–12d* composed of *1*

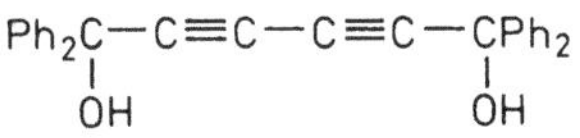

1

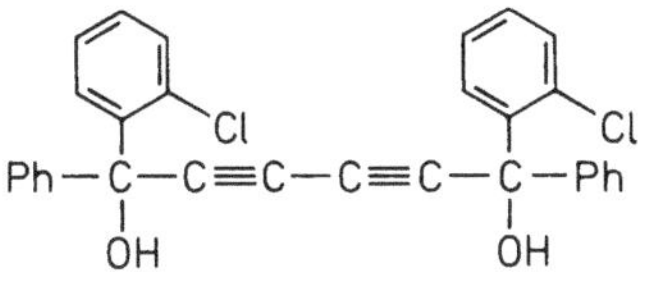

2

a (−)-form ; *b* (+)-form

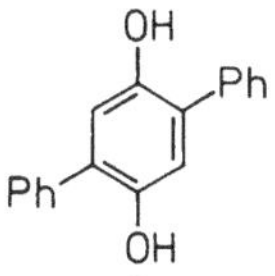

3

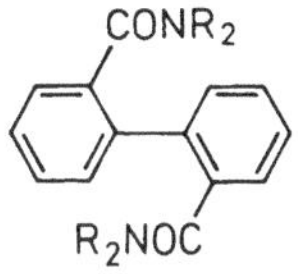

R = cyclohexyl

4

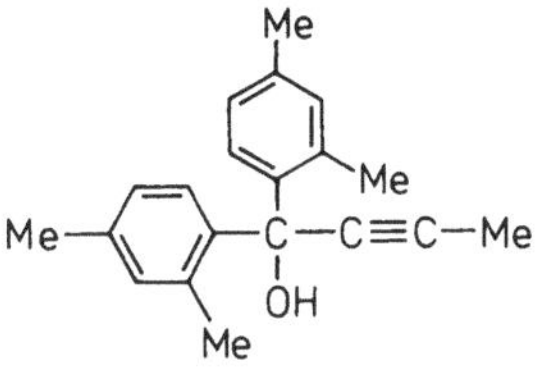

5

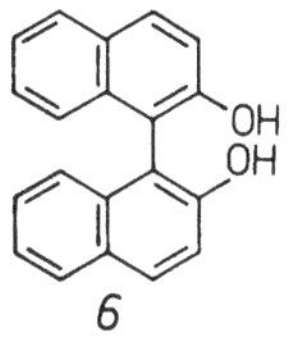

6

a (−)-form

b (+)-form

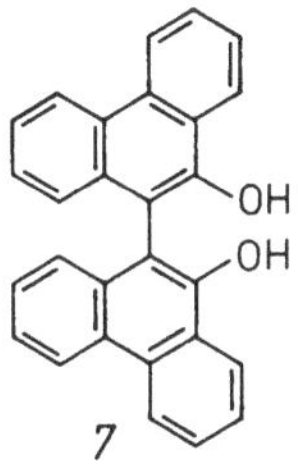

7

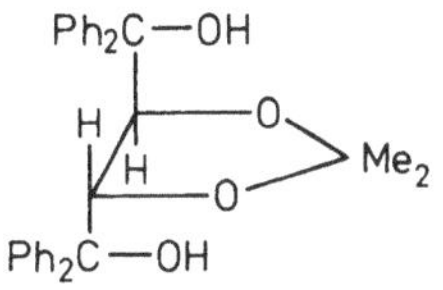

8

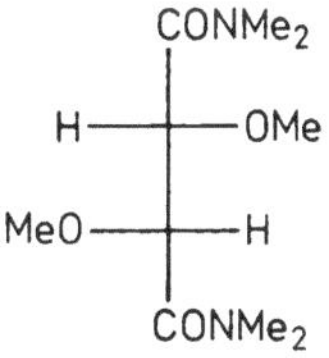

9

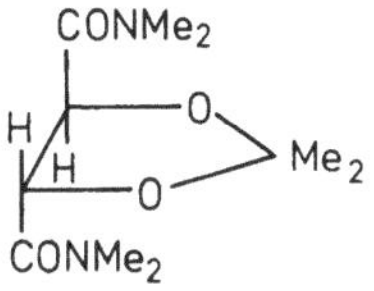

10

a (−)-form ; *b* (+)-form

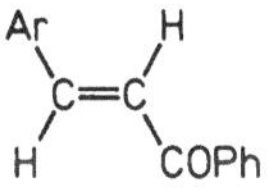

11

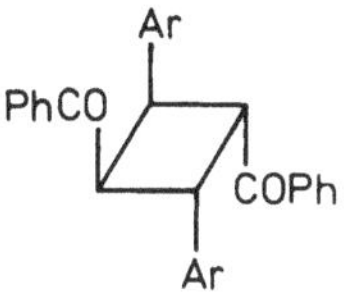

13

syn-head-to-tail

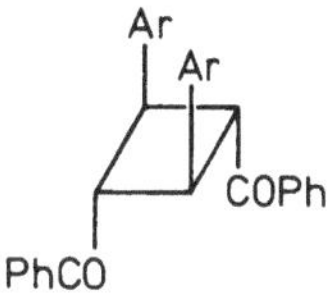

14

anti-head-to-tail

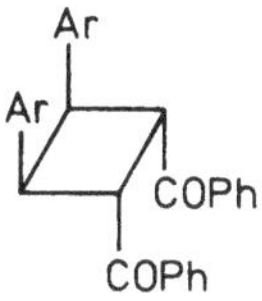

15

syn-head-to-head

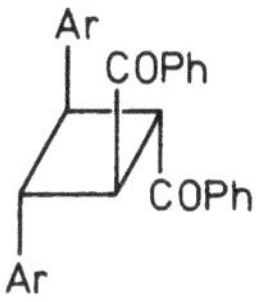

16

anti-head-to-head

a Ar=Ph
b Ar=o−Me−C_6H_4
c Ar=o−MeO−C_6H_4
d Ar=2−Naphthyl
e Ar=o−Cl−C_6H_4
f Ar=2-Furyl
g Ar=2-Thienyl

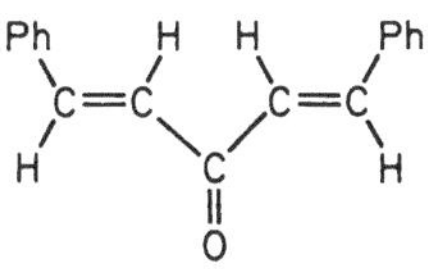

17

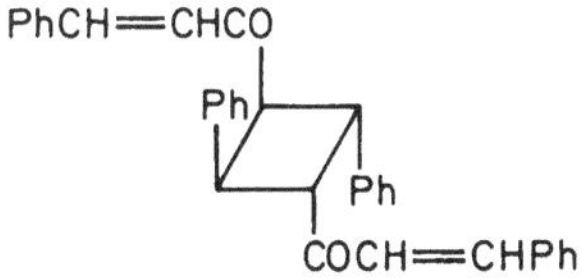

19

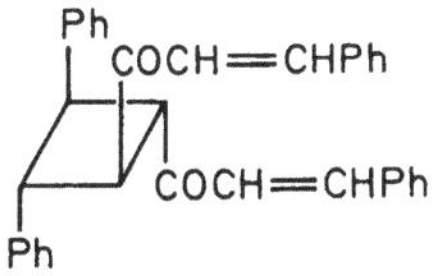

20

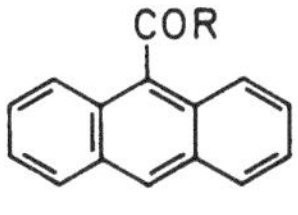

21

a R = H *b* R = Me

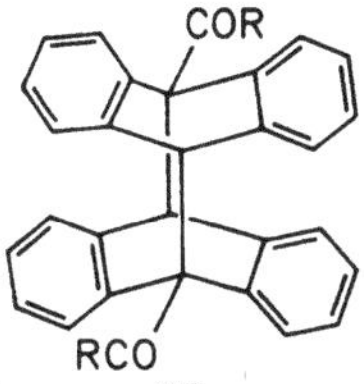

23

Table 1. Crystal forms, melting points, and νOH values of inclusion complexes with chalcones and analogues as guest molecules

Host	Guest	Inclusion complex[a]			
		No.	Crystal form	mp (°C)	νOH (cm^{-1})[b]
1	*11a*	*12a*	pale yellow prisms	87–89	3300
1	*11b*	*12b*	pale yellow prisms	110–112	3310
1	*11c*	*12c*	yellow prisms	184–185	3300
1	*11d*	*12d*	yellow prisms	121–123	3300
1	*11e*	*12e*	pale yellow prisms	133–135	3330
1	*11f*	*12f*	yellow prisms	101–102	3380
1	*11g*	*12g*	yellow prisms	80–81	3400
3	*11a*	*20*	yellow prisms	96–98	3300
3	*17*	*18*	yellow prisms	149–151	3260
1	*21a*	*22a*	yellow prisms	150–152	3300
1	*21b*	*22b*	yellow prisms	107–109	3340

[a] Molar ratio of host and guest is 1:2 in all complexes. [b] All IR spectra were measured in Nujol mull.

Table 2. Reaction time, products, and yields of photocycloaddition reactions of inclusion complexes involving *11*, *17*, and *21* in the solid state

Complex[a]	Reaction time (h)	Product	
		No.	Yield (%)
12a	6	*13a*	90
12b	6	*13b*	85
12c	1.5	*13c*	88
12d	1	*13d*	82
12e	24	—[b]	—
12f	24	—[b]	—
12g	24	—[b]	—
20	6	*13a*	85
18	6	*19*	70
22a	8	*23a*	86
22b	1.5	*23b*	87

[a] Specification in Table 1. [b] Reaction did not occur.

and *11b–d* (Table 1) gave *13b–d*, respectively, in the yields shown in Table 2. The reason for the well controlled reaction can be interpreted as follows: in the complexes with *1* (*12*), two molecules of *11* are packed close together in a position which should give syn-head-to-tail dimers *13*. In order to clarify the reason, an X-ray crystal analysis of *12a* was undertaken [9]. The packing diagram, mutual relation, and geometrical parameters of the reaction centers are shown in Figs. 1 and 2, respectively. It is obvious that hydrogen bonding between the hydroxyl groups of *1* and the carbonyl group of *11a* plays an important role to pack *11a* close together in the complex (Fig. 1). The double bonds are parallel and the distance between two of them is short enough (3.862 Å)

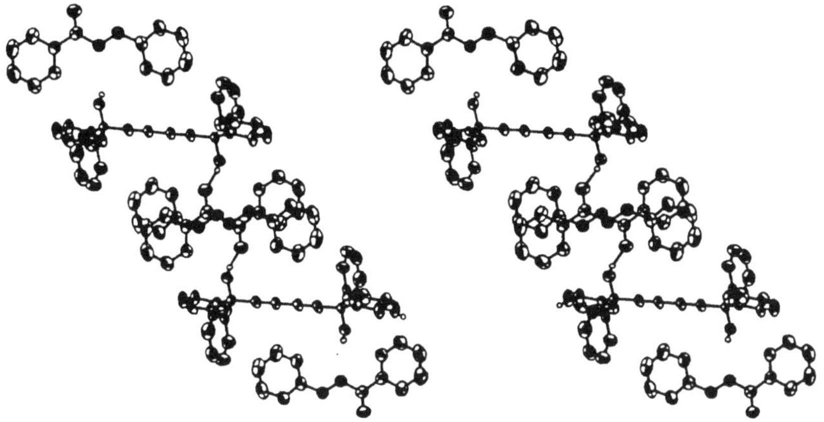

Fig. 1. Stereoscopic view of the two reacting molecules of *11a* in the crystalline host-guest complex *12a*

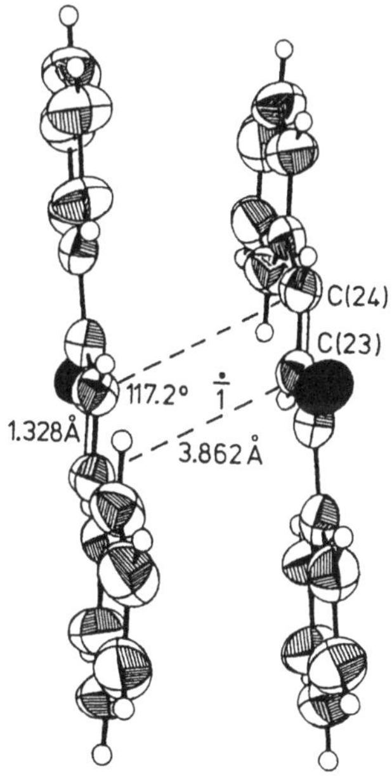

Fig. 2. Mutual relation and geometrical parameters of the reacting centers of the pair of guest molecules *11a*

to react easily. This arrangement of *12a* molecules enables the photodimerization to give the syn-head-to-tail product *13a*, but not the other isomers (*14–16*).

Interestingly, however, *12e–g* are photoinactive. Packing of *11e–g* in their inclusion complexes is probably not suitable, i.e. the distance between the double bonds in each complex is not short enough for photodimerization. Hydrogen bonding in *12e–g* which shows νOH absorptions at higher frequencies than 3330 cm^{-1} is relatively weaker than that in *12a–d* which shows νOH at lower frequencies than 3310 cm^{-1} (Table 1). The above data suggest that the stronger the hydrogen bonding, the closer the packing of *11* molecules in the inclusion complex.

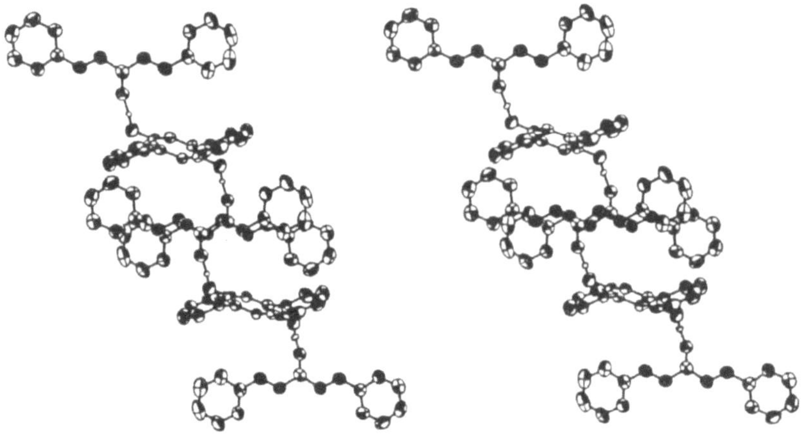

Fig. 3. Stereoscopic view of the two reacting molecules of *17* in the crystalline host-guest complex *18*

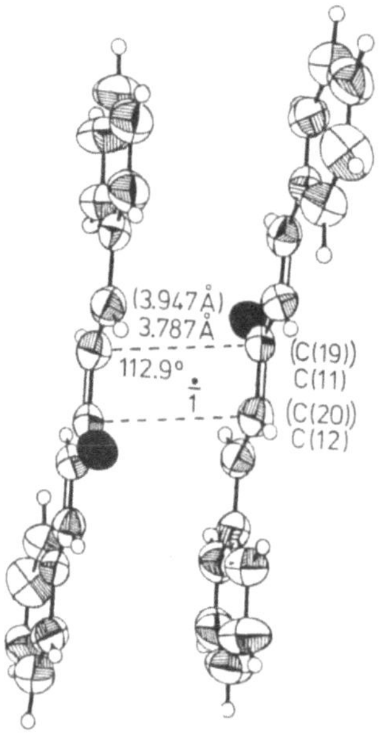

Fig. 4. Mutual relation and geometrical parameters of the reacting centers of the pair of guest molecules *17*

Benzylideneacetone (*17*) is also photoinactive in the solid state, but irradiation in solution for 90 h gives anti-head-to-head dimer *20* in 30% yield [10]. However, when the 1:2 inclusion complex *18* composed of 2,6-diphenylhydroquinone (*3*) and *17* is irradiated in the solid state, syn-head-to-tail dimer *19* was obtained in 70% yield (Table 2) [8]. The νOH absorption of *18* at 3260 cm^{-1} suggests the presence of a strong hydrogen bond between *3* and *17* in the complex (Table 1). X-Ray crystal structural study of *18* shows that hydrogen bonding between *3* and *17* makes the packing tight (Fig. 3) and the distance between the double bonds of *17* shorten (3.787 Å) (Fig. 4) [9]. It is also clear that the arrangement of molecules of *17* in *18* can give the syn-head-to-

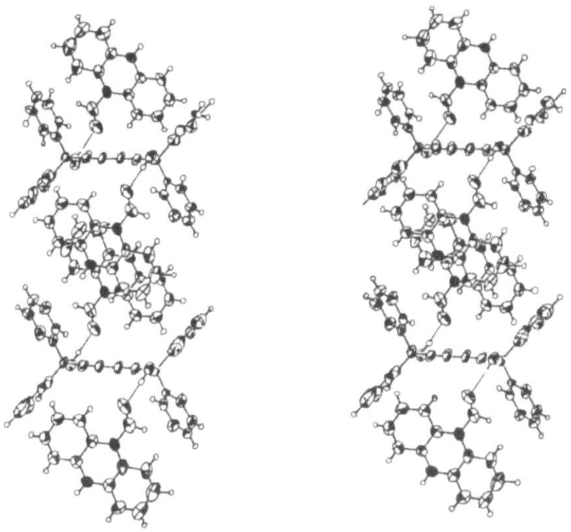

Fig. 5. Stereoscopic drawing of the packing arrangement in *22a* down the plane of the reacting centers (marked by filled ellipsoids)

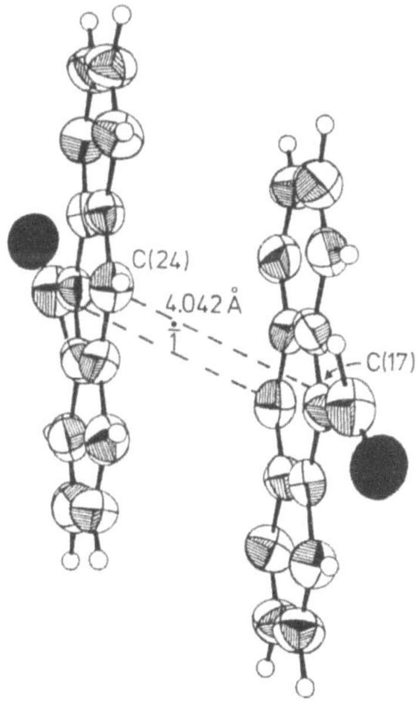

Fig. 6. Relevant geometric parameters involving the reacting centers in *22a*

tail product *19* only by the photodimerization reaction (Fig. 3). In addition, the hydroquinone host compound *3* is available to a photodimerization of *11*. For example, irradiation of the 1:2 inclusion complex *20* composed of *3* and *11a* gave *13a* in 85% yield (Table 2)[8].

Although an irradiation of 9-formylanthracene (*21a*) in solution for 24 h gives its anti-photocycloaddition product *23a* in a low yield[11] irradiation of the 1:2 inclusion complex *22a* composed of *1* and *21a* (Table 1) in the solid state for 8 h gave *23a* in 86% yield [8]. X-Ray crystal structural study of *22a* disclosed that two molecules of *21a* are arranged between two host molecules by forming a hydrogen bond in the direction which gives anti-dimer *23a* by photodimerization (Fig. 5) and that the distance between the two reaction centers of *21a* is short enough to react readily (4.042 Å) (Fig. 6) [12]. Similarly, irradiation of the 1:2 complex *22b* composed of *1* and *21b* (Table 1) in the solid state for 1.5 h gave *23b* in 87% yield (Table 2) [13].

4 Control and Acceleration of Photoreactions via Freeze of an Equilibrium by Complexation and Irradiation in the Solid State

Since 2-pyridone (*24a*, Table 3) exists as an equilibrium mixture with 2-hydroxypyridine (*25a*) [14], it is difficult to isolate *24a* in a pure state. However, the complexation method using inclusion hosts is applicable for the isolation of keto-form *24a* in a pure state. For example, *24a* was isolated by inclusion complexation (1:2 complexes) with *1* (*26a*) (Table 3) and with *3* (*27a*) (mp 151–153 °C), respectively. Both *26a* and *27a* do not contain the enol form (*25a*). The structure of *26a* was studied by X-ray crystal analysis (Figs. 7, 8) [12]. Inclusion of the keto-form is understandable, because *1* and *3* form more stable hydrogen bonds with the carbonyl oxygen of the

Table 3. Melting points and IR spectra of the 1:2 inclusion complexes *26* formed of *1* with 2-pyridones *24*

2-Pyridone (*24*)					Inclusion complex			
No.	R_3	R_4	R_5	R_6	No.	mp (°C)	IR (cm^{-1})[a]	
							νOH	νCO
24a	H	H	H	H	*26a*	176–177	3240	1641
24b	Me	H	H	H	*26b*	176	3280	1635
24c	H	Me	H	H	*26c*	196–197	3220	1640
24d	H	H	Me	H	*26d*	201–202	3200	1653
24e	H	H	H	Me	*26e*	194–195	3270	1620
24f	Me	H	Me	H	*26f*	192–196	3230	1618
24g	H	Me	H	Me	*26g*	208–212	3260	1635
24h	H	H	Me	Me	*26h*	220–222	3070	1644
24i	H	Me	Me	Me	*26i*	216–217	3250	1640
24j	H	Me	$-(CH_2)_3-$		*26j*	208	3250	1622
24k	H	Me	Me	Et	*26k*	166–168	3250	1641
24l	H	OMe	H	H	*26l*	177–178	3170	1638
24m	H	H	H	OMe	*26m*	181–182	3320	1640
24n	H	H	H	Cl	*26n*	150	3360	1632

[a] IR spectra were measured by a combination of KBr disk and hexachlorobutadiene mull methods.

keto-form rather than with the hydroxyl oxygen of the enol-form. A strong hydrogen bond has been observed in *26a* (Table 3) and in *27a* (νOH 2900 cm^{-1}).

The freeze of the equilibrium by complexation with *1* is applicable to various 2-pyridone derivatives such as those summarized in Table 3 [15]. IR spectra show that all complexes consist of the keto-form only (*24*).

The selective inclusion of the keto-form *24* can be used for an efficient dimerization reaction of *24*. Moreover, since molecules of *24* are packed regularly in a host-guest complex, regio- and stereoselective intermolecular reaction of *24* is expected. Irradiation of the 1:2 inclusion complex *26a* composed of *1* and *24a* in the solid state for 6 h gave the trans-anti-dimer *28a* in 76% yield. The efficient reaction is in contrast to that in solution which gives *28a* in 40% yield after an irradiation for 72 h [16]. In solution, collision between molecules of *24a* is prevented in the presence of *25a*. X-Ray crystal structural study of *26a* shows that two molecules of *24a* are arranged between host molecules in positions which give the trans-anti-dimer *28a* by dimerization and the distance between the reaction centers is very close (3.837 Å) (Figs. 7, 8) [12].

Although *26b–n* (Table 3) and *27a* are photoinactive [15], irradiation in the solid state of complexes *30* of N-methylpyridones *29* (Table 4) gives the corresponding

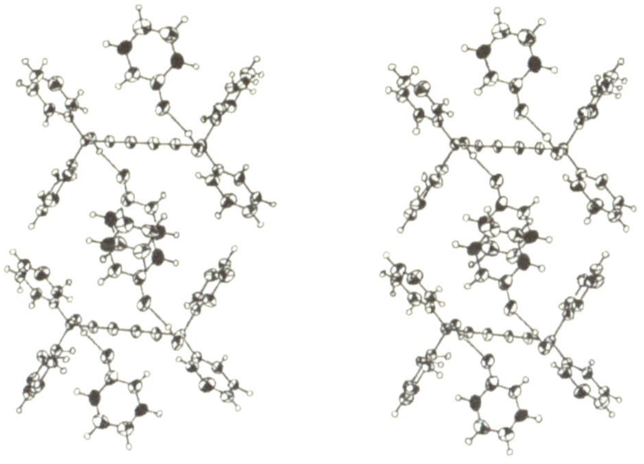

Fig. 7. Stereoscopic drawing of the packing arrangement in *26a* down the plane of the reacting centers (marked by filled ellipsoids)

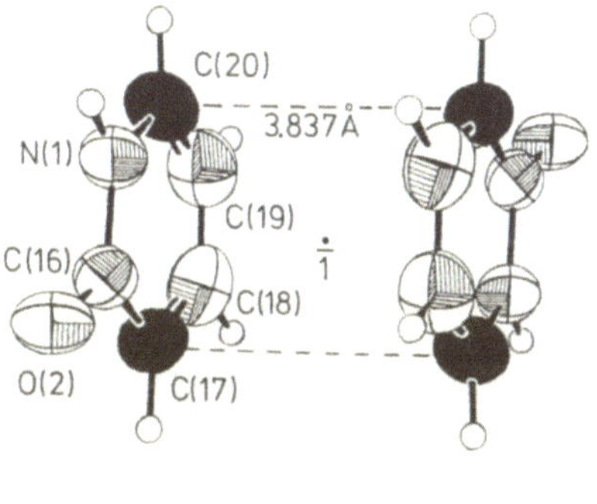

Fig. 8. Relevant geometric parameters involving the reacting centers in *26a*

Table 4. Molar ratios, melting points, and IR spectra of inclusion complexes *30* formed of *1* with N-methylpyridone *29*

N-Methylpyridone (*29*)					Inclusion complex				
No.	R_3	R_4	R_5	R_6	Molar ratio			IR (cm^{-1})[a]	
					No.	(*1*:*29*)	mp (°C)	νOH	νCO
29a	H	H	H	H	*30a*	1:2	137	3120	1648
29b	Me	H	H	H	*30b*	1:2	117–118	3180	1642
29c	H	Me	H	H	*30c*	1:2	162	3180	1660
29d	H	H	Me	H	*30d*	1:1	111	3270, 3170	1654
29e	H	H	H	Me	*30e*	1:2	145	3170	1650
29f	Me	H	Me	H	*30f*	1:2	115–116	3330	1635
29g	H	Me	H	Me	*30g*	1:2	139–140	3140	1645
29h	H	H	Me	Me	*30h*	1:2	161	3130	1642
29i	H	Me	Me	Me	*30i*	1:1	143–144	3210	1642
29j	H	Me	$-(CH_2)_3-$		*30j*	1:2	146–147	3150	1650
29k	H	Me	Me	Et	*30k*	1:1	159	3200	1640
29l	H	OMe	H	H	*30l*	1:1	128	3210	1644
29m	H	H	H	OMe	*30m*	1:2	123	3230	1650
29n	H	H	H	Cl	*30n*	1:2	113	3310	1655

[a] IR Spectra were measured by a combination of KBr disk and hexachlorobutadiene mull methods.

Table 5. Yields of photoreaction products of N-methylpyridones *29* in the crystalline complex with *1* and in ethanol solution

N-Methylpyridone	Product No.	Yield of photoreaction product (%)[a]	
		Crystalline complex	Ethanol solution[c]
29a	*31a*	32[b]	30 (24)
29b	*31b*	19	30 (10)
29c	*31c*	85[b]	21 (15)
29d	*31d*	0	41 (27)
29e	*31e*	74[b]	9
29f	*31f*	1[b]	40 (26)
29g	*31g*	74[b]	3
29h	*31h*	91[b]	37 (22)
29i	*31i*	9	19 (16)
29j	*31j*	94[b]	81 (17)
29m	*31m*	84[b]	0
29n	*31n*	0	0

[a] The irradiation times were 6 h for complexes and 100 h for ethanol solution. [b] Obtained as a complex with *1*. [c] The value in parentheses gives the yield of the other photoproducts.

trans-anti-dimers *31* in the yields shown in Table 5 [15]. Photodimerization of *29a–n* in solution is also summarized in Table 5 [15]. Reactions in solution result in a mixture of products. It is clear, the reactions of the complexes with *1* in the solid state are much more efficient and selective.

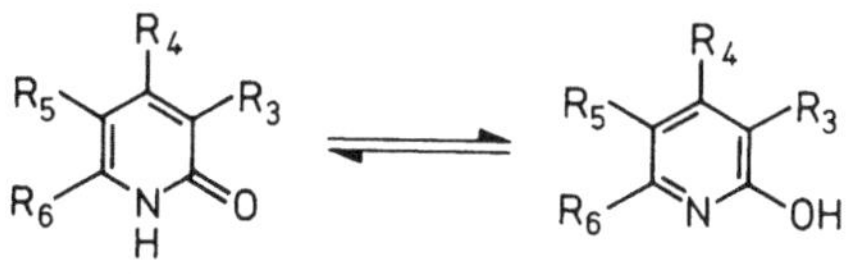

24 25

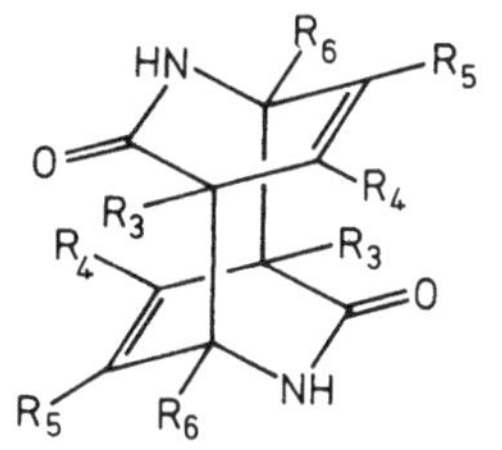

28

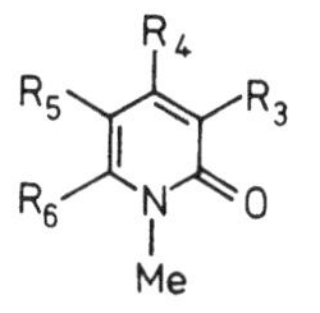

29

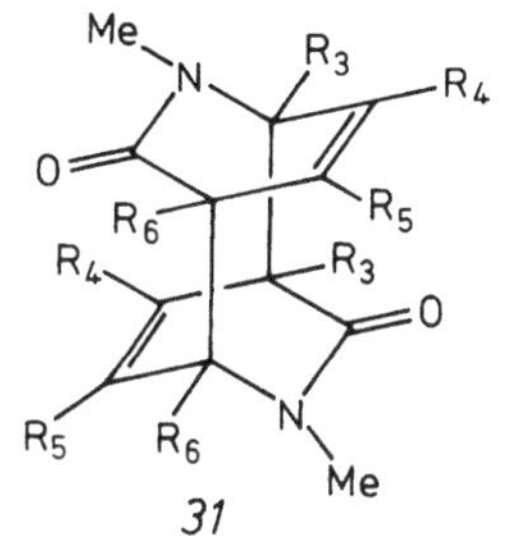

31

The freeze of equilibrium by the complexation method is also applicable to some other compounds. 2-Mercapto substituted tropone *32* has been reported to exist as an equilibrium mixture of 2-mercaptotropone (*32a*) and 2-hydroxytropothione (*32b*), and the latter is predominant both in solution [17] and in the solid state [18]. The equilibrium is frozen and the former was isolated by inclusion complexation with *1*. When a solution of *1* and *32* in petrol ether was kept, a 1 : 1 complex *33* composed of *1* and *32a* was obtained in 90% yield as orange prisms of mp 101 to 103 °C [19]. The structure of *32a* in *33* was elucidated by IR spectroscopy which showed νSH at 2482 cm^{-1}; *33* gave also absorptions of a strongly hydrogen-bonded hydroxyl group at 3270 cm^{-1}.

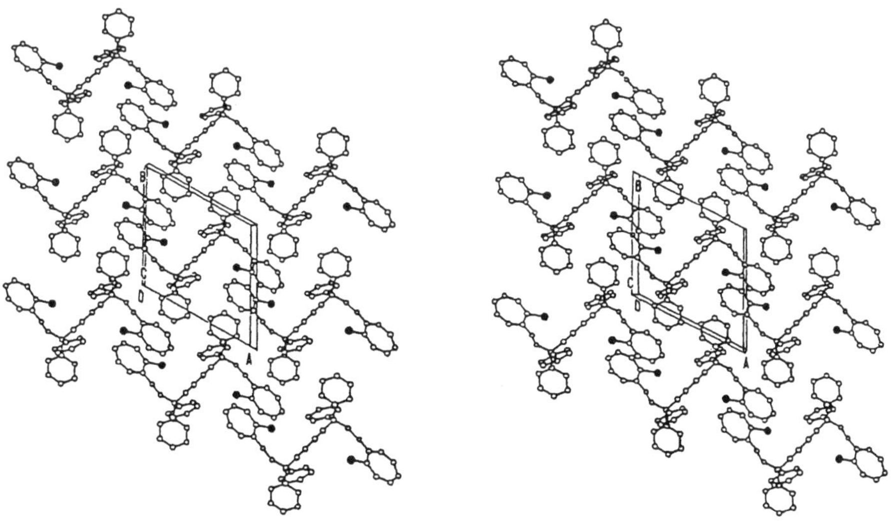

Fig. 9. Stereoscopic drawing of the crystal structure of *33* along the c-axis. Thin lines represent hydrogen bonds; S atoms as bold dots

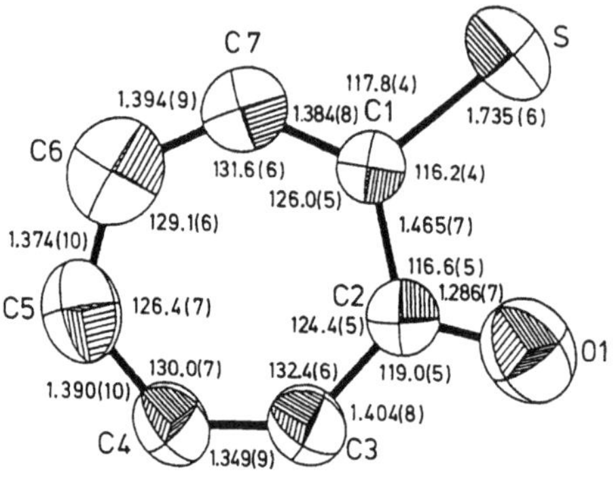

Fig. 10. Bond lengths and angles of *32a* in *33* with numbering of the atoms

The data suggest that the hydroxyl groups of *1* are involved in hydrogen bonds to carbonyl oxygens of *32a*, as is shown in Scheme 1. Moreover, although *32* is easily oxidized to disulfide *34* in solution and in the solid state, *32a* in *33* is stable.

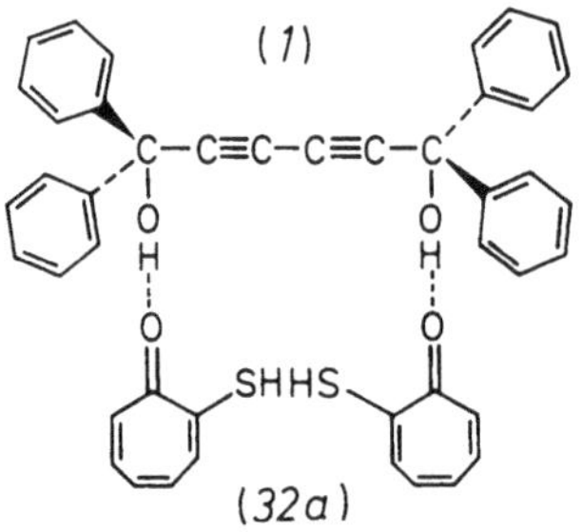

33

Scheme 1

In order to study the structure of *32a* in the inclusion complex in detail, and X-ray crystal structural study of *33* was carried out [19]. The packing diagram of *33* shows that two molecules of *32a* are packed in anti-position to each other and the distance between the two seven-membered rings (3.49 Å) is short enough to cause a π-π interaction (Fig. 9). The stacking of *32a* in *33* is comparable to that of graphite (3.5 Å). Furthermore, *32a* in *33* has a delocalized structure as shown in Fig. 10. Hence, the delocalized structure and the π-π interaction are probably the reason for the characteristic UV spectrum of *32a* which shows absorptions in the longer wave-length region. It is

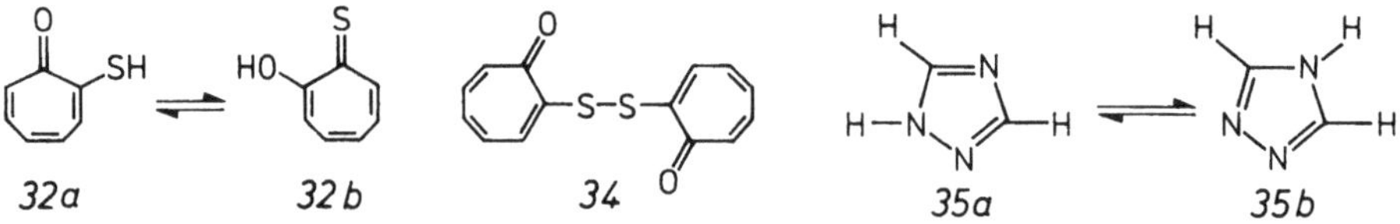

32a *32b* *34* *35a* *35b*

also clear that the stability of *32a* in *33* is due to the packing of *32a* molecules in anti-direction to each other.

1,2,4-Triazole *35* also exists as an equilibrium mixture of the two tautomers, 1,2,4-

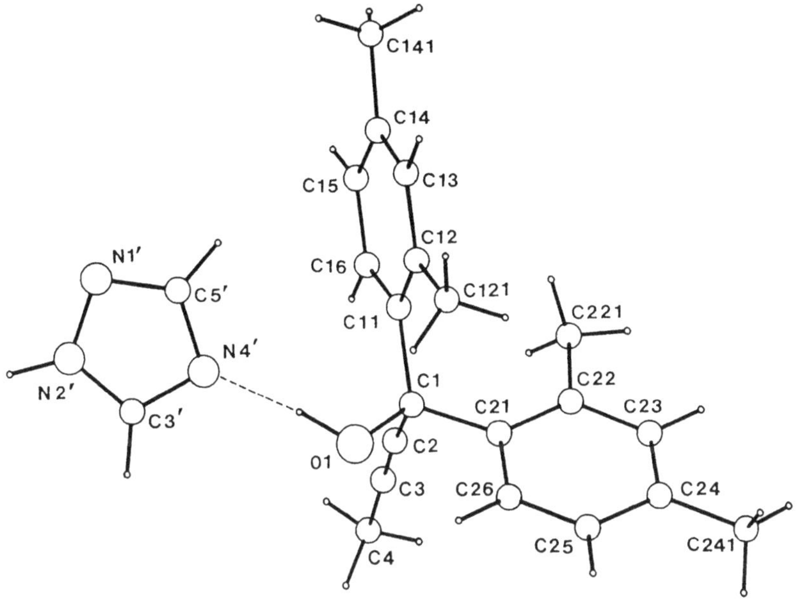

Fig. 11. Perspective view of *36* with numbering of the atoms

triazacyclopenta-2,4-diene (*35a*) and 1,2,4-triazacyclopenta-2,5-diene (*35b*). As a result it is difficult to separate one tautomer [20,21]. The former tautomer *35a*, however, was isolated in a pure state as a 1 : 1 inclusion complex (*36*) with host compound *5*. The complex *36* was obtained in 88% yield by keeping a solution of *5* and *35* in methanol (colorless prisms of mp 101 to 102 °C). A perspective view of *36* based on X-ray analysis (Fig. 11) shows that *35a* is included and the closest contact is a hydrogen bond between the hydroxyl group of *5* and the nitrogen at N4′. The distance of the hydrogen bond was found to be 2.697(4) Å [22].

Cycloocta-2,4,6-trien-1-one exists as an equilibrium mixture of the stable form *37* (*37a* and *37b*) and the unstable form *38* in a 95:5 ratio at 20 °C [23]. Conversion between the two optical conformers *37a* and *37b* is fast around room temperature with the activation energy of 11.9 kcal/mole [23]. It is interesting to freeze the equilibrium in one optical conformer. By inclusion complexation with *2a*, the flipping equilibrium of *37* was frozen in one conformer as a 1:2 complex (*39*) [24]. The IR spectrum of *39* does not show any carbonyl absorption which is assignable to *38*. In order to learn whether one enantiomer (*37a* or *37b*) is included selectively in *39* or a racemic mixture (*37a* and *37b*), photoconversion of *37* in *39* to bicyclo[4.2.0]octa-4,7-dien-2-one (*40*) was carried out. When *39* was irradiated in the solid state by a high-pressure Hg-lamp for 168 h, 50% conversion occurred and (—)-*40* was obtained in 28% yield [$[\alpha]_D$ −69.0° (*c* 0.12, $CHCl_3$)] [24]. Although the optical purity of this product was not determined, it is clear that one of the enantiomers *37a* or *37b* is included selectively, or at least predominantly. Photoreaction of *37* in pentane solution has been reported to give racemic *40* in 30% yield after irradiation for 21 days [25]. Therefore, it is demonstrated that the photoreaction of *37* in *39* proceeds not only enantioselectively but also much more efficiently.

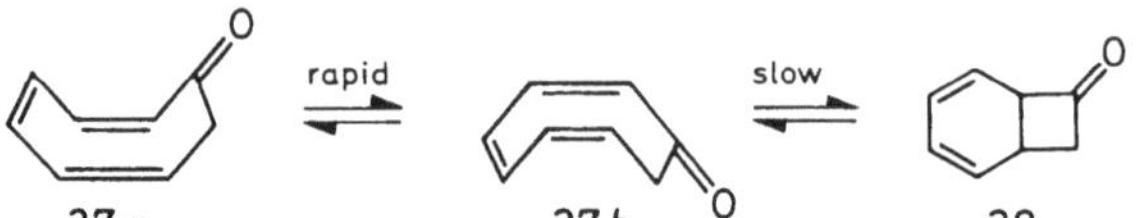

37a *37b* *38*

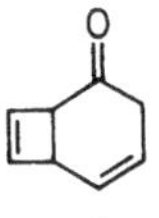

40

5 Control and Acceleration of Photoreactions via Host-Guest Complex Formation by a Solid-Solid Reaction and Irradiation in the Solid State

We found that host-guest complexes are formed by solid-solid reaction of host and guest compounds [26]. The solid-solid reaction can be effected either by grinding both components using an agate mortar and pestle, or agitating using a test-tube shaker, and in some cases just by keeping a mixture of host and guest compounds at room temperature [26]. For example, when a mixture of finely powdered *1* and an equimolar amount of finely powdered benzophenone is agitated using a test-tube shaker for 0.2 h at room temperature, a 1:1 inclusion complex of both components was obtained in quantitative yield [26]. Its IR spectrum is identical to that of an authentic sample prepared by complexation in solution. By the same method, inclusion complexes with some other guest compounds were also obtained (Table 6) [26]. Interestingly, freezing of the equilibrium of 2-pyridone (*24a*) and 2-hydroxypyridine (*25a*) occurs even by inclusion complexation in the solid state. Shaking of an equilibrium mixture of *24a* and *25a* with *1* in the solid state gave a 1:2 complex *26a* composed of *1* and *24a* (see Table 6). This complexation method is also applicable to other host compounds [26].

When the complex formation by solid-solid reaction is combined with a photoreaction in the solid state, a stereoselective photoreaction can be carried out continu-

Table 6. Host-guest inclusion complex formation by solid-solid reaction

Host compound	Guest compound	Reaction time (h)[a]	Inclusion complex	
			Host:guest	Mp (°C)
1	Ph_2CO	0.2	1:1	96
1	Chalcone (*11a*)	1.0	1:2	87–98
1	2-Pyridone (*24a*)	48	1:2	177–182
1	p-$Me_2NC_6H_4CHO$	24	1:2	148–149

[a] Agitation using a test-tube shaker at room temperature.

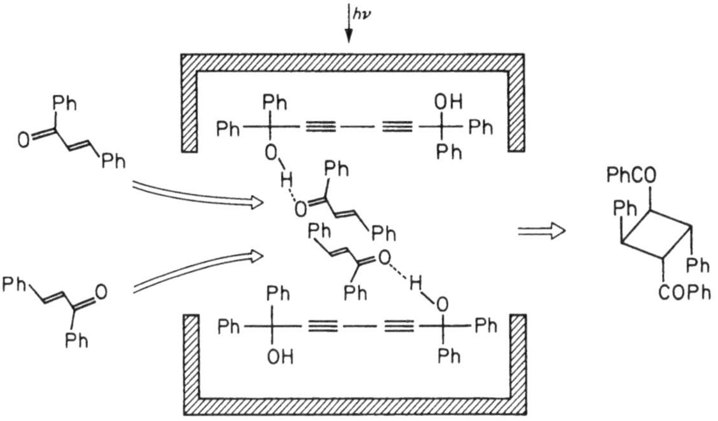

Scheme 2

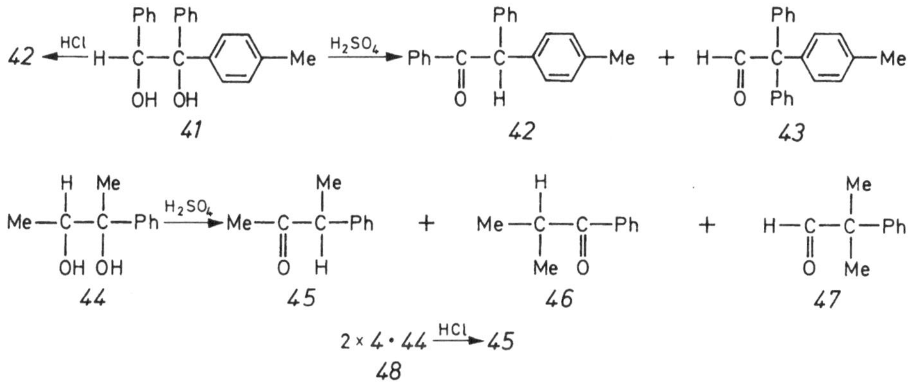

Scheme 3

ously. In other words, the host compound can be used catalytically. By irradiation of 1:1 and 1:2 mixtures of *1* and chalcone *11a* in the solid state with occasional mixing with an agate mortar and pestle for 10 and 40 h, respectively, the syn-head-to-tail dimer *13a* was obtained selectively in 80 and 82% yields, respectively. These results suggest the formation of a 1:2 complex between *1* and *11a* previous to the photodimerization reaction, since plain *11a* does not give *13a* by irradiation in the solid state [5].

Furthermore, irradiation of a 1:4 mixture of *1* and *11a* under the same conditions as above for 72 h gave *13a* in 87% yield. This result shows that the host compound *1* was used almost twice like a catalyst. This is illustrated in Scheme 2. By mixing *1* and *11a*, the 1:2 inclusion complex is formed, and irradiation of the complex gives *1* and *13a*. Further mixing of the recovered *1* and *11a* forms a new complex which upon irradiation gives *13a* again.

Not only the photoreaction but also a pinacol rearrangement (Scheme 3) and a m-chloroperbenzoic acid oxidation (Table 7) occurred in the solid state, selectively and acceleratively.

Table 7. Oxidation of ketones with m-chloroperbenzoic acid (*49*) in the solid state and in chloroform solution[a]

Ketone	Reaction time (h)	Product	Yield (%) in	
			Solid state	Chloroform
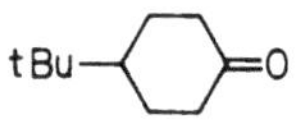	0.5	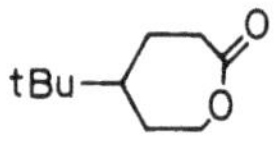	95	94
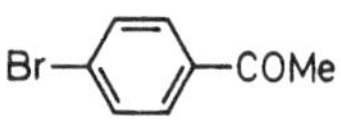	120	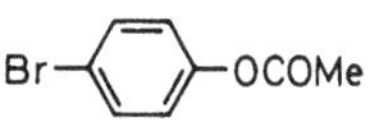	64	50
$PhCOCH_2Ph$	24	$PhCOOCH_2Ph$	97	46
PhCOPh	24	PhCOOPh	85	13
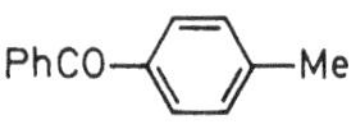	24	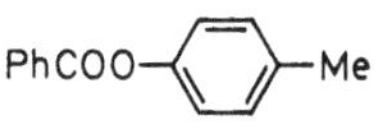	50	12
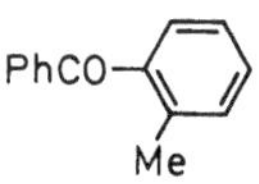	96	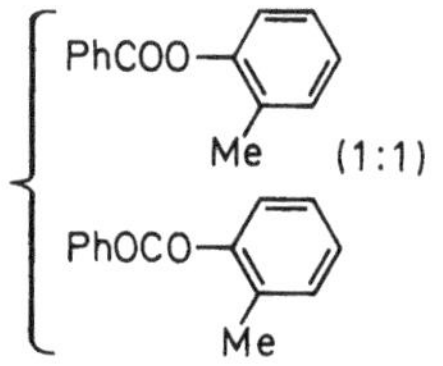	39	6

[a] All reactions were carried out by keeping a reaction mixture or solution at room temperature, and by using a twice molar amount of *49* to ketone. Reactions in chloroform were carried out at a concentration of ketone (1 g) in chloroform (50 ml).

Heating of a mixture of pinacol *41* and 33% sulfuric acid under reflux for 2 h gave two pinacolones *42* and *43* in 80 and 20% yields, respectively. However, when hydrogen chloride gas was passed at room temperature over finely powdered *41* for 3 h, *42* was obtained selectively in 90% yield. The treatment of *44* with sulfuric acid as above gave a complex mixture of reaction products, *45*, *46*, and *47* in 48, 29, and 5% yields, respectively [27]. The oily pinacol *44* formed a 1:2 complex (*48*) with the host compound *4* as colorless crystals. The treatment of finely powdered *48* with hydrogen chloride gas under the same conditions as above gave *45* selectively in 44% yield [27].

When an optically active host compound is used instead of *4*, formation of an optically active pinacolone is expected by an enantioselective pinacol rearrangement in the solid state.

The oxidation reaction of ketones by m-chloroperbenzoic acid (*49*) in the solid state was found to proceed faster than in solution. For example, when a mixture of finely powdered benzophenone and two molar amounts of finely powdered *49* was kept at room temperature for 24 h, phenyl benzoate was obtained in 85% yield (Table 7) [28]. In contrast, the same reaction in chloroform for 24 h gave the same product only in 13% yield (Table 7). Some other examples of the oxidation reaction in the solid state are summarized in Table 7, although the oxidation reaction of the two oily methyl-substituted benzophenones is not a real solid-solid reaction but a liquid-solid reaction. In all cases, however, the oxidation in the absence of solvent is much faster than that in the presence of solvent. Solvent molecules probably prevent an approach of the reagent to the reactant and retard the reaction.

We do hope that the new findings about solid-solid reactions become a starting point for reconsidering and carefully reinvestigating organic reactions which have long been studied in solvents and the belief that most reactions only occur in solution.

6 Control of Enantioselective Photoreactions in Inclusion Complexes with Chiral Host Compounds

An enantioselective photoreaction of a guest compound is expected when an inclusion complex of the guest with an optically active host compound is irradiated in the solid state.

When a solution of *2a* and an equimolar amount of 2-methoxytropone (*50a*) or 2-ethoxytropone (*50b*) in benzene-hexane (1 : 1) was allowed to stand for 12 h at room temperature, a 1 : 1 inclusion complex *51a* (mp 69–71 °C) or *51b* (mp 135–137 °C) was obtained quantitatively as colorless needles. Irradiation of powdered *51a* by a high-pressure Hg-lamp at room temperature for 72 h gave (1*S*,5*R*)-(—)-1-methoxy-bicyclo[3.2.0]hepta-3,6-dien-2-one (*52a*, Scheme 4) of 100% ee (enantiomeric excess) in 11% yield together with (*S*)-(+)-methyl 4-oxocyclopent-2-ene acetate (*56a*, Scheme 5) of 91% ee (26% yield) [29, 30]. Similar irradiation of *51b* for 83 h gave (1*S*, 5*R*)-(—)-*52b* of 100% ee (12% yield) and (*S*)-(+)-*56b* of 72% ee (14% yield). The absolute configurations of *52* and *56* have been studied [31].

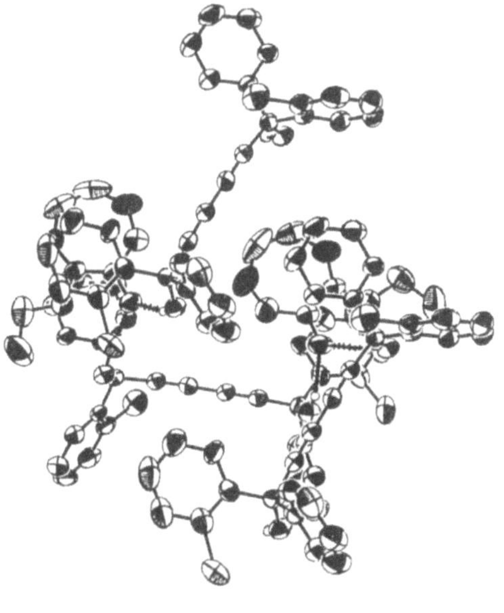

Fig. 12. Stereoscopic view of the crystalline host-guest complex *51b*

The enantioselective photoreaction of *50* to *52* in the complex (*51*) can be interpreted as follows: disrotatory [2 + 2]-photoreaction of *50* in *51* occurs only in the A direction according to a steric hindrance of the o-chlorophenyl group of *2a* (Scheme 4)[29,30]. This interpretation was determined as reasonable by X-ray crystal structural study of *51b* (Fig. 12)[32].

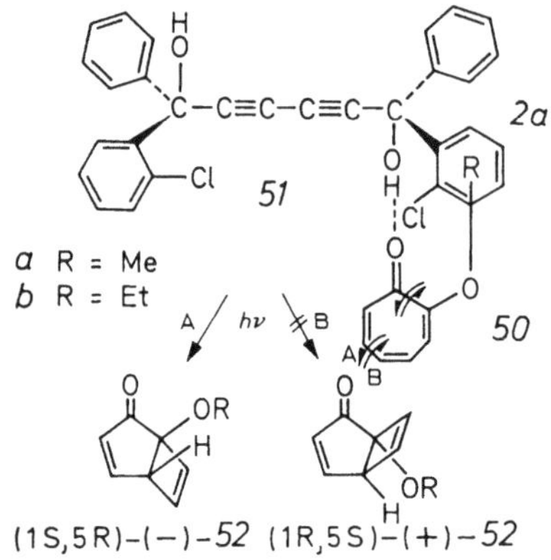

Scheme 4

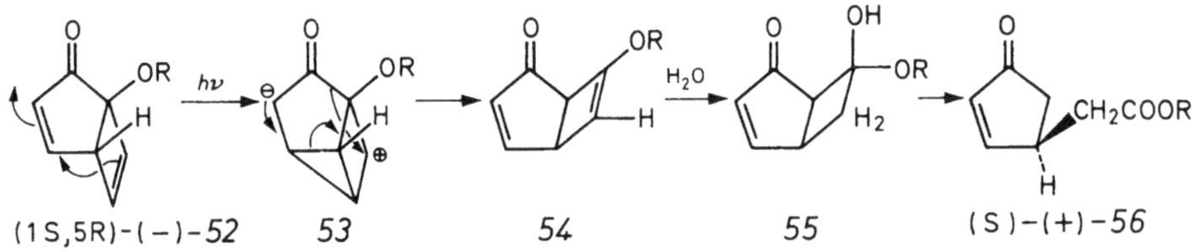

Scheme 5

Formation of *56a* of 91% ee and *56b* of 72% ee shows that the conversion of *52* to *56* (Scheme 5) proceeds with relatively low enantioselectivity. This is probably due to a small amount of water contaminant in the complex, because the irradiation of *52* of 100% ee and *52b* of 100% ee in 2% aqueous methanol gave *56a* of 45% ee and *56b* of 35% ee, respectively. It was also disclosed that this low enantioselective conversion of *52* into *56* is due to a photochemical racemization of *56* via reversible enolization. Irradiation of a solution of *56a* of 97% ee and *56b* of 72% ee in 2% aqueous methanol for 4 h gave *56a* of 34% ee and *56b* of 27% ee, respectively. However, the racemization occurred very slowly by irradiation in a dry methanol solution. Therefore, these results would support the view that the photochemical course from *52* to *56* does not contain any racemization step, as previously postulated[33].

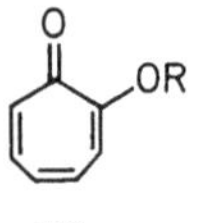

50
a R = Me
b R = Et

When *8a* is used instead of *2a* for complexation with *50a*, two kinds of 1:1 inclusion complexes *57* were obtained: one (*57a*) is photoreactive and the other (*57b*) is photostable. Thus when a solution of *8a* and an equimolar amount of *50a* in benzene was kept at room temperature for 10 h, a 1:1 inclusion complex between both (*57a*) was obtained as colorless fine needles of mp 110 to 123 °C which upon irradiation in the solid state for 49 h gave (1*R*,5*S*)-(+)-*52a* of 3.6% ee in 30% yield [13]. However, recrystallization of *57a* from benzene gave a different kind of 1:1 inclusion complex (*57b*) as colorless needles of mp 123–125 °C which is photostable. Photocyclization of *50a* in *57b* is probably prevented by a steric hindrance. The different behavior of *57a* and *57b* may be clarified in future by X-ray structural studies.

Irradiation of cycloocta-2,4-dien-1-one (*58*) in pentane gives a racemic photodimer, anti-tricyclo[8.6.0.$0^{2,9}$]hexadeca-7,11-diene-3,16-dione (*60*) in 10% yield along with polymeric materials [34]. Efficient and enantioselective photodimerization of *58* was achieved by irradiation of the 2:1 inclusion complex *59* formed between *2a* and *58* [13]. When a solution of *2a* and an equimolar amount of *58* in ether-hexane (1:1) was kept at room temperature for 12 h, *59* was obtained as colorless needles of mp 105 to 108 °C. Irradiation of *59* in the solid state for 48 h gave (—)-*60* of 78% ee in 55% yield.

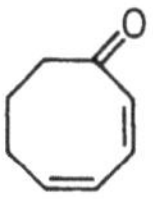
58

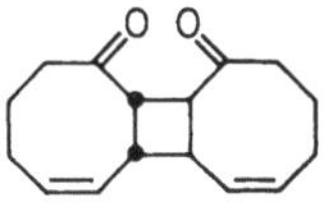
60

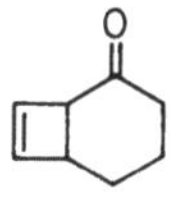
62

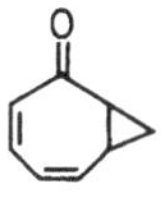
63

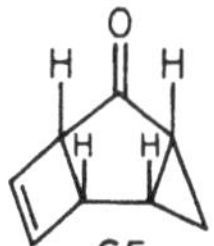
65

Since one unit of *59* contains one molecule of *58* and two molecules of *2a*, at least two of these units should take part in the photodimerization of *58*. In other words, four molecules of *2a* are concerned in each photodimerization of *58*. It is not clear at present whether this has something to do with the relatively high enantioselectivity and efficiency of the photodimerization or not. In future, an X-ray crystal structural study of *59* is required to clarify this question. As in the case of *37*, *58* exists as an equilibrium mixture of two flipping optical conformers. We may also learn from an X-ray analysis which optical conformer is included in *59*.

Interestingly, irradiation for 6 h of the 1:1 complex *61* composed of *8a* and *58* which had been prepared by keeping a solution of *8a* and an equimolar amount of *58* in benzene-hexane at room temperature for 12 h (colorless needles of mp 133 to 136 °C) gave the optically active photocyclization product (—)-bicyclo[4.2.0]oct-7-en-2-one (*62*) [$[\alpha]_D$ —60.6° (*c* 0.18, $CHCl_3$)] in quantitative yield [13].

The photocyclization reaction is in contrast to the photodimerization of *58* in *59*. In the case of *61*, the photodimerization of *58* is probably prevented by steric hindrance. It is almost certain that one optical conformer of *58* is included in *61*, but a real proof of this fact requires an X-ray structural study in the future. However, the formation of the optically active *62* is valuable, because photoreaction of *58* in solution does not give any intramolecular photocyclization product [34].

It has been reported that irradiation of bicyclo[5.1.0]octa-3,5-dien-2-one (*63*) in methanol leads to a mixture of racemic tricyclo[4.2.0.$0^{3,5}$]oct-7-en-2-one (*65*), *37*, and cyclohepta-1,3,5-triene [35]. Control of the reaction was also tried in expectation of the formation of optically active *65*. A 1:2 inclusion complex *64* composed of *2a* and *63* was prepared as colorless prisms of mp 75 to 78 °C by keeping a solution of *2a* and a four molar amount of racemic *63* in ether-hexane (1:1) at room temperature for 12 h. However, a pair of the two optical isomers of *63* was included in *64*, and irradiation of *64* in the solid state for 24 h gave racemic *65* in 71 % yield [13]. Although the reaction control to get single product *65* was achieved, the stereo control to get optically active *65* failed. Complexation of *8a* with racemic *63* gave the 1:2 complex *66* composed of *8a* and optically active (+)-*63*, from which, however, (+)-*63* of only 10 % ee was isolated. Therefore, irradiation of *66* yielded (+)-*65* of 10 % ee.

Inclusion complexation of racemic *63* with the meso isomer of *2* was also examined. The host compound included racemic *63* as did *2a*, and photoreaction of the complex in the solid state gave racemic *65* again [23].

Nevertheless, optically pure *65* could be obtained by complexation with *2a*. For example, when a solution of racemic *65* (0.3 g, 2.5 mmol) and *2a* (0.3 g, 0.62 mmol) in ether-hexane (1:1) (5 ml) was kept at room temperature for 12 h, a 1:2 inclusion complex of *2a* and (+)-*65* was obtained as colorless needles (0.4 g, 88 % yield, mp 98 to 100 °C). Five recrystallizations of the complex from ether-hexane (1:1) gave a 1:2 complex of *2a* and optically pure (+)-*65* (0.12 g, 27 % yield), which upon heating in vacuo gave optically pure (+)-*65* [0.03 g, 20 % yield, $[\alpha]_D$ +164° (*c* 0.2, $CHCl_3$)] [13].

Optically active oxaziridines are useful reagents for the enantioselective oxidation of olefins [37–39]. The following three preparative methods to make this reagent available have been reported: enantioselective oxidation of an imine by (—)-peroxycamphoric acid [37, 38], photocyclization of a nitrone which has a chiral substituent [39], and photocyclization of a nitrone in an optically active solvent [39]. However, an

Table 8. Inclusion complexes *68*, their photoreactions, and photoproducts *69*

	Inclusion complex (*68*)			Irradiation time (h)	Product (*69*)			
	Ar	R	mp (°C)		Yield (%)	$[\alpha]_D$ (°)		% ee
a	Ph	iPr	124–126	24	56	+40.7	(0.31)[a]	—
b	Ph	tBu	135–136	36	41	+8.2	(0.55)	9.5
c	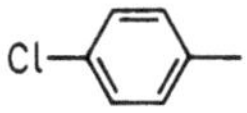	tBu	112–115	32	74	+16.0	(0.61)	30
d	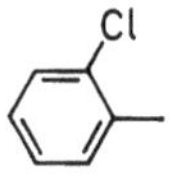	tBu	108–110	12	51	−4.3	(0.83)	100
e	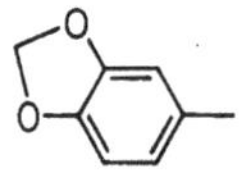	iPr	95–103	16	63	+20.0	(0.83)	28
f	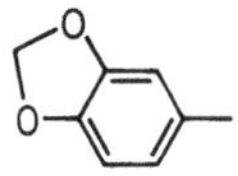	tBu	104–111	21	52	+78.5	(0.38)	94
g	Ph	iPrMeCH	126–128	24	40	+53.9	(0.50)	100

[a] (c, $CHCl_3$).

optically active oxaziridine of more than 30% ee has not been obtained. None the less, enantioselective photocyclization of nitrones occurred in the respective inclusion complexes with *2a*, thus optically pure oxaziridines were obtained.

The 1:1 inclusion complexes *68* composed of *2a* and nitrones *67* were prepared by keeping a solution of *2a* and an equimolar amount of *67* in benzene-hexane (1:1) at room temperature for 12 h [40]. Melting points of the complexes *68* are shown in Table 8. Irradiation of *68* in the solid state gave optically active oxaziridines *69*. Irradiation time, yields and optical purity of the products are summarized in Table 8 [40]. Enantioselectivity in the formation of *67d*, *67f*, and *67g* is high, but that of *69b*, *69c*, and *69e* is low. This suggests a distinct influence coming from the substituents.

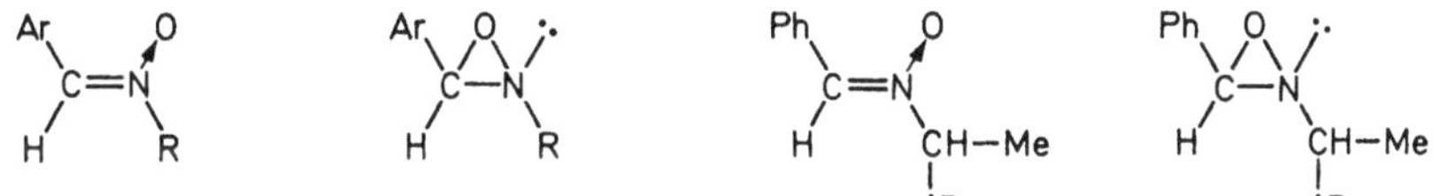

67 69 67g 69g

In the case of *67g* which has a chiral alkyl group, optically pure *67g* was included at the complexation process with *2a*, and (—)-*67g* of 100% ee [$[\alpha]_D$ —66.8° (*c* 0.22, $CHCl_3$)] was obtained. Irradiation of the 1:1 inclusion complex of *2a* and (—)-*67g* of 100% ee gave *69g* of 100% ee which has three optically pure chiral centers [40]. This is not the result of a chiral induction by the optically active alkyl group, since irradiation of *67g* of 100% ee in benzene gave *69g* of only 12% de (diastereomeric excess).

The optically resolved host compounds *6b* and *7b* also forms 1:1 inclusion complexes with *67b* (mp 132 to 134 °C) and *67c* (mp 178 to 180 °C), respectively. However, both complexes are photostable [13].

Development of new synthetic routes to optically active β-lactam derivatives is still an attractive problem in organic chemistry. As a synthetic approach to penicillin derivatives, photocyclization of α-oxoamides *70* to β-lactams has long been studied [41, 42]. This reaction (Scheme 6), however, results in a complex mixture of racemic cis- and trans-isomers of β-lactams *72* and of oxazolidin-4-ones *73*, since the reaction proceeds via a zwitterionic intermediate *71* [43]. Of these isomers, only the optically

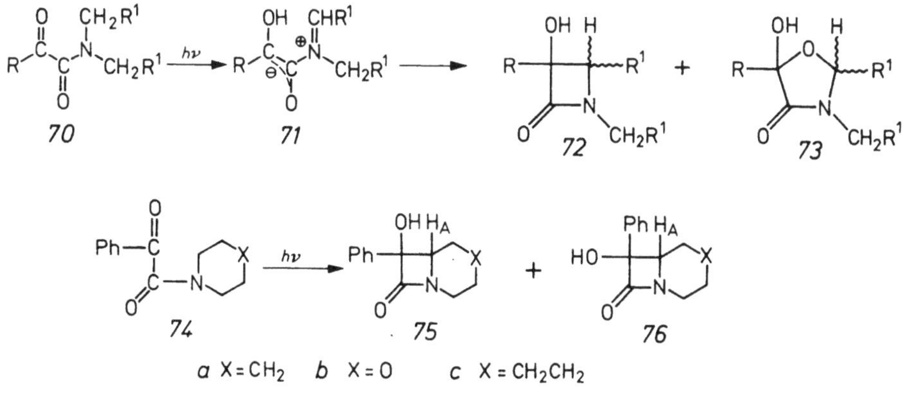

Scheme 6

active β-lactams are the useful compounds. Some severe controls of the photocyclization of α-oxoamides are necessary to obtain optically active β-lactams. Some attempts at such control have been tried. Irradiation of *70* in the solid state gave *72* selectively [44], but this method is not applicable to α-oxoamides derived from a cyclic amine such as *74a* [44]. Furthermore, no stereoselective control was achieved by this method, thus resulting in a mixture of cis- and trans-isomers of *72* [44]. Of course, no efficient enantioselective control of the photocyclization has yet been achieved, although 15% ee of *72* has been obtained by irradiation of *70* in an inclusion complex with deoxycholic acid [45].

On the other hand, stereo- and enantioselective photocyclization reactions of N-benzoylformylpiperidine (*74a*), N-benzoylformylmorpholine (*74b*), and N-benzoylformylhexamethyleneimine (*74c*) were achieved by irradiation of the respective inclu-

Table 9. Inclusion complexes of guest molecules *74a–c* with host compounds *1*, *2a*, and *4*

Guest	Host[a]					
	1	mp (°C)	*4*	mp (°C)	*2a*	mp (°C)
74a	*77*	98–101	*78*	nc[b]	*79*	75–79[c]
74b	*80*	94–96	*81*	nc	*82*	65–70
74c	*83*	113–115	*84*	nc	*85*	84–87

[a] Host:guest ratios are 1:1 except in the case of *80* and *85* (1:2); *79* consists of *2a*, *74a*, and benzene in a 1:1:1 ratio. [b] nc means not clear. [c] Measured in a sealed capillary.

Table 10. Solid-state photocyclization of oxoamides to β-lactams using inclusion complexation

Complex[a]	Irradiation time (h)	Yield[b] (%)	Product [[α]$_D$ value (°)[c] and optical purity (% ee)[d]]	Ratio of *75*:*76*
77	55	41	*75a* and *76a*	77:23
78	30	73	*75a* —	100:0
79	100	67	(−)−*75a* and (−)−*76a*[e] (−62.4, 62.5) (−62.4, 95)	
80	40	38	*75b* and *76b*	77:23
81[f]	50	—	— —	— —
82	50	49	(−)−*75b* and (−)−*76b* (−107.8, 55.8) (−48.7)[g]	78:22
83	25	82	*75c* and *76c*	91:9
84	10	74	*75c* —	100:0
85	30	67	(+)−*75c* and *76*[h] (+22.6, 11.2) (0, 0)	49:51

[a] Specification in Table 9. [b] Yield of a mixture of *75* and *76* which was separated from the crude reaction product by a silica gel chromatography. [c] All [α]$_D$ values were measured in $CHCl_3$ at *c* 1.0. [d] Optical purity was determined by HPLC on Chiralcel. [e] Since optically active *75a* and *76a* could not be separated, it is not clear whether both enantiomers are (−)-ones or not. Therefore, both are tentatively shown as (−)-enantiomers. Since [α]$_D$ value of each enantiomer is also not clear, [α]$_D$ value of the mixture is shown. [f] Compound is inert to irradiation. [g] Optical purity was not determined. [h] When an acetone solution of the mixture of (+)−*75c* and *76c* was kept, racemic *76c* crystallized out, mp 135–137 °C.

sion complexes with the host compounds *1*, *2a*, and *4*. The nine different inclusion complexes prepared from the three α-oxoamides and the three host compounds are specified in Table 9 [30].

On irradiation, all complexes except *81* gave β-lactam derivatives. Irradiation time, yields, and ratios of products are summarized in Table 10 [30]. In all irradiations, the β-lactam derivatives *75a–c* and *76a–c* were produced exclusively. The reason for the efficient control in the inclusion complexes is not clear. A plausible interpretation is that the crystal lattices of the inclusion complexes are too compact to produce oxazolidin-4-ones (cf. *73*) which have a five-membered ring, compared to the β-lactams which contain a smaller four-membered ring.

A perfect stereochemical control of the photocyclization of *74a* and *74c* to cis-β-lactams *75a* and *75c*, respectively, was achieved by the irradiation of their inclusion complexes with *4* (Table 10). In contrast, the stereochemical control in the inclusion complexes with *1* and *2a* is relatively inefficient. A plausible interpretation for the difference is as follows: in the complex with *1* and *2a*, there is a hydrogen bond between the carbonyl oxygen of the amide group of *74* and the hydroxyl group of the host compound as depicted in formula *86* of Scheme 7. It has been deduced that a hydrogen bond is formed to the carbonyl oxygen of the amide group but not to the benzoyl group [12]. The zwitterionic intermediate [43] is stabilized by a similar hydrogen bond as shown in formula *87* of Scheme 7. This stabilization makes an interconversion between *87a* and *87b* easy through the cis-trans isomerization, and the cyclization proceeds less stereoselectively to afford a mixture of *75* and *76*. No such stabilization is expected for the zwitterionic intermediate formed in the inclusion complex with *4* which provides no hydroxyl group. In this case, the intermediate cyclizes immediately after generation and gives *75* selectively. In fact, the photocyclization reaction of *74* in the inclusion complex with *4* proceeds much faster than it does in the complexes with *1* or *2a*.

The conversion of *87a* into *87b* by a rotation about a single bond needs a lot of space which might be available in solution but not in the crystalline complex. This rotation might be severely hindered especially in the complex with *4*.

Enantioselective photocyclization of *74* occurred efficiently in the inclusion complex with *2a*. In particular, the selectivity is very high in the case of *74a*. However, control is inefficient in the 1:2 complex *85* composed of *2a* and *74c*. The host:guest ratio probably depends on the packing of the components in the crystal. The packing is

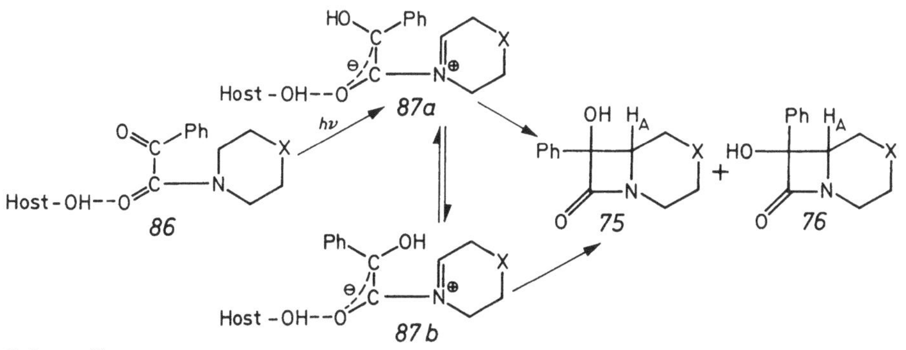

Scheme 7

related to the selectivity of the reaction in the solid state. Therefore, the selectivity would be influenced by the ratio. In the case of the 1 : 1 inclusion complex *79* composed of *2a* and *74a* containing one mole of benzene, this benzene molecule probably works as an important spacer in the crystal exercising an influence on the photocyclization reaction.

In order to control both the stereochemistry and enantioisomerism of the photocyclization product, design of a good host compound is necessary. It seems adequate to design an optically active 2,2-biphenyldicarboxamide derivative, since *4* is very useful for the control of stereochemistry of the photocyclization of *74*.

Since *4* has a high inclusion ability for a wide variety of organic compounds [46], it might be useful for various reaction controls. For example, *4* was found to be useful for a selective synthesis of the β-lactam *89* from N,N-diisopropylacetylformamide (*88*). Irradiation of *88* in benzene solution and in neat liquid has been reported to give the oxazolidinone derivative *90* exclusively in 86 and 65% yields, respectively, but not any *89* [44]. Control of the photocyclization of *88* has been attempted by irradiation of *88* in the solid state at −78 °C. However, the irradiation gave a mixture of *89* (31%) and *90* (29%) [44].

Contrarily, irradiation of the 1:1 inclusion complex between *4* and *88* for 40 h in the solid state gave *89* selectively in 60% yield [47]. By the similar irradiation of the 1:1 inclusion complex between *4* and N,N-dimethylacetylformamide (*91*) in the solid state, β-lactam derivative *92* was obtained as the sole product in 56% yield [47].

The most successful enantiocontrol of photocyclization of a α-oxoamide to a β-lactam in an inclusion complex is represented by the photoreaction of the 1 : 1 complex *94* composed of *2a* and N,N-dimethylbenzoylformamide (*93*) which gives β-lactam *95* of 100% ee in 90% chemical yield (Scheme 8) [32]. The experimental detail is as follows: when a solution of *2a* and an equimolar amount of *93* in ether-petrol ether (1 : 1) was kept at room temperature for 24 h, a corresponding 1 : 1 inclusion complex *94* was obtained as colorless prisms in 75% yield, mp 126 to 127 °C. Powdered *94* was irradiated by a 400 W high-pressure Hg-lamp at room temperature for 27 h, occasionally grinding the complex with a pestle and agate mortar. The reaction mixture was chromatographed on silica gel using benzene-ethyl acetate as solvent to give (−)-*95* of 100% ee in 90% chemical yield [32].

MeCOCON(iPr)$_2$

88

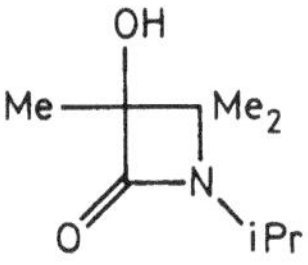

89

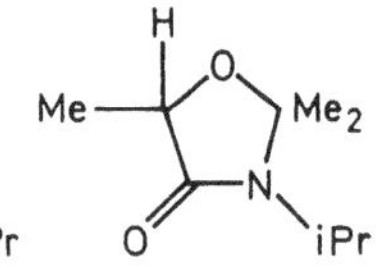

90

MeCOCONMe$_2$

91

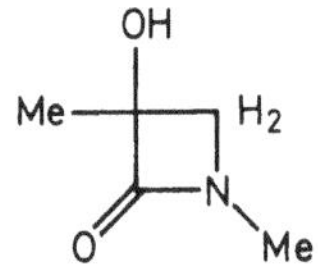

92

X-Ray crystal structural studies [32] (Fig. 13 and Scheme 8 which refers to the crystal structure) showed that one molecule of *93* is held in a fixed conformation determined by two hydrogen bonds and by neighboring host molecules which prevent free rotation about the CO—CO single bond in *93*. Free rotation about this bond would enable the production of the two possible enantiomers. The fixed conformation of the guest molecule by the chiral host molecule causes the least molecular motion during the photocyclization reaction and the high enantioselectivity.

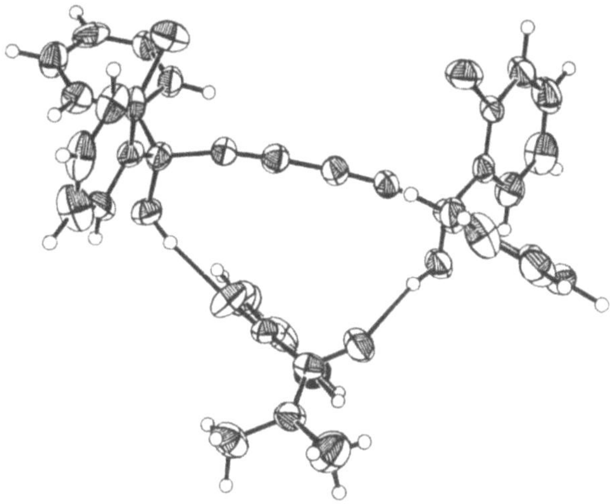

Fig. 13. Stereoscopic view of the crystalline host-guest complex *94*

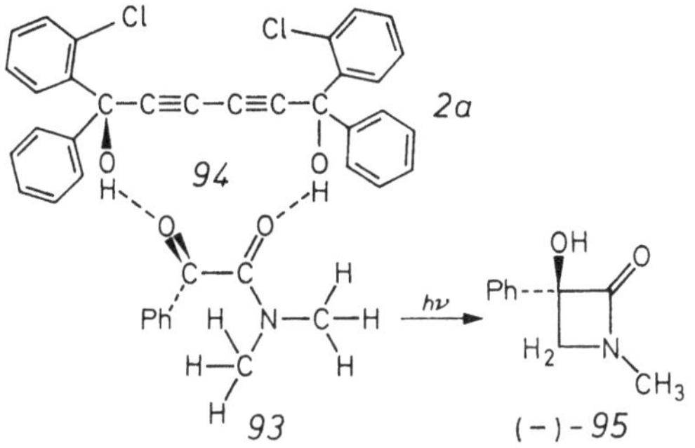

Scheme 8

7 Control of Enantioselective Photoreactions in the Absence of a Chiral Source

The most exciting enantioselective photochemical conversion of a α-oxoamide to a β-lactam has been found in the case of N,N-diisopropylbenzoylformamide (*96*) which gives β-lactam *97*. In the photocyclization of plain *96* in the solid state, optically active β-lactam *97* of high optical purity was obtained in high chemical yield. Thus no optically active host compound is necessary for the enantioselective reaction [48].

Recrystallization of *96* from benzene afforded colorless prisms. That each crystal is chiral was shown by photochemical conversion into the optically active *97*. Crystals of *96* which gave (+)- and (—)-*97* on photocyclization have been tentatively identified as (+)- and (—)-crystals of *96*, respectively. Large amounts of the (+)- and (—)-crystals of *96* can easily be prepared by seeding with finely powdered (+)- and (—)-

crystals, respectively, during recrystallization of *96*. Seeding with (—)-crystals during recrystallization of (+)-crystals gives (—)-crystals and vice versa.

Irradiation of (+)-crystals of *96* with a 400 W high-pressure Hg-lamp, with occasional grinding with an agate mortar and pestle for 40 h at room temperature gave (+)-*97* of 93% ee in 74% yield. Irradiation of (—)-crystals of *96* under the same conditions gave (—)-*97* of 93% ee in 75% yield [48]. Purification to 100% ee can easily be achieved by recrystallization from benzene. Although the photochemical conversion of *96* into *97* on irradiation in the solid state has been reported, enantioselectivity of the reaction has not been discussed [44].

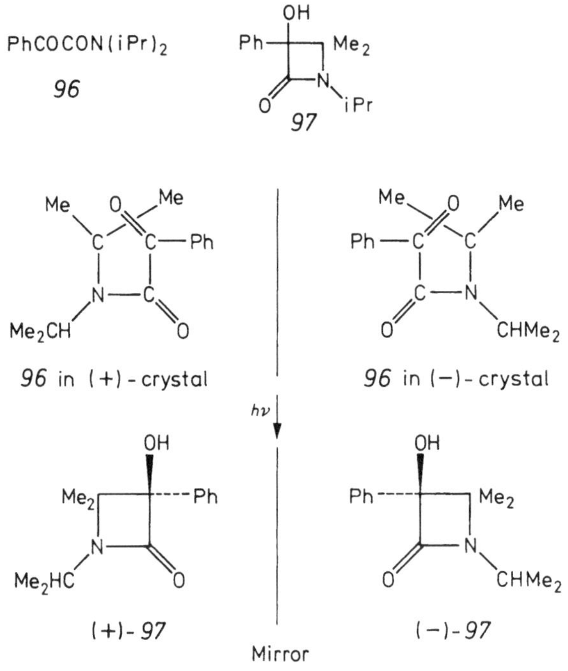

Scheme 9

X-Ray crystal structural analysis of a (+)-crystal of *96* shows that molecules of *96* are arranged so as to be chiral in the crystalline lattice (Fig. 14). Scheme 9 which is depicted by referring to Fig. 14 indicates the reason why each chiral crystal of *96* gives the corresponding chiral *97*.

The above results are valuable in that an optically active compound is produced in bulk from achiral material. Only a few successful examples of photochemical conversion of achiral into chiral material in the absence of a chiral source have been reported hitherto [49], and in these cases the conversion was carried out on a fragment of a chiral crystal. In our case, chiral crystals are available in bulk, and mass production of the chiral compound is possible.

Moreover, the present data may throw some light on the generation of optically active amino acids on the Earth [50,51]. Photocyclization of *96* proceeds efficiently in

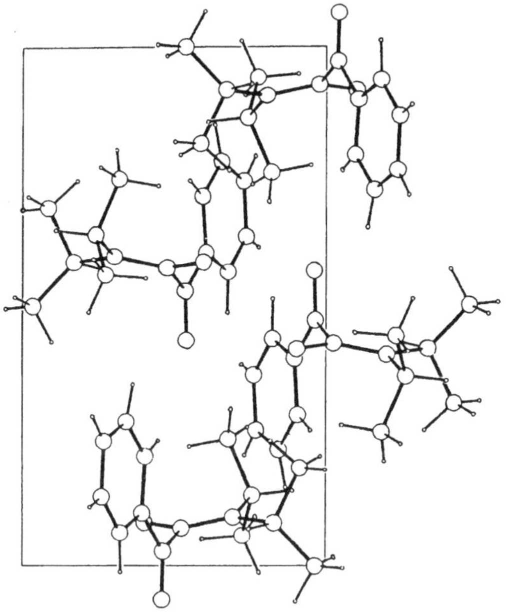

Fig. 14. Packing diagram of *96* in its (+)-chiral crystal

sunlight in the solid state and hydrolysis of the optically active *97* gives an optically active β-amino acid.

8 Acknowledgements

The author is grateful to his coworkers whose names appear in the references, especially Dr. Koichi Tanaka, for their many contributions. The author would like to thank Professors Menahem Kaftory of Technion Israel Institute of Technology, Takaji Fujiwara of Shimane University, and Yuji Oohashi of Ochanomizu Womens University for their valuable X-ray analyses of many host-guest inclusion complexes.

9 References

1. Toda F (1987) Top. Curr. Chem. 140: 43
2. Tanaka K (1988) Tetrahedron Lett. 29:551
3. Toda F, Tanaka K, Nassimbeni LR, Niven ML (in press) Chem. Lett.
4. Lutz RE, Jordan RH (1950) J. Am. Chem. Soc. 72: 4090
5. Stobbe H, Bremer K (1929) J. Prakt. Chem. 123: 1
6. Rabinovich D: J. Chem. Soc. B 1970: 11; Ohkuma K, Kashino S, Haisa M (1973) Bull. Chem. Soc. Jpn. 46: 627
7. Tanaka K, Toda F: J. Chem. Soc., Chem. Commun. 1983: 593
8. Tanaka K, Toda F: Nippon Kagaku Kaishi 1984: 141
9. Kaftory M, Tanaka K, Toda F (1985) J. Org. Chem. 50: 2154
10. Recknwald GW, Pitts JN, Letsinger RL (1953) J. Am. Chem. Soc. 75: 3028

11. Greene FD, Misrock SL, Wolf JR Jr (1955) J. Am. Chem. Soc. 77: 3852
12. Kaftory M (1987) Tetrahedron 43: 1503
13. Toda F, Tanaka K: unpublished data
14. Stefaniak L (1976) Tetrahedron 32: 1065
15. Kuzuya M, Noguchi A, Yokota N, Okuda T, Toda F, Tanaka K: Nippon Kagaku Kaishi 1986: 1746
16. Taylor EC, Kan RO (1963) J. Am. Chem. Soc. 85: 776
17. Nozoe T, Sato M, Matsui K: Proc. Japan Acad. (1953) 28: 407 (1953) Sci. Rep. Tohoku Univ. Ser. I 37: 211 (1952) Nippon Kagaku Kaishi 73: 781; Mroi T (1959) Nippon Kagaku Kaishi 80: 185
18. Ikegami Y (1959) Infrared Absorption Spectra, Nankōdō, vol 8 p 33
19. Toda F, Tanaka K, Asao T, Ikegami Y, Tanaka N, Hamada K, Fujiwara T: Chem. Lett. 1988: 509
20. Goldstein P, Ladell J, Abowitz G (1969) Acta Crystallogr., Sect. B 25: 135
21. Bolton K, Brown RD, Burden FR, Mishra A: J. Chem. Soc., Chem. Commun. 1971: 873
22. Toda F, Tanaka K, Elguero J, Nassimbeni L, Niven M: Chem. Lett. 1987: 2317
23. Ganter C, Pokras SM, Roberts JD (1966) J. Am. Chem. Soc. 88: 4235
24. Toda F, Tanaka K, Oda M (1988) Tetrahedron Lett. 29: 653
25. Buchi G, Burgess EM (1961) J. Am. Chem. Soc. 84: 3104
26. Toda F, Tanaka K, Sekikawa A: J. Chem. Soc., Chem. Commun. 1987: 279
27. Toda F, Shigemasa T, Takumi H (in press) Chem. Lett.
28. Toda F, Yagi M (in press) J. Chem. Soc., Chem. Commun.
29. Toda F, Tanaka K: J. Chem. Soc., Chem. Commun. 1986: 1429
30. Toda F, Tanaka K, Yagi M (1987) Tetrahedron 43: 1495
31. Zandomeneghi M, Cavazza M, Festa C, Pietra F (1983) J. Am. Chem. Soc. 105: 1839
32. Kaftory M, Tanaka K, Toda F (in press) J. Org. Chem.
33. Dauben WG, Koch K, Smith SL, Chapman OL (1963) J. Am. Chem. Soc. 85: 2616
34. Cantrell TS, Solomaon JS (1970) J. Am. Chem. Soc. 92: 4656
35. Paquette LA, Cox O (1967) J. Am. Chem. Soc. 89: 5633
36. Leclercq M, Jacques J (1985) C. R. Acad. Sc. Paris 301: 1231
37. Forni A, Moretti I, Torre G (1987) J. Chem. Soc., Chem. Perkin Trans. 2, 1987: 699
38. Davis FA, McCauley JP Jr Chattopadhyay S, Harakal ME, Towson JC, Watson WH, Tavanaiepour I (1987) J. Am. Chem. Soc. 109: 3370
39. Boyd DR, Campbell RM, Coulter PB, Grimshaw J, Neill DC, Jennings MB: J. Chem. Soc., Perkin Trans. 1, 1985: 849
40. Toda F, Tanaka K: Chem. Lett. 1987: 2283
41. Åkermark B, Johansson N-G, Sjöberg B: Tetrahedron Lett. 1969: 371
42. Henery-Logan KR, Chen CG: Tetrahedron Lett. 1973: 1103
43. Aoyama H, Sakamoto M, Kuwahara K, Yoshida K, Omote Y (1983) J. Am. Chem. Soc. 105: 1958
44. Aoyama H, Hasegawa T, Omote Y (1979) J. Am. Chem. Soc. 101: 5343
45. Aoyama H, Miyazaki K, Sakamoto M, Omote Y: J. Chem. Soc., Chem. Commun. 1983: 333
46. Toda F, Kai A, Tagami Y, Mak TCW: Chem. Lett. 1987: 1393
47. Toda F, Yagi M (in press) Chem. Lett.
48. Toda F, Yagi M, Soda S: J. Chem. Soc., Chem. Commun. 1987: 1413
49. Evans SV, Marcia-Garibay M, Omkaram N, Scheffer JR, Trotter J, Wireko F (1986) J. Am. Chem. Soc. 108: 5648 and references cited therein
50. Green BS, Lahav M, Rabinovich D (1979) Acc. Chem. Res. 12: 191
51. Addadi L, Lahav M (1979) in: Walker DC (ed) Origins of optical activity in nature. Elsevier, New York, chapt 14

Author Index Volumes 101–149

Contents of Vols. 50–100 see Vol. 100
Author and Subject Index Vols. 26–50 see Vol. 50

The volume numbers are printed in italics

ISBN 3-05-500613-5

Bei Fragen zur Produktsicherheit wenden Sie sich bitte an:
If you have any questions regarding product safety,
please contact:

Walter de Gruyter GmbH
Genthiner Straße 13
10785 Berlin
productsafety@degruyterbrill.com